THE CELL SURFACE:
ITS MOLECULAR ROLE
IN MORPHOGENESIS

THE CELL SURFACE:
ITS MOLECULAR ROLE
IN MORPHOGENESIS

A. S. G. CURTIS

LOGOS PRESS
ACADEMIC PRESS

Published by
LOGOS PRESS LIMITED
in association with
ELEK BOOKS LIMITED
2 All Saints Street, London, N.1

Distributed by
ACADEMIC PRESS INC.
111 Fifth Avenue, New York, N.Y. 10003
and
ACADEMIC PRESS INC. (LONDON) LIMITED
Berkeley Square House, Berkeley Square, London, W.1

Library of Congress Catalog Card Number 67–21470

PRINTED IN GREAT BRITAIN BY
UNWIN BROTHERS LIMITED
WOKING AND LONDON

Contents

CONTENTS

INTRODUCTION

"There is something fascinating about science. One gets such wholesale returns of conjecture out of such a trifling investment of fact."

Mark Twain.

This book is about a branch of biology which is emerging from the realms where conjecture holds sway to a solid foundation of experimental fact. But fascinating though the conjecture has been, the investments of fact have recently begun to accumulate to a degree where it seemed apposite to prepare this review and analysis.

Until recently it would only have been possible to write this book in one or other of two ways: either as a survey of the subject arranged in historical order, or by presenting the final product of embryogenesis and attempting to analyse it in terms of possible functions of the cell surface. However, we now have sufficient information about the molecular structure of surfaces to attempt to consider what the results of a given molecular structure would be; first in terms of the behaviour of individual cells, then in the behaviour of populations of cells, and finally in terms of the building of whole embryos. I intend to present the subject in this way.

I hope that the book will show that each level of interpretation rests on all preceding ones and that the molecular approach to the subject is the most meaningful and the most hopeful experimentally. However, it will also appear that at present many of the links between each level of the subject are exceedingly tenuous and it is one of the objects of this book to reveal these deficiencies. But by using this approach it will be shown that a great deal of the conjecture which the subject has collected is redundant.

If, as I very much hope, the explanation of biological phenomena at all levels becomes amenable to the integrated approach which molecular biology offers, this book will be rapidly outdated. The book can however be viewed as an ephemeral attempt to follow through the complete biological implications inherent in the basic design of part of the organism at the molecular level. This aim is what we must succeed in if we are to develop a satisfactory biology. Perhaps the most satisfying proof of our success will be to build working cells and organisms.

In preparing this book I have not invariably quoted the earliest paper by any given worker on a subject. When a scientist has published a number of substantially identical papers, it may be found that only the most accessible of these papers will be quoted. This paper will not necessarily be the first. In addition I have made it my invariable rule not to refer to published abstracts or to theses. Abstracts are almost always so devoid of experimental details and results that no appraisal can be made of them; consequently they have almost no value other than as a certain rather tenuous claim to priority of publication. Theses, though contrarily of immense wealth of fact, are so inaccessible to the majority of workers that they cannot count as scientific literature. I apologise to any worker who feels that this is unfair but it should be pointed out that application of these rules has considerably lessened the amount of literature to be quoted. I offer my particular apologies to any scientist whose work I have inadvertently ignored through my ignorance.

I would like to thank Dr. A. C. Allison, B. B. Boycott, Esq., Dr. D. C. Dumonde, Dr. A. Glauert, Dr. J. E. M. Heaysman, Dr. J. Lucy, Dr. G. Selman, Professor M. S. Steinberg and Professor L. Wolpert for their advice and criticism. I am deeply indebted to my colleagues in the Department of Zoology, University College London and in particular to Professor M. Abercrombie.

University College London   A. S. G. CURTIS
December 1966

CHAPTER 1

The Composition and Structure of the Cell Surface

INTRODUCTION

The fact that we term units of protoplasm "cells" implies that they are separated from the surrounding medium by a membrane. This membrane must lie outside all or the greater part of the cell if it is to be isolated from the external phase, and it is for this reason that we term the membrane "cell surface" or plasma membrane. Of the four main classes of organic molecule in the cell, only the lipid molecules have the properties of low dielectric constant and of having non-polar portions which are required so that extensive membranes shall be formed. These two properties also specify that the membranes shall be of low permeability to solutes, which allows the composition of the intracellular medium to be substantially different from that of the external medium, and of sufficient electrical resistance to permit the cell to develop transmembrane potentials. This book is not concerned with these two functions of the cell surface, permeation and electrogenesis, partly because of the huge literature associated with these two subjects, but mainly because there is a second very different set of cell surface functions connected with multicellularity and morphogenesis. Permeation and electrogenesis are phenomena which can appear in the isolated cell; this book will treat those cell surface properties which control the appearance or loss of multicellularity, and the development of form in these multicellular animals: in other words, interactions between the surface of one cell and another.

A basic requirement of the cell surface membrane is that it shall be structurally fairly stable, for it has to persist throughout the lifetime of a cell, though continuous repair may be possible. Surface films whether single, bi- or multilayer are most stable when their surface tension is close to zero. For cells which lie in aqueous media we should replace the term *surface* tension by *interfacial* tension. If the surface pressure is so high that the interfacial tension would have a negative value, the films are highly unstable, tending to pass into violent expansion; similarly with too low a surface pressure, the high value of interfacial tension tends to lead to membrane contraction (Adam, 1941). Additionally it is highly desirable that the interfacial tension in the interests of stability be rapidly variable in value in response to local stretchings or compressions, so that the membrane can accommodate these deformations by a rise or fall in interfacial tension respectively. For these reasons, we would expect the cell surface to be designed with these properties so that it possesses the required stability. Does the chemical nature of the cell surface and its structure fulfil these requirements? What concomitant properties can be designed into a cell surface which affect other biological properties, while such requirements are met? This chapter will consider these problems at the level of structure and composition, and will allow us to define the surface.

Before discussing this in detail it should be remarked that by 1938 (Harvey and Danielli) our views of cell surface structure, which were derived from studies of permeation, haemolysis, the effect of anaesthetics on cells, electrical measurements and the study of model membrane systems, suggested that the membrane was basically lipoid in nature and probably had the structure of a bi-molecular layer. Harvey and Danielli made these brilliant deductions without being (at that time) able to visualise the membrane directly. The advent of electron microscopy allowed us to see that the cell surface was extremely thin, 100Å, hence the name plasmalemma. The interpretation of electron micrographs will be discussed later.

Readers may feel that there is some ambiguity about describing a certain chemical grouping as "in" or "on" the surface. I would suggest the following definition: that the surface can be divided into (*a*) a region of low dielectric constant and high electrical resistance, and (*b*) outwards of this (on both sides of the surface) a region of high dielectric constant and low resistance. Groupings in the outermost regions are to be termed "on" the surface, because these groups are able to react as groups actually on the edge of or in the outer aqueous phase. The groups "in" the surface are those in the region of low dielectric constant. Nevertheless the low dielectric constant core of the membrane affects reactions "on" the surface to a certain degree and the polar part of the surface will control reactions inside the membrane to some extent. Molecules and particles adsorbed on to the surface may provide new surface components.

THE COMPOSITION OF THE PLASMALEMMA

Bulk analyses and isolation of the cell surface

In this section the results of various attempts to characterise the components of the cell surface will be described. Either the surface has been extracted and then analysed, or else enzymological, immunochemical, histochemical and electrophoretic techniques have been used to identify components of the plasmalemma *in situ*. In the second section a general survey will be made of the structure of the plasmalemma. This will include the findings of electron microscopy and X-ray diffraction studies as well as the structural conclusions which can be made from our knowledge of chemical composition.

Attempts to analyse the composition of the surface require the development of a suitable means of stripping the cell surface off the cytoplasm, and the discovery of tests which will show whether or not it is the surface which has been isolated. At present neither the techniques of isolation nor those of characterisation have been developed far. Two structures, myelin and the red cell ghost, have provided the most frequently used material for isolation of plasmalemma components. The red cell or erythrocyte ghost was introduced for use earlier than myelin, but has in fact proved to be a most unsatisfactory material. At one time it was believed that the ghost (the membranous shell left after haemolysis of the erythrocyte) provided an isolate of pure plasmalemmal material. Gorter and Grendel (1925) made acetone extracts from red cells and the extracts were then spread at air–water interfaces. The area of lipid in these films, when related to the number of cells extracted, appeared to be just sufficient to cover each cell with a bimolecular layer of lipid. In fact these results are purely fortuitous because it was not stated at what surface pressure the films were spread, in consequence of which, it is impossible to make any comparison with the plasmalemma *in situ* on the cell, which is probably at high surface pressure. Moreover it is not known how to compare molecular areas of this lipid at air–water and oil–water interfaces, so that even if the surface pressure of the spread film had been recorded no useful conclusion could have been made. Subsequently (Ponder, 1949) it was found that acetone does not extract ghosts completely. It was also discovered by Schmitt, *et al.* (1938) using polarisation studies, that the ghost appeared to contain a considerable thickness of lipid. Waugh and Schmitt (1940) observed that the dried ghost membranes (stroma) are 120–215 Å thick. These results were confirmed by electron microscopy (see Ponder, 1949). It is curious that Gorter and Grendel's work formed a main basis of the bimolecular lipid leaflet theory of cell surface structure, a theory which has reached much confirmation from careful research work. For these reasons both lipid analysis (de Gier and van Deenen, 1961) and protein estimations (Ponder, 1949; Maddy and Malcolm, 1965; Poulik and Lauf, 1965) of ghosts, which purport to represent plasmalemma composition, should be viewed with caution. It may of course be that stromal lipid is equivalent to plasmalemmal lipid but this is uncertain. Gier and van Deenen's data show that the lipid composition of the erythrocytes of various species is very variable.

The use of myelin as a source of cell surface material is based on the work of Geren (1954) and Robertson (1958a,b) who showed that myelin is composed of Schwann cell surfaces, any one cell surface being wound in a spiral round the axon (see later). Unfortunately at present, all myelin analyses are based on analyses of bulk tissue such as white matter from the brain, which of course means that contributions from nuclear and other cytoplasmic materials appear in the analyses. Boell and Nachmansohn (1940) stripped the myelin off giant fibres from the squid but they did not analyse their material. Wolfe (1962, 1964) summarises a considerable number of analyses of mammalian myelin. He publishes an analysis of the lipids of white matter as follows:

cholesterol	164	millimoles/kg dry weight.
sphingomyelin	78	,,
cerebrosides	169	,,
cephalins	199	,,
lecithin (phosphatidyl choline)	103	,,

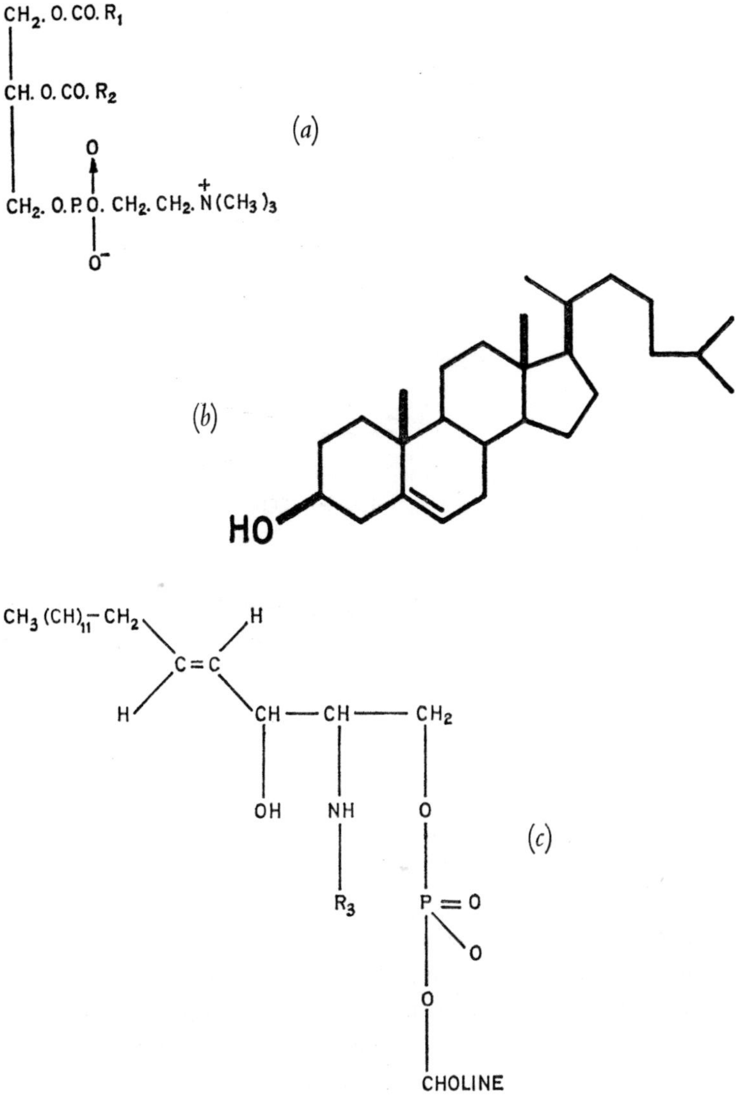

$$CH_2.O.CO.R_1$$

$$CH.O.CO.R_2$$

(a)

$$CH_2.O.P.O.CH_2.CH_2.\overset{+}{N}(CH_3)_3$$
$$O^-$$

(b)

HO

$$CH_3(CH)_{11}-CH_2$$

$$C=C$$

$$H$$

$$CH-CH---CH_2$$

$$OH \quad NH \quad O$$

$$R_3 \quad P=O$$

$$O$$

$$O$$

(c)

CHOLINE

Figure 1. Formulae and general shape of main lipid components of plasmalemmae. (a) Lecithin (phosphatidyl choline); R_1 and R_2 may be oleyl, palmityl, steryl or lineolyl. Phosphatidyl serine and ethanolamine have similar structures. (b) Cholesterol. (c) Sphingomyelin; R_3 is C_{15} or C_{16} fatty acid; in the cerebrosides phosphoryl choline is replaced by galactose.

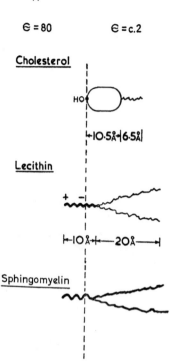

Figure 1—*continued*

(*d*) General shape and arrangement of these molecules at a water-lipid interface; water to left, lipid to right.

The structural formulae of these components are given in Figure 1: sphingomyelin and cerebrosides are sphingosine-containing lipids; in the former a long chain fatty acid is substituted on the amino-group of sphingosine and choline phosphate on the outermost adjacent hydroxyl group; in the latter the phosphoryl choline is replaced by galactose. Phosphatidyl serine and phosphatidyl ethanol-amine have also been described as components of myelin (Robertson, 1960a). Nussbaum *et al.* (1963) found twice as much cephalin as lecithin in adult rat brain myelin. They also found that cholesterol formed 16·7 per cent of all lipids and sphingomyelin some 1 per cent.

Vandenheuvel (1963) published the following data on myelin derived from several sources.

	% Total lipid	Saturated and unsaturated fatty acids according to chain length					
		Weight % of lipid type					
		14	16	18	20	22	24 and 26
Phosphatidyl	20	4·2	29	63	0·7	0·6	—
Sphingolipid	40	0·6	3·8	23·8	7·4	15·1	49·1
Cholesterol	40	—	50·7	—	—	—	49·1

In addition two classes of protein-lipid compound are found. Le Baron and Folch (1956) described proteolipids and phosphatido-peptides from brain myelin. Finean et al. (1956) found almost no proteolipid in sciatic nerve; does this suggest tissue differences in surface composition? The decomposition products of proteolipid and phosphatidopeptides had been previously named neurokeratin. Lipid compounds in myelin with beta-D-N-acetyl-neuraminic acid (sialic acid) have been termed mucolipids or gangliosides (Wolfe, 1962). Nevertheless we are still not quite certain whether these components are those of the surface, let alone whether other types of cells have surfaces of the same composition. Richardson et al. (1963) described a protein fraction from mitochondria as a structural substance, and it is possible that a similar structural component might exist in the cell surface.

Recently a number of attempts have been made to isolate the cell surface from cell types other than the erythrocyte (Neville, 1960; Bruemmer and Thomas, 1957; Levin and Thomas, 1961; Thomas and Levin 1962; Kono and Colwick, 1961; McCollester, 1962; Miller and Crane, 1961 and Woodin and Wienkeke, 1966) in which unfortunately little progress has been made in identifying the frac-tions obtained as cell surface, other than from microscopy and their lipoprotein composition. The evidence derived from studies on erythrocyte ghosts (see for example Poulik and Lauf, 1965; Wolfe, 1964; Maddy and Malcolm, 1965) is vitiated by the fact that such ghost preparations have invariably contained non-plasmalemmal material. O'Neill and Wolpert (1961) prepared surface fractions from *Amoeba proteus*, the identification was confirmed by good electron micrographs, which also showed the mucoid filamentous coat on one side of the membrane (this coat is found on the membrane in life). They also used fluorescent antibody staining to detect the presence of the coat. Herzenberg and Herzenberg (1961), working on the localisation of the H-2 antigens, H_2^d and H_2^k, in mouse liver cells, using a hemagglutination technique, found that these were associated with a surface membrane fraction. However it is possible that the antigenic marker becomes associated with this fraction during extraction, and that identification of the purity of the membrane fraction is unsatisfactory.

Despite the fact that the preceding two groups of workers have used the elegant method of antigen markers to identify the cell membrane fraction, there is still room for great development in methods of identifying the membrane. Although we strongly suspect that many antigens such as blood group substances are located on or near the cell surface, it is not clearly established that they may not also occur in the cytoplasm (see later in this chapter). Other criteria have been used to identify membrane isolates, for instance Wallach and Ullrey (1962) broke Ehrlich ascites cells open by embolism and identified membrane fractions by their ATPase activity, also observed in suspensions of whole cells and thus presumably surface located. They also argued that the presence of sialic acid and phospholipid in the fraction suggested that it was derived from membrane. Again Kamat and Wallach (1965) used the presence of ATPase and certain surface (or supposedly surface) antigens to recognise the plasmalemma fraction from cell homogenates. Neifakh *et al.* (1965) used the presence of ATPase activity and the nature of cytochromes present to show that the membrane fraction was not contaminated with mitochondria. Such a conclusion depends on the assumption that all the mitochondrial components remain together during fractionation, so that there is no possibility of mitochondrial lipid being the only mitochondrial component to contaminate the plasma membrane fraction.

It will appear that criteria for the identification and preparation of isolated cell surface for analysis depend upon somewhat circular arguments. Identification is based on the supposed surface location of an antigen, enzyme or other type of molecule. To some extent their surface location can be ascertained from histochemical location of staining or from the kinetics of enzymic reaction, but in every case a certain doubt obtains as to whether these supposed surface molecules are only or even ever located in the surface. Consequently there is the danger that surface fractions are identified from the supposed surface location of a molecule, which itself was previously obtained in a supposed surface fraction by another worker. The best way to avoid this is to use as many criteria as possible as some workers have done, but many others, e.g. Emmelot and Bos (1962, 1966), Ashworth and Green (1966), O'Brien and Sampson (1965), etc. have assumed that a method of preparation or one criterion is sufficient to identify plasmalemma preparations. These studies have not as yet revealed anything significant about plasmalemma composition save that Neifakh *et al.* (1965) claim to have found an acto-myosin-like protein in plasma membrane fractions from liver cells.

Other criteria such as the absence of non-membranous cytoplasmic or nuclear components, reactions to various reagents, and enzyme analyses are required, and possibly most satisfying of all the reconstitution of a functioning membrane. Electron microscopy may not always be a suitable means of identifying the surface fraction because a properly extracted and pure surface fraction may have undergone phase change during extraction. The present methods of cell surface separation being unsatisfactory, we shall have to rely at present on myelin analyses for our knowledge. Further evidence on the chemical nature of the plasmalemma must be derived from more indirect studies. This also involves the question of the so-called "surface coats" because if coats are present as part of the cell they represent the outermost part of the cell, and in consequence analyses of the plasmalemma alone would not represent chemical analyses of the surface; "surface coats" are discussed later in this chapter.

Enzymology of the cell surface

Apart from the elegant studies of Rothstein and his co-workers (see Rothstein, 1954) on yeast cell surface enzymes, very little work

has been carried out on this topic. With the development of histochemical techniques in electron microscopy, a few reports on surface-
located enzymes have appeared (Marchesi *et al.* 1964); de Thé *et al.*
(1963) found an ATPase on HeLa cells and described the presence of
nucleoside phosphatase on the surface of glial cells in the retina;
Epstein and Holt (1963) also located an ATPase-like enzyme on the
surfaces of cells infected with herpes virus. Kamat and Wallach
(1965), Wallach and Ullrey (1962), Emmelot and Bos (1962) and
Turkington (1962) all found ATPase in cell fractions which they
suggest contain surface membrane. A number of reports have
suggested a surface location for one enzyme or another without
providing any clear evidence for this point of view.

Antigenicity

Immunological reactions ought to provide a precise means of
recognising the nature of at least part of the surface components, and
of discovering similarities or disclosing differences between the
surfaces of different cells. If it were conclusively shown that a certain
antigen occurred only on the surface, then its recognition by
immune reactions would provide an excellent means of identifying
plasmalemma during extraction procedures. A variety of immune
reactions have been used to identify surface antigens, but it is possible
to object to some of these techniques as not certainly locating the
antigen on the surface. Immune fluorescent staining of unfixed cells,
whether carried out with a fluorescent antibody conjugate against
the antigen, or by treatment of the cells with an antibody followed
by a fluorescent antiglobulin conjugate, does not locate the antigen
site precisely, because the precise position of the dye uptake cannot
be resolved down to the plasmalemma level. Möller (1961) carried
out ingenious experiments and recognised a "ring" staining of cells in
which a given antibody showed specificity for one cell type (H_2. see
later, page 14). This reaction appeared as a ring of staining around
the periphery of the cell. Since lipid solvent treatment before the
antibody was added led to a general diffuse unspecific staining of
the cell, Möller suggested that the ring reaction was a surface staining
which only occurred on intact cells. It seems to me that the claims
of this technique are very probably correct but that they are not yet

fully established. This test cannot of course rule out the occurrence of the same antigens both within the cytoplasm and on the membrane, for the membrane may prevent entry of the conjugate into the intact cell, or into its deeper levels. Hiramoto *et al.* (1960) described findings which are similar to Möller's for the HeLa cells.

Two main classes of test almost certainly reveal surface antigens. These are the agglutination tests including the immune adherence test and the anti-globulin (Coomb's) test, which are presumed to act by antigen on the cell surface combining with the antibody (and in the latter reaction with a second antibody to the first) to form chemical bonding links between cell and substrate or cell and cell, since the antibodies are bivalent.

Similarly other agglutination reactions are widely thought to act by combination of an antibody with a surface antigen. Those reactions in which lysis of the cell occurs (see Chapter 2), perhaps due to a resulting instability of the plasmalemma, or in which cytotoxicity results, may very well be due to action at the cell surface but proof is lacking.

It is of course possible that immune adherence and the antiglobulin reaction etc. are due to the combination of antibody with sites deep in the plasmalemma, e.g. in pinocytotic pits, building up and projecting beyond the surface. Some types of antibody molecules are large enough to penetrate some hundreds of angstrom units beneath the plasmalemma and yet leave their "tails" at the surface. Antibodies might just conceivably lyse holes in the membrane by a skeletonisation reaction (see Sobotka, 1956) before they reached the antigen. However, two observations argue against this interpretation. First, some blood group antigens, e.g. the B_2 antigen in the brown trout (Sanders and Wright, 1962), or the incomplete Rh group, are only revealed as antigenic after the cells have been treated with papain or trypsin. These antigens would thus seem to be embedded in the cell. Consequently those antigens which react without pretreatment would seem to be on the surface. Second, Coombs *et al.* (1951) found an inagglutinable (by the antiglobulin test) class of cell which became agglutinable if alternate layers of globulin and antiglobulin were built up on the cells. They suggested that antigen sites in this class of cell were inside or beneath the cell surface and although accessible to the antibody could not participate in the agglutination reaction. If this is so, cells which react directly in the simple test would seem to bear antigens on the surface. However, the finding (Kelus *et al.* 1959) that ficin treatment of HeLa cells considerably enhanced their reactivity in a mixed agglutination test suggests that much of the antigen is not exposed on the surface.

Electron microscopical evidence on the location of antigens has been provided by Morgan *et al.* (1961) and by Easton *et al.* (1962). Easton and his co-workers conjugated antibodies with ferritin which is electron dense. They found that after thirty minutes' incubation, mammalian cells showed antibody molecules in large numbers apparently located on the surface. After further incubation the antibody entered the cell by pinocytosis. Similarly Morgan *et al.* used ferritin conjugated antibody to reveal the location of influenza virus, which can be considered as an antigen, at the cell surface. It would be of great value if further studies were made.

Since few developmental studies of surface antigens have been made, we shall have to consider data on the composition of the cell surface derived from studies of adults.

However, before discussing the nature of surface antigens it should be pointed out that some antigens, though located at times on the surface, are so loosely and reversibly bound to the surface that they can hardly be considered as typical components of a given cell's plasmalemma. The Lewis blood group substances in man (Sneath and Sneath, 1959) and the J antigen in cattle appear to be reversibly adsorbed by erythrocytes. Pereira and Fiqueirdo (1958) found that a variety of adenovirus antigens were attached to the surface. In addition Kodani (1962) found that human amnion cells in culture could adsorb blood group substances from the medium. Hamburger *et al.* (1963) and Anderson and Walford (1960) found that various serum proteins could become very firmly attached to cells and affect their antigenic properties. Neter (1963) described the adsorption of antigens and antigen/antibody complexes to the cell surface.

More is known of the chemistry of the blood group substances (located on the surface by a variety of immunological techniques) than other metazoan surface antigens, though of course the bacterial surface antigens are perhaps still better known. Morgan and Watkins (1959) and Morgan (1963) have summarised our knowledge of mammalian, in particular human, blood group substances. They are mucopolysaccharides of very variable molecular weight, a range of from 2×10^5 to 3×10^6 being quoted, but apparently substances of identical blood group type can have a range of molecular weights. They all contain L-fucose, D-galactose, N-acetylglucosamine and N-acetylneuraminic acid (sialic acid). The immunological specificities of human blood groups are carried by the following structures.

SPECIFIC GROUPS

Blood
Group

A   O-α-D-N-acetylgalactosaminoyl-(1>3)-O-β-D-galactosyl
     -(1>4)-N-acetyl-D-glucosaminoyl
B   3-O-α-D-galactosyl-(1>3)-D-galactosyl.
H   O-L-fucosyl and O-β-D-N-acetylglucosaminoyl
Le^a 3-O-β-D-galactosyl⟍N-acetyl-glucosaminoyl
        4-O-α-L-fucosyl⟋

The aminoacid moiety is resistant to trypsin but not to papain. Klenk and Uhlenbruck (1960) obtained evidence that the related M and N group substances on the surface of erythrocytes were susceptible to destruction by receptor destroying enzyme (RDEase or neuraminidase), and thus that their specificity depended on the presence of sialic (N-acetylneuraminic acid). Similarly Dodd *et al.* (1963) found that crude N-acetylneuraminic acid preparations inhibit the activity of anti-Rhesus$_0$D antibody, which suggests that the Rh$_0$D blood group either contains or is related in structure to N-acetylneuraminic acid. It is instructive to compare this work with that of Grubb (1955), who found that there were probably 2000 Rh$_0$D receptors on each red cell. But using electrophoretic methods Cook, Heard and Seaman (1960, 1961) showed that 10^{10} red cells release from the electrophoretic surface (see Chapter 2) 40 microgram of N-acetylneuraminic acid which would correspond to c. 10^7 sites per cell, if there is one N-acetylneuraminic acid molecule per Rh_0D substance molecule. This comparison suggests that much of the acid is not located in Rh_0D substance on the surface. As will be seen later, although much of the N-acetyl neuraminic acid may be extracted from the inner regions of the plasmalemma or even from the inner cytoplasm, so much is detected on the surface by electrophoretic methods that it is surprising that immunochemical tests have not provided evidence of greater quantities of it. It is also surprising that electrophoretic methods have not provided evidence of the other blood group substance sugars, but is this for want of trying?

Blood group substances have been chiefly studied on erythrocytes and in secretions, but are they present on other cells? Human A and B antigens have been detected on the surface of leucocytes (Glynn and Holborow, 1959). The same authors used fluorescent conjugates to study the distribution of the various blood groups in various human tissues. They found that A and B were present in or on most tissues but that the H substance was absent from many non-epi- or endo-thelial tissues. Kelus et al. (1959) used mixed agglutination tests to show that ficin-treated HeLa cells carried the H, Mn and J^n antigens, but lacked A, B, Rh, CD. The original patient, source of HeLa cells, was O,Rh+. Szulman (1964) using immunofluorescent staining investigated the distribution of A,B and H antigens on the surface of human embryonic cells. He found that they were initially present on all epithelia and endothelia, except liver adrenal and nerve cells. At birth they were absent from all tissues except blood cells, and stratified endo- and epithelia which retain them.

The histocompatibility antigens, (H_2) of mice are the next best known and identified surface antigens of cells although their chemical nature is at the time of writing little known. Möller (1961) and Möller and Möller (1962), using a most elegant fluorescent conjugate staining method and cytotoxicity tests, concluded that part at least of the antigen system substances were located on or in the surface, although Basch and Stetson (1962) detected these antigens intracellularly. Davies and Hutchison (1961) detected variations in H_2 antigenic activity from tissue to tissue in C57B1/6 mice.

A large amount of preliminary work has been done on the surface location of other unnamed or classified antigens. This has mostly been done in the attempt to develop methods of specifically identifying a given cell type. Hiramoto *et al.* (1960) using fluorescent conjugates detected surface antigens on HeLa cells. Ben-Or and Doljanski (1960) used single cell suspensions from C57BL mice and showed that there was some degree of specificity of tissue and species of cellular antigens but no attempt to locate the site of the antigen was carried out. Spiegel (1954) demonstrated a similar species specificity of antibodies which were cytotoxic to only one species of sponge cell (incidentally these antibodies prevented, rather obviously, cell aggregation in the appropriate species). Brand and Syverton (1960) using hemagglutination reactions, which suggest a surface location of antigen, showed that human EE and ERK-1 cells had components in common with rhesus monkey erythrocytes but not with rhesus kidney cells. Kite and Merchant (1961) located antigens on the surface of the mouse L fibroblast using agglutination and complement fixation reactions and showed that these cells had the same antigens as mouse LL(-M) cells but there was some difference from human epithelial cells. Periodate treatment, which affects carbohydrates, destroyed the antigenicity of the cells, so did lipase. These studies suggest that there really are surface antigenic differences between the different cells of the same species, implying differences in surface composition due to differentiation.

Few studies of the developmental appearance and differentiation of surface antigens have yet been made. Möller (1963) studied the H_2 isoantigens of newborn and embryonic mice from the 13th day of embryonic life. In 13–14 day old embryos some antigenicity was present in or on liver cells when transplantation survival tests were carried out, but whereas H^b, unlike H_2^a antigens, could be detected by cytotoxicity tests from the day of birth, the H_2^a antigen could not be detected till some days after birth. Möller points out that these results may either indicate the appearance of different antigens at different times or differing sensitivities of the tests at various ages. Saison and Ingram (1962) reported that the A and O blood group antigens in pigs were not adsorbed to the red cells till 6 or 7 days after birth, though present in the blood all the time. Owen (1962) found that in the rat the C and E blood group substances appeared on the 10th day of foetal life, the D antigen at 16–18 days, but the A only 32 days postnatally. Szulman (1964) made the most interesting observation that the A,B and M antigens in human embryos are initially present in the surface of all epithelial and endothelial cells except the liver, adrenal and nerves, but before birth are lost from all except stratified epithelia and endothelia. Nace (1963) reported an antigen affecting cell survival in protein containing media which might possibly be present on the surface of frog neurula cells.

Electrophoretic evidence

The presence of charged groups at the cell surface causes the cells to move in an electric field. Electrophoresis (see Chapter 2) allows measurement of the number of surface charges per cell. Study of the surface charges at various pH and ionic conditions and before or after treatment with enzymes and other specific reagents which inactivate charged groups, has been used to characterise the chemical nature of these charged groups on the cell surface. It is theoretically possible to continue such investigations by producing charges on normally uncharged groups of the surface by specific reactions.

The electrokinetic properties of the surface of a cell vary with pH and ionic strength. The surface density and type of charged groups, the adsorption of small ions to the surface, changes in the thickness of the double layer, adsorption of macromolecules, elution of membrane components and rearrangement of surface components may all vary with pH and ionic strength and will affect the electrophoretic mobility of a cell. Consequently the behaviour of charge with pH and ionic strength affords only a crude means of identifying the nature of the charged groups.

Specific ion adsorption if present will suggest the presence of certain components on the surface, or its absence will indicate other features of the surface. This adsorption of ions in the Stern layer will, if the ions are cations and if the concentration of cations is increased, lead to suppression of negative charge on the surface (zero point of charge), and later to the appearance of positive charge on the surface (reversal of charge). The concentration of a given ion required to bring about charge reversal decreases with increasing valency of the ion, but also within a valency group the exact order depends upon the specific chemistry of the surface charged groups. This order is termed the *reversal of charge spectrum*, and varies in its details according to whether the surface carries carboxyl, phosphate or sulphate groups (see Douglas and Parker, 1957; Douglas and Shaw, 1958). Hence the reversal of charge spectrum provides a rough means of identifying the surface groups. The most specific analytical method is provided by treating the surface with specific reagents, enzymes etc. which will either destroy, unmask or create charged groupings whose presence or absence can be determined electrophoretically.

Bangham and Pethica (1961) published pH/mobility (i.e. mobility as a function of charge) curves for a variety of cell types. The erythrocyte like all other cells so far investigated has a negative charge at physiological pH values. Some cell types, e.g. the polymorphonuclear leucocyte and the erythrocyte, lack a true isoelectric point (Bangham and Pethica, 1958). The erythrocyte does not exhibit an isoelectric point down to pH $1 \cdot 8$, but many other cells, e.g. liver (Bangham and Pethica, 1961), including bacteria, show true isoelectric points between pH $3 \cdot 0$ and $4 \cdot 5$; but a few cells, e.g. lymphocytes, have lower isoelectric points down to pH $2 \cdot 2$. The form of the pH-mobility curve can be used roughly to estimate the pK values of surface components but cannot be used as a precise analytical test. Carboxyl groups have a pK of *c*. $2 \cdot 0$ to $4 \cdot 8$ and hence surfaces fall within the carboxyl range, but values of $3 \cdot 5$ to $4 \cdot 0$ are more likely for most carboxyl compounds. Mouse liver cells have a pH-mobility curve which levels off near pH $4 \cdot 0$, suggesting a pK of c. $4 \cdot 0$ and the presence of carboxyl groups on the surface. Bangham *et al.* (1962) found that the Ehrlich ascites cell had an isoelectric point between $3 \cdot 0$ and $4 \cdot 0$ but the isoelectric point of the Klein lymphosarcoma 537 was between $3 \cdot 0$ and $3 \cdot 5$.

If amine groups are present on the surface it would be expected that a true isoelectric point would be found and that the pH-mobility curve would show almost no change in mobility from c. pH $5 \cdot 5$ to $10 \cdot 0$ (Bangham and Pethica, 1961). This is found (see above) for many cells but not for erythrocytes. Further confirmation for the absence of positive charged groups on the erythrocyte has been obtained by two techniques. Heard and Seaman (1960) showed that halide and thiocyanate anions are equivalently adsorbed onto erythrocytes over a wide range of concentrations, which argues that there are few or no positive charges on the surface. However somewhat contradictory results were obtained with phosphate ions. Bangham *et al.* (1962) observed different adsorption of I^- and Cl^- to the S 37 tumour cell, which suggests the presence of positive charges on the surface. Heard and Seaman (1961) treated erythrocytes with acetaldehyde and found that this treatment had little effect on the surface charge, since acetaldehyde would react with $-NH_2$ groups they concluded that few positive charges could be present on the erythrocyte surface. Eisenberg *et al.* (1962) found that the charge density of liver cells rose on heating them above 60°C.

The cation charge reversal spectra are perhaps of greater analytical value. Bangham and Pethica (1958) compared spectra found for cells with those for surfaces of known composition, and suggested that sheep lymphocytes have surfaces similar to lecithin. Bangham and Pethica (1961) continued this work using mouse lymphocytes and Ehrlich ascites cells for which they suggested that a cephalin-like phosphatide provides the charge. Uranyl ions are particularly effective in reversing charge on these cells which further suggests that phosphate groups are borne on the surface. Against this interpretation it is possible that the charged groups involved in electrophoresis may lie inside the membrane at least to some degree so that a protein layer might lie outside the lipid all the same. Sheep polymorphonuclear leucocytes require a higher concentration of uranyl ions to bring about charge reversal than lymphocytes do, and this fact and their general charge reversal spectrum suggests that they have carboxyl surfaces (Wilkins *et al.*, 1962a,b).

Unfortunately the use of charge reversal spectra for identification of the surface groupings can be criticised on the following grounds. First, model systems are insufficiently investigated at present for precise comparisons to be made with cell surfaces, if indeed these are possible. Second, the presence of low concentrations of ions such as phosphate or protein around the cells may raise the concentration of particular cations required for charge reversal.

The use of specific reagents including enzymes to study surfaces composition is well illustrated by the work of Haydon and Seaman

(1962). They investigated how much of the charge on erythrocytes and *Escherichia coli* cells was due to structural components of the surface, e.g. carboxyl groups and how much was due to adsorption or desorption of electrolyte ions. They also considered the possibility that the surface might have an appreciable depth of charged groups, all or some of which would be detected by electrophoretic measurements. To do this they titrated the cells with the cation, methylene blue, and measured the resulting change in surface charge electrophoretically at various ionic strengths. It was thought that methylene blue binds specifically with carboxyl groups, though there was a slight uncertainty about the stoichiometry of the reaction. Plots of reduction in surface charge due to methylene blue binding against increasing concentration of methylene blue for various electrolyte concentrations gave a series of curves meeting at a point representing infinite cation concentration at which the binding sites would be saturated. The reduction in charge at that point represented that part of the total surface charge due to carboxyl groups. Since previous work with the erythrocyte suggested that the only class of charged groups on the erythrocyte surface was carboxyl groups, they could calculate that part of the charge under any given electrolyte conditions which was due to electrolyte adsorption or desorption by simple subtraction. Since the same number of binding sites were found over a wide range of electrolyte concentrations the density, i.e. concentration of carboxyl sites, of the material of the outer part of the membrane is probably constant over the whole of the outer 20 Å of the membrane. This was confirmed by making measurements at different ionic strengths so that the plane of the zeta potential lies at varying distances from the non-conducting lipid part of the membrane. This exceedingly careful work, which ought to be copied for other cell types, was carried out by Haydon and Seaman after they and their co-workers had investigated the effect of trypsin on the erythrocyte surface.

Trypsinisation of the erythrocyte usually lowers the surface charge density, though chick red cells have their charge density raised by this treatment (Seaman and Uhlenbruck, 1962). The electrophoretic mobility of the human erythrocyte is temperature independent which suggests that ion adsorption phenomena do not bulk large in accounting for changes in charge density due to trypsin treatment. Heard and Seaman (1960) and Seaman and Heard (1960) investigated the effect of trypsin on the surface charge of the ery-throcyte with great care. Although trypsin is strongly positively charged at pH 7 there is no evidence that its adsorption on the cell is responsible for the lowering in charge density, because there is no evidence that a specific anion binding on to the trypsin takes place as a result. Moreover di-isopropylfluorophosphate (DFP) inactivated trypsin does not affect surface charge, whereas if it acted by simple adsorption it would still be expected to affect charge. Further evidence against such a possibility is derived from the observation that the "chemical peptidase", N-bromosuccinimide produces the same reduction in charge. Although trypsin might act by allowing reorientation of the surface structure without loss of ionogenic material, Seaman and Heard consider that this is unlikely, because crosslinking of a type unbreakable by trypsin (e.g. S—S bonds) would impede penetration of trypsin far into the surface. Never-theless it is known that trypsin will enter tissue cells, possibly of course by pinocytosis, but Heard, Seaman and Simon-Reuss (1961) found no effect of trypsin on the surface charge of tissue cells. Seaman and Heard obtained no evidence that phosphate was responsible for any of the charge on the erythrocyte.

Cook, Heard and Seaman (1960) were able to detect a sialo-mucopeptide (sialic acid is more correctly known as N-acetyl-neuraminic acid) in the supernatant after red cells had been treated with trypsin. Later Cook et al. (1961) treated red cells with neur-aminidase, an enzyme which liberates acylated neuraminic acids by cleavage of an α-glycosidic linkage. They showed that N-acetyl-neuraminic acid was released from the cells by this treatment and that the release was accompanied by a fall in surface charge density. The reduction in charge density of c. 2500 esu.cm^2 found would be brought about by the removal of some 40 microgram of N-acetyl-neuraminic acid per 10^{10} cells. In fact a yield of twice as much acid was obtained. The authors were able as a result of previous work (see above) to dismiss the possibility that the changes in surface charge density could be due to the unmasking of cationic groups or to the adsorption of the enzyme or impurities. This large yield of N-acetylneuraminic acid might be explained on the grounds that the use of the Gouy-Chapman equation for calculating surface charge ignores those charged groups which lack counter ions owing to the space occupied by components of the surface structure in a penetrable surface (see Haydon, 1961). Cook, Heard and Seaman considered that the discrepancy was too large to be accounted for entirely by Haydon's correction. They suggested that some of the acid must be liberated from portions of the structure deep to the plane of electro-kinetic shear. Rather surprisingly Haydon and Seaman (1962) found that neuraminidase treatment of the erythrocyte did not affect the uptake of methylene blue, and this suggests that the surface carboxyl groups are mainly not due to N-acetylneuraminic acid. Wallach and Eylar (1961) measured the amount of N-acetylneuraminic acid released from Ehrlich ascites cells by trypsin and found that it was five times greater than that which could be accounted for by the reduction in charge density; they considered that much of the acid must be located inside the cytoplasm. Haydon and Seaman were able to measure the average density of carboxyl groups on the human erythrocyte surface and found that they were about 50 Å apart. If this is the distribution, the ionic atmospheres of each group will not overlap at appreciable ionic strengths (greater than 0·04), and in consequence rigorous use of the usual electrophoretic equations will be precluded.

This type of work has unfortunately been almost confined to the erythrocyte and various bacteria. At the time of writing work on protozoan or metazoan cells other than red blood cells is of pre-liminary nature. I have already described work using pH-mobility curves and charge reversal spectra. Ruhenstroth-Bauer et al. (1962) treated HeLa and mouse Ehrlich ascites cells as well as other tumour cells with neuraminidase. This treatment reduces the surface charge density. Forrester et al. (1962) used hamster fibroblasts and their polyoma transformants; on treatment with neuraminidase both types of cells lost many of their surface charges, but the tumour (polyoma) type still had a higher charge density than the normal cell type. They suggested that some 20 per cent of the charge was due to phosphate in view of the effects of calcium in reducing charge, but it is also possible that the calcium was chelated with carboxyl groups. These results suggest in a preliminary manner that much of the charge of mammalian tissue cells may be due to N-acetyl-neuraminic acid, but the results of Wallach and Eylar raise some doubts, as does work by Cook *et al.* (1962) on the Ehrlich ascites cell which showed that more sialic acid was released by trypsinisation than could be accounted for by changes in surface charge density, and that this was partly due at least to unexplained happenings during the washing of the cells. Simon-Reuss *et al.* (1964) found no correlation between the amount of sialic acid released by neuraminidase and the surface charge of various mammalian cells. It is of considerable interest that Gasic and Baydak (1962) found that adsorption of muco-polysaccharides from the medium on to cells could easily occur. Vassar (1963) found that neuraminidase reduced surface charge of a variety of human cells.

More extensive work on bacteria (see Davies *et al.* 1956; Adams and Rideal, 1959; Haydon and Seaman, 1962) has shown that amino-sugars may be responsible for the charge on some bacteria, and that differential anion adsorption occurs in others, suggesting the presence of positive charges. James *et al.* (1963) have treated bacteria with specific amino-acid decarboxylases to identify the contribution of various groups to the charge. This work on the erythrocyte and on bacteria indicates the care which should be applied to the interpretation of electrokinetic data and which has not so far characterised work on tissue cells. It is essential to be able to evaluate the degree of specific ion adsorption, the charge density at various levels in the surface, the degree of folding of the surface which affects measured charge, the possibility of surface rearrangement and the presence of charges of both signs.

There remains one final problem. If the cell surface is a polar-isable electrode (see Chapter 3) these investigations of the surface would be more suspect because metabolic processes would produce current flows which would charge or discharge the surface with accompanying changes in electrophoretic behaviour.

DEFINITION OF A SURFACE STRUCTURE

At this point we cannot attempt to integrate information on the chemical and structural nature of cell surfaces. There is much

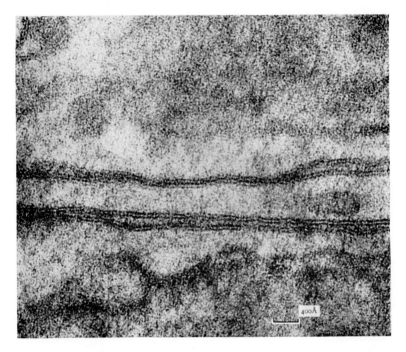

1.1 The plasmalemma in section. 4 bimolecular lipid leaflets can be seen in this electron micrograph of Schwann cells. Permanganate fixation. The upper two plasmalemmae are separated by a gap of approximately 100 Å width, whereas the lower two are in close contact in the middle of the picture. By courtesy of Dr. E. G. Gray.

1.2 The plasmalemma in the surface view. Carbon replica of two cells; they join on the right-hand side of the picture. Irregular bumps are probably mitochondria underneath the membrane. Fine roughness probably represents protein molecules adsorbed from the medium. By courtesy of Dr. G. C. Easty.

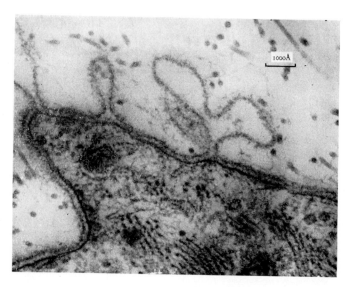

1.3 Basement membranes in atrophic muscle. A rather tenuous basement membrane is thrown into folds and stands some 200 Å off the plasmalemma. Osmic fixation. By courtesy of Professor R. Miledi.

evidence that the plasmalemma is the effective outer covering of the cell, both from morphological studies and from techniques such as immunology and electrophoresis. Permeation studies and electrical properties (see Chapter 2 for a very brief synopsis) also suggest the plasmalemma as the true outer surface of the cell. In future chapters further confirmation of this point of view for phenomena such as adhesion will be presented. Evidence of a layer outside the plasmalemma which is integrally part of the surface is tenuous.

However as will be appreciated we cannot talk of the cell surface in the way in which a physical chemist would talk of one. Not only does the plasmalemma have depth but even the outer surface of the plasmalemma appears to demonstrate a "depth" in reactions and phenomena which are classically thought of as true-surface phenomena. Electrophoretic observations show that the charged groups on the surface of at least one cell type are not arranged in a plane but rather are arranged in a layer at least 20 Å thick. Of course even the most "ideal" surfaces in practice have a certain depth in which various phenomena occur at various levels but the surface of the plasmalemma appears to be a much deeper structure than is normally considered in surface studies.

It is surprising how little reliable evidence is available on the chemical composition of the plasma membrane. The preceding sections have provided an outline of possible composition. There is fairly good evidence that lipid is present in the plasmalemma, but the evidence for protein is much less satisfactory. The elegant work on myelin described later, and the similarities in the electron microscope image of myelin and other plasmalemmae and various artificial lipid containing systems, are good evidence that the cell surface contains lipid. But we cannot define which lipids are present with certainty even in myelin. The low capacitance of the membrane (see Chapter 2) suggests that lipid is present but, as Kirkwood (1955) has shown, this could also be due to protein. No evidence as to the nature of the lipid in other plasmalemmae is available at present but we can say that there are probably differences between cells of different types in view of differences in electrical resistance and in the types of mucopolysaccharides on the cell surface.

Despite the confident statement in many papers that the mem-

B

brane bears protein monolayers on one of or both of its surfaces, there is at present no evidence other than the demonstration of a few enzymes, the somewhat questionable identification of protein components of myelin, and the electron microscopy of protein monolayers on artificial lipid membranes, that protein is present on the cell surface. This may seem surprising, but neither electrophoretic, immunological nor staining studies have confirmed the presence of protein. On the other hand, there are several simple techniques such as electrophoresis after specific enzyme attack of proteins (rather than peptide chains in mucopolysaccharides) that might be tried. However, it is much more certain that various mucopolysaccharide compounds are present on the surface, appearing in both electrophoretic and immunochemical tests. Both these suggest the widespread occurrence of the amino-sugar N-acetyl-neuraminic acid on the surface of many cells. Electrophoretic and immunological investigations also indicate that the surface chemistry varies widely from cell type to cell type. But we are unable to state what groupings are responsible for these differences.

The next problem is to translate the composition of the surface into its structure. Obviously lipids will tend to form association colloids because of their low dielectric constants, but a considerable number of structures are possible (see Lucy and Glauert, 1964). Harvey and Danielli (1938) suggested that the plasmalemma was a bimolecular lipid leaflet in which the polar ends of certain lipids limited the vertical extent of the structure to two layers. However the inclusion of polar lipids allows a variety of micellar structures to be formed. In the next section we shall examine which structures might be found in plasmalemmae. It seems very likely that other components of the surface, in particular amino-sugars, may affect surface structure, but nothing appears to be known about this. Again we do not really know whether different cells have plasmalemmae of different lipid composition and whether this would affect structure. However it will be possible to develop the idea that plasmalemmae contain lipids in order to explain their structure, and in turn to account for the mechanism of cell adhesion and more tentatively for details of cell behaviour. Finally it may be possible to link morphogenetic movements to the lipid composition of the surfaces of the cells involved, through features of cell behaviour. A somewhat similar though still more incomplete story will be developed for the sialic acid components of the surface. Unfortunately we cannot yet even attempt similar treatments for details of the exact lipids or other components of the plasmalemmae involved.

THE STRUCTURE OF THE PLASMALEMMA

Evidence from electron microscopy and X-ray diffraction studies

CRITIQUE OF TECHNIQUES OF ELECTRON MICROSCOPY

At present our interpretation of plasmalemma structure depends largely on the results of electron microscopy and to a lesser extent on those of X-ray diffraction and optical microscopy. Unfortunately too few studies of fixation and other preparative mechanisms have yet been made for us to be wholly sure about the interpretation of electron micrographs. We are faced with two separate problems: first, which chemical groupings show definite staining properties with electron dense stains so that the various components of the surface can be mapped to their correct positions; second, whether specimen preparation may not have lead to a rearrangement of structure, i.e. to a phase change in either the lipid or other surface components. During fixation with osmium tetroxide there is probably oxidation of surface groups, reduction of the fixative with microflocculation to give a degree of negative (resist) staining. Reduced fixative may migrate to flocculate in particular charge regions. Similar considerations probably apply in the case of other types of fixative or stain containing metal ions.

Before discussing the structure of the plasmalemma and considering the value of various types of evidence I shall briefly describe the standard appearance of the plasmalemma in section as viewed by electron microscopy. With fixation by osmium tetroxide single or double densely staining lines represent the plasmalemma (Plate I.1). The line or pairs of lines have been reported by different authors to be of different thicknesses (see Elbers, 1964 for a survey of reported values), both for one cell type and between cell types. When two staining lines appear they are usually each c. 20–30 Å thick and are separated by a 30–40 Å non-staining core. With per-manganate fixation a more constant picture is observed, Robertson (1958b) finding a pair of 20 Å thick dense lines separated by a 35 Å core, though Sjorstrand and Elfvin (1962) using polyester embedding described an asymmetrical structure composed of an outer 25 Å dense band and an inner 35 Å dense band separated by a 30 Å core. This discrepancy appears to be due to the different embedding media used. These electron micrographs of the plasmalemma apparently provide further support for the Harvey–Danielli model hypothesis that the plasmalemma is composed of a bimolecular lipid leaflet.

The question of what components of the surface are stained by permanganate or osmium tetroxide is still a little unclear. Rie mersma and Booij (1962) showed that OsO_4 binds to lecithin with a molar ratio of $1 \cdot 4$, which corresponds closely to the average of $0 \cdot 33$ double bonds per lecithin molecule (impure lecithin). This correspondence cannot however be taken as proof, as Elbers (1964) does, that osmium tetroxide reacts solely with the double bonds in lipids. Hayes *et al.* (1963) concluded that osmium tetroxide reacted mainly with the unsaturated lipids of serum lipoproteins, in view of the binding ratio. More conclusive evidence has been derived from the staining of systems of known composition. Mercer (1957) found that protein or lipid systems stained equally densely. Stockenius (1960) reported that uranyl linoleate shows lamellar banding in electron micro-graphs, and suggested that this appearance could only be due to uranium being attached to terminal carboxyl groups. Fixation of uranyl linoleate with osmium tetroxide did not alter the banding. He concluded that this result meant that OsO_4 could not be reacting with the double bonds in the centres of the linoleate molecules for, if it had, additional bands would have appeared in the centres of the linoleate lamellae. Similarly Trurnit and Schidlovsky (1961) found staining by OsO_4 of saturated lipids. Stockenius (1962, 1963) suggested that osmium collects around the hydrophilic regions of the phospholipid molecules in hexagonal phase soaps. Although these observations appear to suggest that osmium staining represents the hydrophilic ends of lipid molecules a note of caution should be introduced. Riemersma and Booij (1962) found that the reaction product of OsO_4 and lecithin could be split into a lecithin-diol and a lower osmium oxide: after this splitting the lecithin-diol could be restained. He concluded that OsO_4 reacts with double bonds to form an osmium oxide sol which then migrates and binds coulombically to a positive polar end group. However these results might as well be interpreted as proving the binding of OsO_4 reduced by other groupings located at the polar ends of the molecules. If OsO_4 had reacted with double bonds it is surprising that the lecithin could be restained. If however osmium staining can migrate within the plasmalemma interpretation of the structure becomes more difficult. Stockenius (1959) also observed that when protein monolayers of globin are adsorbed on to myelin figures then the thickness of the stained bands increases by some 7 Å from 18 to 25 Å. Does this increase represent reaction with a protein monolayer or increased thickness of charged groupings to adsorb osmium oxide sols?

Potassium permanganate reacts with a wide variety of organic molecules to produce colloidal MnO_2. Elbers (1964) suggests that the plasmalemma may be destabilised by permanganate fixation but the evidence for this conclusion seems a little unclear. Robertson (1960a,b) suggested that the protein, mucoprotein or mucopolysaccharide adsorbed on the lipid layers is of different character on the inner and outer surfaces of the plasma membrane because there are slight differences in staining.

Further evidence that osmium reacts and stains at the polar ends of the molecules can be derived from Finean's work (1959) on the saturated synthetic phospholipid, distearoyl phosphatidylserine. This shows an X-ray spacing of 63 Å before and after fixation with OsO_4 and after embedding, though the electron microscope gave a periodicity of c. 50 Å. In general it may be remarked that both in studies of synthetic lipids and in studies of myelin, given X-ray periodicities correspond to smaller periodicities as measured from electron micrographs. Elbers (1964) attributes this to drying effects in the electron microscope.

During the preparation of specimens for electron microscopy fixation is succeeded by dehydration and embedding in media of low ionic strength and very high viscosity. These changes may result in alterations in lipid structure. Luzzati and Husson (1962) examined a number of lipid-water systems by X-ray diffraction techniques. Lamellar stuctures were only found when the water content was less than 40 per cent. The phase diagram for human brain phospholipid showed that the transition boundary between hexagonal and lamellar phases lies close to the temperature and water content of lipids in living tissue. This work by Luzzati and Husson raises the possibility that irreversible structural changes of state may occur during fixation and the subsequent dehydration as specimens are prepared for embedding. However Stockenius (1962) showed that lipid and phospholipid phases prepared for electron microscopy at constant temperature showed the same structure in electron micrographs as could be deduced from X-ray diffraction data. He admits doubt as to whether the practice of using fixatives at 0°C may not have allowed the transformation of hexagonal to lamellar structures. Joy and Finean (1963) found that myelin shows irreversible structural changes on cooling below —3°C. Further evidence for the possibility of such changes arises from the findings by Elworthy and McIntosh (1964) that lecithin structure changes from hexagonal micelles to bimolecular leaflets as the dielectric constant of the immersion medium falls below 25. The dielectric constants of various dehydration media are below 25.

EVIDENCE FOR A BIMOLECULAR LEAFLET STRUCTURE

Is the plasmalemma in life based on a lamellar or a hexagonal phase lipid structure? Although a considerable variety of staining patterns of the plasmalemma of various cell types have been described (see Elbers, 1964) in general it can be said that the plasmalemma appears as one or more staining bands. When two or more bands appear the bands are parallel. Robertson (1958b, 1960a,b) gave the thickness of the plasmalemma as c. 75 Å, there being two outer dense lines c. 20 Å thick separated by a lightly staining core of 35 Å using $KMnO_4$ staining. It is not clear whether the differences in plasmalemma appearance found for various cell types are real or due to variable techniques. Robertson (1958b) suggested that plasmalemma structure is constant for all cell types, in which case the elucidation of plasmalemma structure from myelin (a description of which follows immediately) would be of general application.

The identification of plasmalemma structure has been approached with considerable success from the study of myelin. Geren(1954) showed that myelin was formed by a Schwann cell wrapping itself spirally around an axon. Since the Schwann cell puts on very many turns, a massive structure is built up, composed as Robertson (1958a, b) confirmed almost entirely of cell surface (Figure 2). This massive

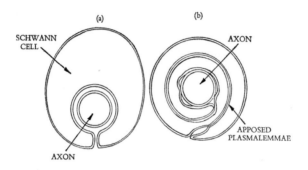

Figure 2. Formation of myelin from a Schwann cell. (*a*) Schwann cell surrounds an axon *a* but its plasmalemma (double line) is separated from the axon by a gap. (*b*) Schwann cell has now wound itself twice around the axon, forming two layers of double cell surfaces. At this stage Schwann cell cytoplasm still lies between the apposed plasmalemmae, but later the cyto-plasm is withdrawn and the apposed cell surfaces come together to form myelin. Sectional view.

structure is suitable for structural studies by polarisation and X-ray diffraction techniques and provides a source of material for analysis of surface components. Consequently myelin is an ideal material in which to correlate chemical, electron microscopical, optical and X-ray diffraction findings. These correlations have been made in particular by Robertson (1958b, 1959) and Finean (1956, 1957). The interpretations themselves are probably based on Harvey and Danielli's (1938) suggestions that the plasmalemma has a bimolec-ular lipid leaflet as its main component. Examining this line of work in greater detail, we find that Bear and Schmitt (1937) suggested from the results of polarisation microscopy that myelin has a radial repeat of c. 180 Å in whole live nerve, and this received further confirmation from measurements by Schmitt *et al.* (1941) using X-ray diffraction techniques. Finean (1956, 1957) was able to repeat

these measurements and suggested that the 180 Å spacing corresponds to the 120-140 Å repeat found in dried or osmic fixed myelin.
A 120 Å repeat was found in osmic fixed myelin by electron microscopy (Fernandez-Moran and Finean, 1957), with alternate "heavy"
and "thin" lines appearing at a c. 60 Å spacing: this result must
incidentally be regarded as one of the earliest pieces of evidence
showing that electron microscopy does visualise true rather than
artefactual structure. Fernandez-Moran (1957) centrifuged myelin at
140,000g and was then able to obtain separation of the lines of osmic
staining into double lines; this observation may possibly in part be
evidence for the existence of a bimolecular lipid leaflet within the
repeat period. But his earlier work (Fernandez-Moran, 1954) in
which he had made X-ray diffraction studies of artificial lipid
bimolecular leaflets, had shown that the dimensions of their repeat
is such that the 120 Å of the myelin repeat should contain four lipid
layers in its thickness and some additional material. The association
of lipid molecules is of such a type that the 120 Å would contain two
bimolecular lipid leaflets; this suggestion had already been made by
Schmitt et al. (1941). However there would have to be a layer or
layers of additional material c. 20 Å to provide the full repeat period.
Unfortunately it has been impossible to derive a full molecular
structure for the plasmalemma by X-ray diffraction, although
Finean (1957) located the position of the phosphate groups to be
60 Å apart radially in this way. For this reason the next two steps of
the interpretation of myelin structure and its identity with plasmalemmal structure have been obtained from electron microscopical and chemical evidence.

Geren (1954) had made the original suggestion that myelin was
formed from the plasmalemma of the Schwann cell surface, but
Robertson (1958a,b, 1959, 1960a,b) worked out the precise way in
which the plasmalemma was formed into myelin and related this to
earlier chemical and X-ray diffraction etc. evidence. Using high
resolution electron microscopy he showed that one repeat layer was
formed from two plasmalemmae opposed back to back, that is to
say with their inner surfaces pressed together. This structure forms
as the inner cytoplasm is withdrawn from between "inner" and
"outer" plasmalemmae of part of a Schwann cell, the "inner"

surface of course lying next to the myelin just formed in the previous "turn" of the Schwann cell round the axon (see Figure 2).

It can now be seen that the myelin repeat is composed of two plasmalemmae (back to back), the inner sides of the two plasma-lemmae fusing into a single line, and that the dimensions of repeat indicate the presence of four lipid layers; hence Robertson concluded that the plasmalemma was constructed of a bimolecular lipid leaflet, covered perhaps with protein or mucoprotein or muco-polysaccharide on either side. Further evidence in favour of this was obtained by Robertson from examination of the staining of myelin and other plasmalemmae. With suitable staining conditions two dense staining lines appear, sometimes at rather variable distances (see Sjorstrand and Elfvin, 1962) but roughly at a separation such that the staining must represent the hydrophilic ends of the mole-cules (see Stockenius, 1960), being twice the length apart of lipid molecules such as lecithin. This pattern of staining is interpreted to indicate the presence of a bimolecular lipid leaflet which previous physicochemical studies (see Harvey and Danielli, 1938) had indicated must have its polar groups arranged at the outside of the leaflet.

Attempts to work out the chemical structure of the myelin plasmalemma have been founded on three observational bases, namely bulk analyses of myelin, some X-ray diffraction data, and physicochemical data from models and the staining of myelin, and on a variety of theoretical bases. The reliability of myelin analyses has been discussed in earlier sections. Finean (1956, 1957) worked out a possible structure on the basis of analyses which showed cholesterol, phospholipid and cerebroside occurring in the ratios 2:2:1. X-ray diffraction data indicated the position of phosphate groups: this model is shown in Figure 3. It should be noted that the phospholipid molecule would have to have a curved end in order to fit into the structure, perhaps associating with cholesterol in this position. Later evidence (Katzman and Wilson, 1961) on the extrac-tibility of lipids from myelin agrees with this model. This model has been criticised by Vandenheuvel (1963) on the grounds that Finean's model does not produce the closeness of molecular packing that would result from intermolecular attraction. Since Vanden-heuvel's paper is wholly theoretical it will be discussed later. It is possible as will appear shortly that these same lipids might be packed in such a way as to give other lipid phases. Despite the great elegance of the work of Finean, Fernandez-Moran and Robertson on plasmalemmal structure it must be remembered that the structure of other plasmalemmae may be different and that no comparable analysis has yet been carried out for them.

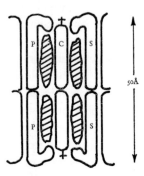

Figure 3.   Finean's model of cell surface structure: lipid components only. P is lecithin and S sphingomyelin. Cholesterol is cross-hatched. Choline phosphate is shown by cross on cerebroside C. Bimolecular leaflet structure. (Surface shown in section.)

EVIDENCE FOR MICELLAR STRUCTURES IN THE PLASMALEMMA

The possibility that the true nature of lipid in some or all plasmalemmae is a hexagonal phase structure has already been mentioned. A number of observations of plasmalemmal structure have been made by electron microscopy which depict the globular or hexagonal micelles. Before describing these observations it should be pointed out that apparent globular structures may result from the overlap of structures of other shapes (overlap artefact). Klug et al. (1964) have pointed out this error of interpretation and shown how to overcome it (see also Robertson, 1966). Dourmashkin et al. (1962) described a hexagonal array of 70–90 Å diameter pits in chick liver cell plasmalemmae, after saponin treatment. Bangham and Horne (1962) and Glauert et al. (1962) criticised this description on the grounds that the structure might be an artefact due to the action of saponin. Sjorstrand (1963a,b,c) found minute globular components, some 50 Å in diameter, in the plasmalemmae and mitochondrial membranes of mouse kidney cells (see also Sjorstrand and Elfvin,

1964). Fernandez-Moran (1957) observed a rather similar structure in myelin. Sjorstrand proposed that the membrane might consist of globular lipids with protein molecules between them, which may agree with another finding by Fernandez-Moran (1957) namely that trypsinisation of myelin produced lipid leaflets c. 100 Å long by 60 Å deep by 60 Å broad. Gent et al. (1964) extracted a particle 60 Å long by 60 Å by 40 Å by treating myelin with lysolecithin. Robertson (1963b) described a repeat period of c. 95 Å in a hexagonal array of stained lines in the plasma membrane of the synaptic junctions of the Mauthner cells in the goldfish. Somewhat similar structures c. 50 Å–250 Å have been described by Fernandez-Moran (1954, 1961) and by Lasansky and De Robertis (1960) in retinal rod outer segments. Roots and Johnston (1964) were unable to visualise any bilammellar plasmalemma in electron micrographs of ox brain cells: they suggest that the surface may have a lipid crystalline structure. Di Stefano (1966) described a unit structure roughly $75 \times 75 \times 112$ Å in the plasmalemma of chick fibroblasts, but his electron micrographs were not checked according to the criteria of Klug et al. (1964). Robertson (1966), finding no evidence for a globular structure from X-ray diffraction studies of plasmalemmar stacks in retinal rod outer segments, investigated whether overlap artefacts are responsible for the appearance of globular staining patterns in electron micrographs for plasmalemmae. Permanganate staining forms small granules of stain in the plasmalemma. Robertson argues that since sections for electron microscopy are 500 Å thick, considerable overlap of the images of these granules will occur if the section is slightly tilted. Probably the best evidence yet for the existence of a globular structure in the plasma membrane has come from the work of Blasie et al. (1965). Not only did they find evidence for a globular structure from electron micrographs of the outer segment membranes of the retinal receptors, but low angle X-ray diffraction patterns of partially dried isolated outer segment membranes gave evidence of a globular structure of 40 Å diameter.

OTHER CONSIDERATIONS

Light microscopy has made a small contribution to this subject. Barer (1956) using phase microscopy with immersion media of various refractive indices estimated that the refractive index of the cell surface lay between 1·36 and 1·38 in a variety of cells. Ambrose's technique of contact microscopy (1956) was used by him to measure the refractive index of cell surface of embryonic chick heart fibroblasts (1961). Curtis (1964) used an interferometric technique to investigate the effect of osmic fixation on the refractive index and found that it raised a live value of 1·370 to 1·371; this is not a significant change, but 3M NaCl and 3M sucrose (the latter unpublished) raise the refractive index considerably. These results suggest that in life the membrane contains little water (a result perhaps favouring the lamellar structure) but that it can be permeated by sucrose and NaCl which may act by disrupting its structure. In view of these values for the refractive index of the cell surface it is surprising that artificial lipid membranes appear to have refractive indices of 1·66 (Thompson, 1964), though these very high values are suspect because they predict an improbably high lipid concentration (200 per cent) in the membrane and do not agree with Barer's observations, and were obtained by measurement of the Brewster angle, which is an uncertain technique.

Artificial cell membranes have been prepared by Setala et al. (1960) from cholesterol-gliadin and their surface appearance was studied directly after osmic fixation. These films resemble the surface replicas made by Easty and Mercer (1960) (see later). Mueller et al. (1962) prepared thin phospholipid membranes of phosphatidyl choline and other phospholipids. Reflection interference microscopy was used to measure the thickness of the thin stable film as 61 Å. The membrane had a specific resistance of 1×10^6 ohm.cm^2 but were fairly permeable to water. The refractive index was 1·66; this value is much higher than that for the cell surface, which together with the thickness of lipid (in cells c. 40 Å) raises the suggestion that the model and live structures are not wholly comparable. Thompson (1964) gives further details of this work.

By metal shadowing fixed and dried cells it is possible to prepare replicas of their surfaces. Early work by Hillier and Hoffman (1953) on the erythrocyte surface suggested that it was covered with 100 Å wide plaques. Coman and Anderson (1955) and Nowell and Berwick (1958) reported that replicas of normal and tumour cell surfaces were of different texture when viewed by electron microscopy. Catalano, Nowell, Berwick and Klein (1960) examined the cell surfaces of sublines of the mouse MC1M sarcoma; the sublines are believed to differ in cell adhesiveness, but they found no obvious differences between them. Easty and Mercer (1960) made replicas of hamster kidney (cortical epithelium) cells and their stilboestrol-induced tumorous form. Their replicas, see Plate I:2, appear to have greater resolution than those of other workers. They found that the tumour cells had surfaces different in no obvious manner from normal cells. Small 200 Å particles were observed in the replicas which may represent adsorbed protein molecules. A certain degree of internal structure, nuclei, mitochondria etc. appeared in the replicas, together with a number of filaments, which were probably microvilli, and pore-like structures which may possibly be pores in the surface.

Most analyses of supposed cell surface material have shown the presence of protein and it has often been assumed that this protein is a component of the cell surface and not a contaminant. It was pointed out on p. 9 that at present the art of cell surface isolation is not sufficiently developed for pure cell surface to be obtained. Although myelin is a bulk source of cell surface it is impossible at present to rule out the possibility of its contamination by cytoplasmic components during its extraction. In addition it is impossible to exclude the possibility of a thin layer of cytoplasmic protein being wound into the myelin structure between the plasma membranes. Furthermore, immunological and electrophoretic techniques have failed to reveal the presence of protein on the outer side of the plasmalemma lipid, while these techniques have clearly demonstrated the presence of mucopolysaccharide as a cell surface component. Despite these arguments Vandenheuvel (1963, 1965) has suggested that the cell surface in myelin carries its protein surface monolayer in the β-keratin configuration. He assumes that the amino-acid composition of myelin represents a single protein. The protein content of myelin is just sufficient to cover the cell surface at an average area per amino-acid of $32 \cdot 6$ Å^2. This degree of packing of the protein would produce a fairly high surface tension whereas we know that other cell surfaces at least have low surface tensions. It can now be appreciated that at the present treatments such as that of

Vandenheuvel are very speculative. It is interesting that Vandenheuvel ignores the presence of aminosugar compounds in myelin,
described on page 7. It is still very possible that protein is an in-
tegral part of cell surface structure but at the moment there is very
little evidence to support the idea.

Theories of surface structure and composition

The two main theories of plasmalemma structure have already
been mentioned in the preceding section. These are the bimolecular
lipid leaflet theory and the hexagonal phase lipid theory. At this
point I intend to discuss in more detail the chemical details of these
models and the virtues of each model from a theoretical point of
view. The lipid structure of the bimolecular lipid leaflet theory as
stated by Finean was worked out on myelin. The structure proposed
is based on three main points: the size and shape of the component
molecules in relation to the measured dimensions of the plasmalemma; the position of the phosphate groups; and the expected
orientation of the polar groups of the lipid on the outside of the
leaflet. This last expectation is based on the study of artificial lipid
layers at oil–water interfaces, and presupposes that either (a) the
protein or mucopolysaccharide adsorbed on the surface stabilises
the system by linkage with the polar groups of lipids, or (b) that the
outer mucopolysaccharide etc. is so diffuse that the immediate
surroundings of the edge of the lipid molecules are of high dielectric
constant, maintaining their ionisation. The lateral cohesion of the
lipid molecules may be partly due to hydrogen bonding, but close-
range van der Waals forces probably play a considerable part. In
this scheme the protein or mucopolysaccharide components have
not been given any structural role.

Vandenheuvel (1963) has cogently criticised details of Finean's
model. He studied the various structures that could be built with
exact stereomodels of lipids, and proposed that the most stable
structures would be the ones likely to occur in the membrane (see
Figure 4). He accepts the bimolecular lipid leaflet structure rather
than the hexagonal phase lipid. Vandenheuvel also discussed Finean's
presumption that phospatidyl choline was bent round on top of
the cholesterol to which it was bound by hydrogen bonding.

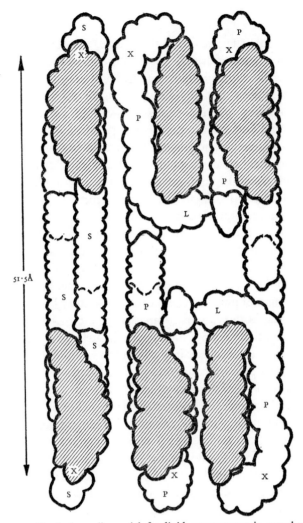

Figure 4. Vandenheuvel's model for lipid arrangement in membranes.
P refers to lecithin and S to sphingomyelin. Cholesterol is cross hatched,
position of phosphate groups shown by crosses. Two lecithin–cholesterol
pairs and one sphingomyelin–cholesterol pair are shown. Note that in
this figure the two members of a lecithin–cholesterol pair have their shorter
axes turned through 90° about the long axis with respect to one another.
Where chains overlap, the extent of the underlying chain is shown by
broken lines. It is also possible that the opposite members of a lecithin–
cholesterol pair have the same orientation but are turned through 90°
(short axes) with reference to the sphingomyelin–cholesterol pair. The side
chain L on the lecithin is shown as a linolenic group; it might also be linoleic
or oleic. (Adapted from Vandenheuvel (1963) by kind permission of Dr.
F. A. Vandenheuvel.)

Willmer (1961) suggested that this model, put forward by Finean, did not agree with measured values of the surface area per molecule in cholesterol-lecithin monolayers (though admittedly these were gaseous state films, not liquid state as in the cell surface). Vandenheuvel considers that the cholesterol-lecithin bond proposed by Finean would be far too weak since it could only be an ion-dipole bond (hydrogen bonding being impossible because of molecular shapes). Vandenheuvel proposes two basic stable structures. These are the cholesterol-sphingomyelin complex and the cholesterol-lecithin complex (see Figure 4). These structures are those which would give maximum van der Waals interaction between the component parts. All phosphatidyl and phosphatidal lipids will fit in the model for the cholesterol-lecithin complex, placing themselves in the lecithin part. The fatty acid components of lecithin, attached to the C_2 atom as unsaturated chains, can be fitted into the model, though of differing chemical natures. It will be seen from the diagrams that both types of complex have a cross-sectional area of c. 100 Å^2 and that they are both approximately 40 Å long.

In fitting the complexes into a bimolecular lipid leaflet structure, Vandenheuvel pays more attention to the details of myelin analyses than Finean. Many of the sphingolipid fatty acids have a chain length of C_{24} and Vandenheuvel points out that such a chain length will allow interdigitation of the cholesterol-sphingolipid molecules to give a bimolecular leaflet 51·5 Å wide, 48 carbon atoms wide. The C_{24} fatty acid lipids would account for 33 per cent of the bimolecular units. Another third of the lipid could be accounted for by the pairing of C_{22} or C_{23} chains with C_{26} or C_{25} respectively. The remaining third would be made of phosphatidyl units, three quarters of which, containing a palmitic or oleic chain on either side, would have a length (taken in pairs) of 51·5 Å. Only a very small proportion of the lipids cannot be fitted to this model. He suggests that under conditions of extreme dehydration such as occur during fixation, a certain amount of collapse of the membrane components (in particular the short chains) will occur. He indicates that protein is located on the surface as zigzag chains attached by ionic links to the lipid. He points out that his model allows the existence of pores, sufficiently large to permit the passage of solvated alkali metal ions.

This model, although mainly at present justified on theoretical grounds, may explain the significance of details of human myelin lipid composition.

Despite the lack of any statement about the details of the muco-polysaccharide or protein structures on the surface, it would be very unwise to dismiss the arrangement of these molecules as random. Matalon and Schulman (1949) and Eley and Hedge (1956) amongst others have shown that proteins have very specific interactions with lipids. The hexagonal phase lipid theory explains the specific adsorption of various proteins rather better than the bimolecular lipid leaflet theory, in which all parts of the lipid surface would be alike for adsorption of proteins.

Although Lucy and Glauert (1964) pointed out that one of the main defects of the bimolecular lipid leaflet theory was that it failed to explain the high permeability of some membranes, Vandenheuvel's theory escapes this criticism. It seems to be that both Finean's model and others could easily allow pore permeation through intermolecular spaces in the leaflet structure.

Kavanau (1963) in a short paper made a number of interesting speculations about a membrane structure in which transformation between several equilibrium states was possible. The theory pictures the membrane as a lipid-lipoprotein complex. In one state, the "open" state (see Figure 5), the membrane consists of hexagonal micelles of lipid 180–200 Å thick, bound to outer and inner protein monolayers. Cytoplasmic matrix is supposed to be present in the gaps between the micelles. Transformation can occur to a "closed" state, in which the micelles merge into at first a discontinuous layer formed of micelles tightly packed side by side with closure of the gap, and then into a true bimolecular leaflet. As this transformation occurs, the membrane lessens in thickness in order that the micelles may expand laterally. In the discontinuous closed state the micelles are termed "bimolecular discs". Kavanau suggests that several factors may be able to bring about transformation. Since he presumes that divalent cations will tend to link the carboxyl groups of pairs of lipid chains; their presence will lead to close packing of the surface into the closed state. Changes at the surface, which might arise from a variety of metabolic processes, will cause displacement of the

divalent cations when the pH falls and thus under these conditions
the membrane will tend to the "open" state. Increasing the concentration of divalent cations will tend to "close" the surface.
Temperature and pressure changes will also affect the structure, as
will alteration in the dielectric constant of the surrounding medium.

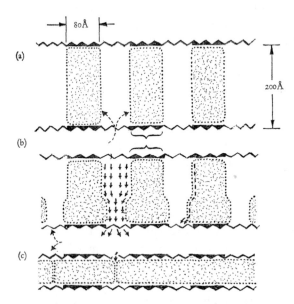

Figure 5. Kavanau's model of cell surface structure. (*a*) Lipid micelles
(fine dots) lie between protein layers (zig-zags). The polar groups of the
lipids are shown as fine dots. Each micelle is some 80 Å in diameter and about
200 Å high. (*b*) The hydrophilic matrix is starting to flow back into the
cytoplasm. (*c*) The matrix has completely left the membrane, which has
now become a bimolecular lipid leaflet with protein monolayers on either
side. (By kind permission of the Editor of *Nature*, from Kavanau (1963)
with modifications.)

Kavanau cited the evidence mentioned earlier concerning the
possible hexagonal phase lipid structure of the membrane in support
of his ideas. The theory was chiefly developed to explain cell movement (see Chapter 5). Although Kavanau's ideas are of great interest,
they are not yet worked out in terms of detailed structure of lipids
etc., nor is there any clear evidence at all of this type of phase-state
transformation occurring in life or even in artificial systems.

The work of Lucy and Glauert (1964) on the structure of lipid systems using electron microscopy has already been mentioned. Here I want to consider the more theoretical aspects of their paper. Their work on artificial lipid mixtures suggested that lecithin-cholesterol micelles of the radial structure shown in Figure 6 occurred in the plasma membrane. The micelles would be arranged in penta- or hexagonal array with pores between the micelles. The precise arrangement of any protein or mucopolysaccharide components is not clear, but Lucy and Glauert suggest that a protein molecule may substitute for a lipid micelle and so become fitted into the hexagonal array. One alternative arrangement is that the pores are filled with extended protein chains, another that (as in Kavanau's scheme) the proteins form a proper monolayer over the surface. All three or any two of these arrangements of protein or mucopolysaccharide might occur in one surface. The thickness of the membrane does not appear to be specified by Lucy and Glauert.

Work on permeability of the cell surface gives a rather different picture of its surface structure. Since this book omits detailed discussion of work on permeation phenomena, such work is only summarised briefly. As has already been mentioned, some workers predicted the existence of pores to explain the permeation of ions through the plasmalemma; the particular point is that evidence indicates that cations up to a certain size (as solvated ions) are able to permeate the membrane with ease, there being a sharp cut-off above a certain ion size (see Solomon, 1961 or Mullins, 1961). This suggests the existence of some geometric restriction factor, explained as being due to pores in the membrane. I do not think that we should necessarily visualise these pores as permanent holes in the membrane, but rather as tracks of short lifetime between the constituent molecules which will not necessarily be visualised by electron microscopy. Chapman's finding that considerable molecular vibration may take place in artificial membranes at room temperature suggests that transient pores may easily form in the cell surface (Chapman, 1966). Other permeation theories (see Danielli, 1954 for a general survey) require the presence of carrier or enzyme molecules in the plasmalemma, often able to move from one side of the membrane to the other.

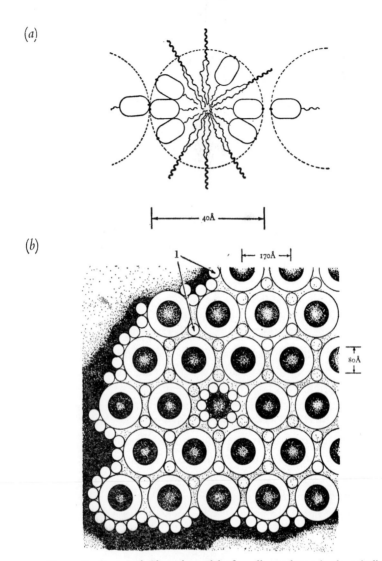

Figure 6. Lucy and Glauert's model of a cell membrane in the micellar state. (a) Diagram of a lecithin-cholesterol micelle, with the same conventions for the two types of molecule as in Figure 1(d). (b) Top view of a micellar structure, composed of lecithin-cholesterol and saponin. Saponin of course does not occur in the plasmalemma, but this diagram gives a general idea of the possible structure of a hexagonal phase lipid structure. White rings are saponin-cholesterol micelles. The black circles are holes 80 Å diameter in membrane. Small circles *l* are lecithin-cholesterol micelles. From Lucy and Glauert (1964) (by kind permission of the authors and the Editor of the *Journal of Molecular Biology*.)

Katchalsky (1961) pointed out that many of these examples of supposed active transport could be simply explained in thermo-dynamical terms without the need to propose specific systems within the membrane doing work to move solute from one side of the membrane to the other. There is some evidence (see Hokin and Hokin, 1960) that phosphatidic acid may be a passive sodium carrier in the membrane. It is hard to see how such structures, in particular the enzymatic ones, could be made to fit in a bimolecular lipid leaflet structure unless one proposes special regions of other structure.

Do these alternative structures provide the necessary membrane stability? The bimolecular lipid leaflet may become highly stabilised if its surface pressure is low enough. Harvey and Danielli (1938) originally considered that adsorbed protein lowers the interfacial tension of the surface lipids to a value which ensures stability. More recently Danielli (unpublished) has suggested that lecithin monolayers at the oil–water interface tend to show very high surface pressures, and that a zero interfacial tension would result from the combination of lecithin and cholesterol in a monolayer. Unfortunately the majority of investigations of the behaviour of lipid and lipid protein monolayers have been at the air–water interface (see van Deenen et al. 1962), in which interesting interactions may be observed, but which does not closely resemble the cellular situation. It is however of considerable interest that the stable bimolecular films prepared by Huang et al. (1964) and Thompson (1964) have very low interfacial tensions, of the same order as that of live cells (see Chapter 2).

At this point it is suitable to discuss whether either of these structures actually exists in the surface. Although there is fairly good evidence that the bimolecular lipid leaflet structure exists in myelin there is less good evidence that it is found in the plasmalemmae of cells other than the special case of myelin. Careful study of electron micrographs of plasmalemmae of several cell types suggests that the bimolecular lipid leaflet structure is present in these cells, though it must be admitted that there are reasons for thinking that fixation and embedding processes alter the structure (see earlier). At present so little is known with accuracy about membrane composition or reaction with fixatives or stains that no deductions about membrane structure can be made from these two lines of evidence. Membrane electrical properties favour the bimolecular lipid leaflet structure when the transmembrane resistance is high, and the hexagonal phase structure when the resistance is low. In conclusion it must be stated, as Korn (1966) does in an excellent review, that the nature of the plasmalemma structure is still unknown to us, though we have some substantial clues. It must be remembered that it is very probable that plasmalemmae structure may vary from cell type to cell type or even between different parts of a cell and perhaps with the metabolic state of a cell.

SURFACE COATS

Holtfreter (1943a) described a "surface coat" on the outermost part of the egg proper of the amphibian, lying directly inside the vitelline membrane. As cleavage occurred, this coat was not destroyed or divided with the cells, but remained as a continuous structure covering the entire outermost cell layer of the embryo during blastula and gastrula stages. The evidence for its existence was visual, and Holtfreter suggested that this coat had the function of co-ordinating cell movements (see Chapter 7). Dollander (1962) reviewed the question of the existence of this coat on amphibian embryos, and from his own electron micrographs was able to demonstrate a very lightly staining layer outside the plasmalemma which passed from one blastomere to the next. He considers that the surface coat described by Holtfreter does not exist. Evidence for its functions will be considered in later chapters. Birks, Katz and Miledi (1959) described a similar enveloping membrane around atrophic muscle cells (Plate I.3). Both these structures are of a type which does not coat the whole of one cell surface, but which instead spreads continuously over parts of several cells. In this respect they resemble basement membranes. Molecularly they may not be in direct contact with the cell surface and the membrane described by Birks et al. is separated from the cells by a gap.

A variety of types of basement membrane structure have been described in the literature and, as will appear, it appears that the term has been frequently used with little thought. One of the classical basement membranes is that of the amphibian skin. Weiss and Ferris (1956) described this as a structure composed of alternate layers of aligned collagen-like fibrils, each layer being in different orientation from that immediately above or below. It appeared that the membrane was separated from the base of the epithelial cells by a space some 600 Å wide, whose nature was not at that time clearly revealed. But Saltpeter and Singer (1960) described a space (the adepidermal space) between cell and basement membrane. Jakus (1956) described the fine structure of Descemet's membrane which lies between the stroma and endothelial cells of the mammalian cornea. This basement membrane appeared to be separated by a gap from the cells. Descemet's membrane is thinner than the basement membrane of the amphibian skin and lacks the layered structure. In the rat kidney, van Breeman et al. (1956) described a gap between the cells which stained as a basement membrane. Farquhar and Palade's result shows (see Chapter 3) that this basement membrane probably has no real existence. Consequently it appears that the term basement membrane covers a variety of structures, but those whose structure is best established appear to be not part of the cell surface but structures passing between or over a number of cells, often, perhaps always, separated from them by a space.

The second type of "coat" described in the literature is the one which is supposed to cover the entire outer surface of a cell outside the plasmalemma, and which does not stretch as a lamina from cell to cell, although some suppose (see later) that it has an adhesive function binding cell to cell laterally. I feel that the term "coat" has been loosely used even in this context, and that some workers really refer to the plasmalemma and associated internal structures when they use it. The evidence for the existence of a layer of material outside the plasmalemma is unsatisfactory at present. In electron micrographs, faint staining is sometimes seen outside the plasmalemma (see Epstein et al. (1964) and many other authors). Epstein was examining the adhesion of viruses to the cell surface and claimed that the virus particles adhered to the coat (represented by the faint staining) and not directly to the plasmalemma. Morgan et al. (1961) who made particularly fine electron micrographs, were unable to see any coat acting in the attachment of viruses, which apparently directly contacted the plasmalemma. The tenuous staining outside the plasma-lemma may represent a coat on the surface, and has been con-sidered as forming the intercellular material of the gap. Thus by defining the coat as part of the cells, two cells in adhesion would be in molecular contact. But arguments (see Chapter 4) against the reality of the gap substance would apply against the existence of such surface coats. An alternative explanation of the staining zone around cells is that it represents either a trace of waste material found in life or a product leached out of the cells during fixation. It is probable that much protein etc. is lost from cells during osmic fixation. Bennett (1963) terms both types of coat the "glycocalyx".

Other work on the coat is more circumstantial and results may in fact refer to properties of the plasmalemma. For example, Gasic and Gasic (1962), Gasic and Baydak (1964), Gasic and Berwick (1963) and Defendi and Gasic (1963) have described Hale staining for muco-polysaccharides at the outside of HeLa and other cells. These deposits are visible by light microscopy (but cannot of course be exactly located at the surface) and by electron microscopy. This staining was abolished by neuraminidase which attacks sialic acid. The iron particles of the Hale stain were visible by electron microscopy as large and sporadically situated bodies attached on one side to the plasmalemma. In consequence, it is difficult to say whether they represent odd globules of coat or are in fact aggregates of stain indicating the position of mucopolysaccharides such as sialic acid in the plasmalemma. But electrophoretic work confirms that the sialic acid lies in or very close to the edge of the plasmalemma. Thus, if we consider that the plasmalemma represents a material of high dielectric constant, to whose surface electrophoretic measurements refer, there is little evidence for a diffuse coat outside the plasma-lemma. Gasic and Baydak (1964) found that tumour cells, which did not show Hale staining, can absorb mucopolysaccharides from the medium and become stained. Pepsin treatment also made the cells stainable in the absence of mucopolysaccharides. These results suggest that staining may sometimes be due either to components of the body fluids adsorbing on the cell or that staining may also occur inside the cell membrane. The excellent work of Gasic and his co-workers should probably be accepted as confirmation of the location of N-acetylneuraminic acid as a component of the plasma-lemma. I feel that in this work there has been a tendency to use the term "coat" when plasmalemma would be more suit-able.

The remaining evidence usually quoted in favour of the existence of surface coats is of a very circumstantial and somewhat unsatis-factory nature. Moscona (1962, 1963a) has evidence of a number of macromolecular substances of mucoid nature found in the medium after cell disaggregation with enzymes or chelating agents (see also Curtis, 1958). A number of observations, discussed in Chapter 4, led Moscona to suggest that they were essential for cell adhesion in reaggregation. These substances are supposed in some unexplained manner to be responsible for the specificity of adhesion observed in reaggregates. Rashly and entirely without good evidence I made a similar suggestion (1958)—it is much more likely that the substance was a product of cell cytolysis; see also Steinberg's (1963a) criticism of Moscona's evidence. It might well be thought that these sub-stances are in fact "coats" which have been stripped off the cells and which must be returned there before adhesion can occur. But since there is no evidence that they are in actual fact involved in adhesion or normally present (see Chapter 4), this supposition is unnecessary.

Jones (1960) suggested that the uptake of the dye, Orange G, which stains the outer parts of Rous sarcoma cells, indicates the existence of tanned protein on the cell surface. Unfortunately light microscopy which he used is insufficient to demonstrate such a localisation of staining unequivocally.

Rosenberg (1960) described an exudate laid down by trypsinised cells on glass surfaces; this might be a coat material but it could equally well be macromolecular material lost from the cell after trypsinisation, which is known to damage the membranes (Hebb and Chu, 1960). Closely associated with this concept is the proposal (P. Weiss, 1945) that exudates (microexudates) control cell move-ment; this is discussed in Chapter 5. Bell (1960) treated amphibian neurula cells with ultrasound and obtained a jelly-like material which he considered to be equivalent to the "surface coat", but which is more probably a product of cell injury.

EXTENSIONS OF THE CELL SURFACE. MICROVILLI

Although the cell surface does not protrude filaments formed solely of plasmalemmae, the cell surface does of course limit the cytoplasm, and hair-like protrusions of the surface are found in microvilli and cilia, whose cores are composed of the inner cytoplasm.

Microvilli have been described from many cells. They protrude from the cell surface like fingers, being formed of a core of cytoplasm surrounded by a normal plasmalemma. At their larger limits they merge into narrow pseudopodia (filipodia), and indeed they may possibly arise by the retraction of a pseudopodium which leaves behind a small structure (the microvillus) which cannot be easily resorbed. The majority of microvilli appear to be between 800 Å and 2000 Å in diameter but in egg formation thinner forms are produced. Wischnitzer (1963) classes these structures into macrovilli (c. 2000 Å dia.) and microvilli (c. 1000 Å dia.), but if his usage were adopted much confusion with earlier reports will result. On adult tissue cells microvilli of c. 2000 Å dia. and 10,000 to 20,000 Å length have been reported (Gey *et al.* 1954; Easty and Mercer, 1960).

Although no proper survey of the occurrence of microvilli in embryonic tissue is yet possible, it is to be remarked that they are found during oogenesis in a large variety of species, and may perhaps occur in all species during egg formation. Colwin, Colwin and Philpott (1957) described microvilli at the surface of the virgin egg of the worm *Hydroides hexagonus*. Endo (1961) found a few microvilli at the surface of the egg of *Clypeaster japonicus*. Humphreys (1964) discovered microvilli on the eggs of *Mytilus edulis* during oogenesis; he suggested that they may be involved in vitelline membrane formation. In the unfertilised eggs of the lamellibranch *Spisula solidissima* according to Rebhun (1962) microvilli of 300–400 Å diameter are found before fertilisation which contribute to vitelline membrane formation. Kemp (1956) described the formation of microvilli during oogenesis in the frog, *Rana pipiens*, and reported that the microvilli disappeared before fertilisation. Wischnitzer (1963) found microvilli on the egg surface in the newt *Triturus viridescens* while it was in the ovary. The follicle cells did not bear microvilli though they did bear a few large "macrovilli". He suggests that the follicle cells may be involved in the formation of the vitelline membrane. The microvilli pierced the zona radiata around the developing egg but did not reach the surface of the follicle cells.

In mammals reports have been confined as yet to man, the rat, the mouse and guinea-pigs. Sotelo and Porter (1959) described microvilli, of 800 Å diameter, in the rat egg before fertilisation, as did Chiquoine (1960) in mice; these villi penetrated the zona pellucida. Odor (1960) found that these microvilli persisted till the formation of the first polar body in the rat, but Izquierdo and Vial (1962) reported that microvilli persisted at least till the eight-cell stage. Blanchette (1961) however reports that the oocyte lacks microvilli. Wartenberg and Stegner (1960) described microvilli on the human egg during oogenesis. Anderson and Beams (1960) observed microvilli on the guinea-pig egg during oogenesis. Mazanec (personal communication) describes the loss of microvilli from the surface of the rat egg by the first cleavage; simultaneously microvilli develop in the region of contact between the first two blastomeres.

I shall not describe the development of microvilli found late in development as cells take up their adult differentiated state. But one example of a very different cell surface structure is worthy of consideration here.

On amoebae filamentous projections outside the plasmalemma have been described (see O'Neill and Wolpert, 1961); these are some 60 Å in diameter and are probably formed of mucoprotein.

Bang (1955) and Blough (1963) have provided evidence that virus-release from infected cells takes place by the formation of microvilli which bud off from the cell. Goldberg and Green (1960) examined the immune cytolysis of cells by electron microscopy and described the break up of the plasmalemma into islands some $0 \cdot 3 \mu$ wide; these islands then formed into spherical structures c. $0 \cdot 3 \mu$ in diameter. This remarkable transformation appears possibly to parallel the formation of cortical granules during oogenesis.

In this chapter we have examined the evidence which reveals the basic structure and chemical nature of the cell surface. At the moment a disproportionate amount of this evidence comes from two abnormal types of cell surface, the red cell surface and myelin, and it is very desirable that future work should reveal the chemistry and structure of the surfaces of other types of cell. It seems reasonably certain that the cell surface is composed of a mixture of phospholipids and cholesterol, and that it probably contains mucopolysaccharides and some enzymes on its outer surface. Despite the insistence of elementary textbooks on the presence of protein on the outer or inner surfaces of the lipid or mixed up with it, there is very little evidence at present that protein is present in the cell surface. Most workers at present favour the idea that the lipid is structured as a bimolecular lipid leaflet but there is some evidence that the lipid is in other phase states, at least in some cell types. In the next chapter the physicochemical behaviour of these cell surface structures will be considered.

From this point we can look as from a watershed. We have accumulated sufficient evidence and hypothesis, inaccurate though much of it probably is, to begin to predict the behaviour of cell surface membranes in biological systems. By comparing predictions based on the different models of membrane structure we have discussed with observed results, we can test the accuracy of our understanding of the cell surface, and indicate those parts of the subject most in need of research. The remainder of the book will be occupied with this task.

CHAPTER 2

Basic Physical and Chemical Properties

The various models of plasmalemma structure discussed in the last chapter will now be used for examining the biology of the cell surface. The very first step taken in this chapter, is to describe the first set of consequences of the structure: these are the basic physical and chemical properties. These control in their turn adhesive and motile properties of cells, basic of course to the third step of our argument, which is the nature of cell behaviour. The lipid of the surface will have a low dielectric constant and high electrical resistance. This provides an ideal system for the generation of a potential difference across the membrane because of differences in charge due to ionic differences on either side. The outer part of the surface will carry charged groups, because amphipolar ions (e.g. amino acids) will tend to accumulate at the lipid–water interface or already be present as part of the structure. Since the ionic parts of these molecules will not be soluble in lipid, an electrostatic field will be found at and near the surface; this falls off from a high potential at the surface very rapidly, because of the excess ions of opposite charge attracted to the vicinity of the surface. The zone in which the potential falls effectively to zero is termed the double layer.

MEMBRANE RESISTANCE, IMPEDANCE AND TRANSMEMBRANE POTENTIAL

The great majority of electrical measurements on the plasmalemma refer of course to adult muscle and nerve cells. Work on adult cells will not be discussed other than in very brief outline, because to do so adequately would distract us from developmental studies. The electrical resistance of the plasmalemma is rather variable, though values of c. $1 \text{ k}\Omega/\text{cm}^2$ equivalent to a specific resistance of 1×10^4 ohm/cm are widely accepted (Mullins, 1961). However, Loewenstein (1966) has shown that in many tissues such as liver or epithelia there exist low resistance pathways through a number of cells (see also Potter *et al.* 1966). In these cases parts of the membranes of two or more cells have low transmembrane resistances, and when the low resistance parts of the membranes of two or more cells are opposite, a pathway is present. The resistance of nerve and muscle membranes varies according to whether the cells are electrically active or not. During discharge or under the influence of hyperpolarising currents, membrane resistance falls to a low level. Membrane resistances in early embryonic cells appear to be variable. Cole and Guttman (1942) measured a membrane resistance of 140 Ω/cm^2 in the newly laid egg of *Rana pipiens*. Lundberg (1955) found a resistance of c. 1500 Ω/cm^2 in the unfertilised egg of *Psammechinus miliaris*, and Kanno and Loewenstein (1963) obtained membrane resistances up to 7510 Ω/cm^2 for the eggs of *Triturus viridescens* (see also Rothschild, 1938).

In the majority of cells the membrane resistance is ohmic : in other words the current/voltage relationship is linear, but in electrically excitable cells, that is, those in which the plasma membrane reacts to electrical stimuli by changing its properties so that electrical discharge occurs, the relationships are non-linear (Grundfest, 1961, 1964), so, as a result the resistance may rise or fall below its resting value. It is widely believed that the changes in resistance with applied current are due to changes in the structure of the plasmalemma. These alterations can occur within a few milliseconds. The changes in ionic permeation and in electrical resistance of the membrane are so large that many authors (Kavanau, 1963; Grundfest, 1964) consider that the bimolecular lipid leaflet theory cannot explain them. Furthermore, there is the unexplained fact that the structure responds to current flow: we need to be able to account for this on any theory of membrane structure.

Measurements of the impedance of cells can be used to determine the capacitance per unit area (see Schwan and Morowitz, 1962 for a recent theoretical analysis). Early work was done on whole cells, either in suspension or singly (Cole and Curtis, 1950). In interpreting these results it was assumed that the cell structure consisted of a shell of one dielectric constant surrounding an inner shell whose dielectric constant was constant and similar to water. Electron micrographs show many intracellular membranes. If these membranes close off regions of the cell or by their shape elongate paths of low resistance they so distort current flow that the impedance rises and false measurements of plasmalemma capacitance result. Surprisingly this effect does not appear to be significant though we cannot be sure until direct transmembrane impedance measurements are made with intracellular electrodes. For the majority of measurements (Cole, 1962), on a wide variety of cells, values of c. $1\,\mu\text{F/cm}^2$ are found whether obtained on whole cells or as by Cole (1949a) with an intracellular electrode in squid axon.

Measurements have usually been made over a fairly small frequency range (0·1 to 10 kc/s), and the membrane capacity rises with frequency, with constant phase angle (Cole, 1949a, 1962). Armed with capacitance measurements it is possible, assuming values of the membrane dielectric constant (from studies of lipids), to calculate the surface thickness from the relation:

$$C_m = \frac{e_r \cdot e_m}{d},$$

where C_m is the capacitance, d the thickness, e_m the dielectric constant and $e_r = 8\cdot84 \times 10^{-14}$. Assuming dielectric constant values for lipid of 3 the membrane would be 33 Å thick for measured values of cell capacitance. However, as Cole (1949b) points out, it is probably more reasonable to use the static dielectric constant (rather than that referring to a certain frequency) which would give greater thickness, say 100 Å, to the membrane.

Few impedance measurements have been made on embryonic cells. Cole and Guttman (1942) measured a membrane capacity of c. 1 $\mu\text{F/cm}^2$ for the frog's egg. Cole (1962) gives a value of c. 0·8 $\mu\text{F/cm}^2$ for *Arbacia* eggs. Hubbard and Rothschild (1939) found spontaneous rhythmic impedance changes in the trout egg after fertilisation which corresponded to a 4% change in capacitance. The changes had two periods, one of 1·5 minutes and one of 6 minutes.

The membrane or resting potential of embryonic cells has very rarely been measured in contrast to the frequent studies on adult cells. Kanno and Loewenstein (1963) found that the oocytes of *Triturus viridescens* and *Xenopus laevis* had membrane potentials of between —10 and —20 mV. Rothschild (1938) was unable to detect membrane potentials in the fertilised egg of *Echinus esculentus*. Rather surprisingly Lundberg (1955) claimed that the unfertilised egg of *Psammechinus miliaris* had a + 5 to + 10 mV membrane potential. No studies appear to have been made on blastula cells or of early stages of developing nerve cells, though it seems certain that this promising line of research will be soon taken up. Measurements of membrane potentials in explanted chick spinal nerve cells from 7- to 9-day embryos by Crain (1956) gave values for the resting potential up to 65 mV with action potentials of 95 mV. Fingl *et al.* (1952) measured potentials in the 3-day chick embryo heart, finding resting potentials of 29·2 and 39·3 mV in the atrium and ventricle respectively and action potentials of 39·2 and 53·5 mV. Despite the lack of measurements it seems probable that most, if not all, embryonic cells have membrane potentials.

SURFACE CHARGE, POTENTIAL AND THE STRUCTURE OF THE DOUBLE LAYER

If charge groups are present on a surface an electrical double layer is formed. It consists of an excess or deficiency of electrons on the surface and an equivalent amount of charge of opposite sign distributed near the interface in the aqueous solution. This charge on the surface gives rise to an electrostatic field in the medium surrounding the cell. The potential of this field drops exponentially away from the surface (see Figure 7). This affects adsorption to the surface and the interaction of cells by adhesion and various other effects. These effects of considerable biological importance are discussed later in this chapter and in Chapter 3. The charge on the surface may either be due to adsorbed amphipolar ions or to the original structure of the surface. Unfortunately it has been difficult to arrive at a satisfactory picture of double layer structure, it being usual to make a number of simplifying assumptions.

Surface charge is usually measured electrophoretically. An electrical field exerts a force on a particle bearing surface charges so that it migrates in the field. It is assumed that the moving particle carries a thin layer of the medium with it. At the outside of this layer there is continual shear between the moving particle with its layer of

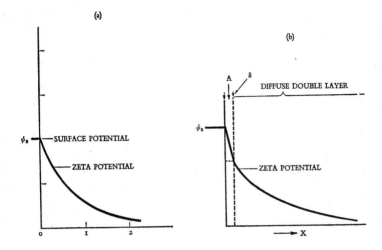

Figure 7. Double layer structure. (a) Potential in the double layer according to the Gouy–Chapman treatment. Ordinate potential; abscissa distance from surface into medium, in units of xx, where $1/x$ is the double layer thickness and x absolute distance. (b) Potential in the double layer according to the Stern treatment. A is the region where ions from the medium do not penetrate because of their finite size, δ is the plane of adsorbed ions. Ordinate and abscissa as in (a).

medium and the bulk medium. The potential at this plane of shear (the slipping plane) is termed the zeta potential (see Figure 7), and is (classically) calculated from the velocity of migration of the particle by the following formula:

$$\text{migration velocity (mobility)} = \frac{\epsilon E \zeta}{6\pi\eta}$$

where field strength is E, the zeta (electrokinetic) potential ζ, the dielectric constant of the medium ϵ, and η the medium viscosity. This formula is based on the Gouy–Chapman theory (see below). In practice the measurements are made in a monovalent 1:1 electrolyte. It is usual to measure mobility at various pH values. Some criticism of this practice (see Pulvertaft and Weiss, 1963) has been made on the grounds that the surface may be irreversibly damaged by such treatments in electrophoresis. This is perhaps exaggerated because it has been shown that the cells are viable after exposure to electrophoretic cell media at normal pH, and that mobilities obtained at

C

pH 7·0 are reproducible after exposure to media at pH 3·0 (Bangham, personal communication). Unfortunately few measurements have been made on embryonic cells, whose density is often (owing to the presence of yolk platelets) so large that it is impossible to suspend them in the conventional type of electrophoresis apparatus.

The presence of cells in an electrical field deforms it, which may be allowed for using the Henry correction. If surface conductance on the cells is appreciable, the electrophoretic mobility will be reduced, the correction for which is rarely applied, even for studies on surfaces of known composition let alone for cells. Temperature also affects the zeta potential (Mehrishi and Seaman, 1966).

The zeta potentials found from electrophoretic mobilities are interpreted in the light of the structure assumed for the double layer, and this can lead to corrections to the zeta potential (see later). The simplest model of the double layer is due to Gouy and Chapman. In this model the charge on the surface is treated as though it was smeared out uniformly. The charge in the aqueous medium is thought of as a population of point-like ions. The solvent only affects the double layer by its dielectric constant. (The charge sets up an electrostatic field in the medium whose potential drops away from the surface.) The potential at the surface is termed the surface potential. It can be shown (see Overbeek, 1952) that the potential drops away over a distance $1/x$, where the approximate relation is

$$x^2 = \frac{4\,\pi e^2 \Sigma n_{io} \, z_i^2}{\epsilon kT}$$

where e is the proton charge, n_{io} the bulk concentration of ions of valency z, ϵ the dielectric constant of the medium and k Boltzmann's constant.

The surface charge density is calculated from the zeta potential on the Gouy–Chapman theory by the following simplified relation:

$$\sigma = 3\cdot52 \times 10^4 \; \mathbf{I} \; \sinh(\zeta/51\cdot3)$$

where $\mathbf{I}$ is the ionic strength, and surface charge σ is related to surface potential ψ_0 thus:

$$\sigma = \frac{\epsilon x}{4\,\pi}\,\psi_0$$

(for low surface potentials ψ_0 <25 mv) where $1/x$ is the double layer thickness.

The magnitude of the surface potential will control a number of biological characters. It will influence surface pressure, tension, the kinetics of surface reactions and adsorption, the molecular configuration of the surface, the plasmalemma permeability, interfacial pH and the formation of cell adhesions. The manner in which it changes with chemical changes in the surface can be used to elucidate surface chemistry. The extreme thinness of the double layer (in aqueous media) implies that a very steep potential gradient exists in it, e.g. for a surface potential of 15 mV in 0·1 M sodium chloride the gradient is 150,000 V/cm. This high value has some interesting biological consequences.

However the Gouy–Chapman theory, though often used, leads to ridiculous conclusions at high ionic concentrations (Overbeek, 1952), because ions have a definite size. It is impossible to pack many ions into the double layer whereas the Gouy–Chapman theory allows it. Stern introduced two corrections: the first is to suppose that the nearest counter-ions (co-ions) are sufficiently large for their centres to be unable to touch the surface. Since the solvated Na ion has a radius of 1·74–3·75 Å depending on the number of water molecules associated with it (Ling, 1962; Mullins, 1961), it can be seen that an appreciable part of the double layer is taken up with the space in which no ionic centres can be present. This space is termed the molecular condenser and part of the potential drop will occur over this and part over the diffuse part of the double layer in which counter ions are present. The consequence of this correction will be to rather increase the double layer thickness compared with the Gouy–Chapman thickness. Stern also suggested that specific adsorption of ions might occur at the inner side of the diffuse part of the double layer, in which case the potential drops more rapidly than would otherwise be expected.

These assumptions are probably not sufficient to explain plasmalemmar behaviour. Haydon (1961) and Seaman and Heard (1960) pointed out that the surface of cells may be a three dimensional matrix of charges penetrable to some extent by counter-ions. Seaman and Heard obtained evidence that this was the situation in the erythrocyte and Haydon developed means of calculating the surface potential of such double layers from electrophoretic measurements. In addition a further set of problems, both those peculiar to biological systems and more general ones, can be proposed. The potential gradient may be so large that dielectric saturation may set in, effectively reducing the dielectric constant below the value for water. Lyklema and Overbeek (1961) pointed out that the dielectric saturation was only 2 per cent for gradients of 5×10^5 V/cm and in consequence there is only likely to be a small error in calculation of surface potentials, even for surfaces in high ionic concentrations, where the gradient might reach this value in the steepest part of the potential/distance from surface curve. However, they point out that the potential gradient may affect the viscosity of the medium close to the cells, and this, though not affecting calculation of surface potential from zeta potentials, does increase the value of zeta potential calculated from a given electrophoretic measurement by some 10 per cent for cells in say 0·5M media. These authors also reinvestigated the assumptions made about the slipping plane in electrophoresis theory and concluded that the concept of a slipping layer should replace that of a plane. The consequence of doing this would be to raise zeta potential values for given mobilities.

In addition the cell surface may be crenated and folded, producing a surface roughness. Not only may the surface double layer be confined to some extent in these crevices, reducing the apparent surface charge but conversely, if the roughness is more macroscopic there will be additional surface area to provide charge and this will lead to overestimate of the density of surface charge. Hunter (1960) produced some evidence that the human erythrocyte had a surface of this latter type. If the surface is penetrable to counter-ions there will tend to be a shielding of surface charges by these counter-ions and a shielding by the outer groups of the inner groups if the surface has depth. Both these effects will lead to under-estimates of surface potential. The cell surface almost certainly bears charges on its inner surface and the diffuse double layers of this and the outer surface may interact if there are fixed charges within the membrane, Haydon (1961). Haydon points out that models of surface structure propose fixed charges within the membrane and thus it seems very probable that the interaction between each double layer does occur with a resulting under-estimate of charge density from zeta-potential measurements.

These difficulties in knowing what type of surface we are dealing with on cells makes uncertain the calculation and derivation of surface potentials and charge density from electrophoretic (electrokinetic) measurements. Despite these points a considerable number of measurements of surface charge of mammalian cells have been made, and in the resulting calculations of zeta and other potentials the simple Gouy–Chapman model was used, with possible under-estimate of surface charge and potential.

It is theoretically possible to carry out colloid titrations of the cell surface to measure the charge. Terayama (1962) titrated ascites hepatoma cells with the polyions clupein sulphate and polyvinyl sulphate to measure the surface charge. It was necessary to assume that these polyions did not enter the cytoplasm in order to interpret these results as measurements of surface charge; yet these measurements give charge densities some six times greater than those obtained from electrophoretic measurements.

However, the majority of measurements of surface charge and potential have been obtained by electrophoretic measurements. Few measurements on normally adherent cell types have been carried out, it being rather too usual to confine one's work to blood cells which can be prepared with ease. A summary of measurements is given in Table I. When measurements have been made on tissue cells it has been usual to disaggregate them chemically or physically, but it is of considerable interest whether these processes may not have altered the surface charge density or caused macromolecules from ruptured cells to adsorb on to the surface of intact cells. The values given in Table I are of course calculated using the Gouy–Chapman theory and may be under-estimates. It will be appreciated that the ionic strength affects the measured mobility and zeta potential but not the surface potential. The values refer to monovalent cation electrolytes. Yet cells live in mixed mono- and di-valent cation electrolytes, but unfortunately no study of the double layer under such conditions appears to have been carried out other than that by Forrester et al. (1962) in brief. In consequence

TABLE I

Electrophoretic Data for Various Cell Types

Cell type	Mobility (μ/sec/$|V|$/cm)	Zeta potential (calculated for $D=80$) (mV)	Ionic strength	Surface potential	Charge density (esu/cm^2)	Author
Blood cells						
Human erythrocyte	1·48	−22·9	0·10	—	5139	Seaman, Kok and Heard, 1962
,, ,,	1·08	−16·7	0·145	—	4424	Seaman and Heard, 1960
,, ,,	1·49	−23·0	0·086	—	3790	Hunter, 1960
Hamster ,,	1·08	−17·0	0·12	−23·0	5000	Ambrose and Easty, 1960
Chimpanzee ,,	1·18	−18·3	0·145	−20·0	4860	
Dog ,,	1·28	−19·8	0·145	−21·8	5289	
Horse ,,	1·16	−18·0	0·145	−19·6	4772	Seaman and Uhlenbruck, 1962
Pig ,,	0·88	−13·6	0·145	−14·7	3582	
Sheep ,,	1·44	−22·3	0·145	−24·6	5975	
Chick ,,	0·82	−12·7	0·145	−13·7	3342	
Sheep leucocyte	0·90	−14·0	0·145	−15·1	3676	Bangham, Pethica and Seaman, 1958
,, lymphocyte	0·97	−15·5	0·145	−16·8	4090	
Embryonic cells						
Arbacia eggs	—	−31·0	0·50	−38·0	16,400	Dan, 1936
Mouse embryonic fibroblasts	0·97	−15·0	0·145	−16·3	3956	Heard, Seaman and Simon-Reuss, 1961
5 day chick heart	0·80	−12·5	0·145	—	2200	Collins, 1966
5 ,, ,, liver	0·90	−14·0	0·145	—	2350	,, ,,

						Reference
Adult tissue cells						
Human colon epithelium	1·23 sd=0·113	−19·2	0·083	−22·7	3882	Vassar, 1963
Rat liver	1·00 se=0·015	−15·5	0·174	−16·6	4500	Ben-Or et al. 1960
Rat liver, regenerating, 12 hr	1·36 se=0·021	−19·5	0·174	−21·4	5709	Ben-Or et al. 1960
72 hr	1·17 sd 11%	−18·3	0·172	—	5302	,, ,, 1962
Hamster kidney	0·48	−7·5	0·12	−8·5	1800	Ambrose and Easty, 1960
,, ,, C13	1·15	−17·9	0·150	—	4495	Forrester et al. 1962
,, ,,	1·02	−16·0	0·150	—	4290	,, ,, 1964
Mouse fibroblasts	0·82	−12·7	0·145	−13·7	3342	Heard et al. 1961
,, ,, L strain	0·88	−13·6	0·145	−14·7	3582	1961
Tumour cells						
Hamster kidney tumour	0·96	−15·0	0·12	−17·0	3700	Ambrose and Easty, 1960
,, ,, type I	1·26	−19·8	0·15	—	5400	Forrester et al. 1964
Mouse MCIM sarcoma	0·96	−15·0	0·12	−17·0	3700	Purdom et al. 1958
,, ,, ascites	1·80	−28·0	0·12	−33·0	7210	,, ,, 1958
Ehrlich ascites	1·0	−15·5	?	—	—	Bangham et al. 1962
Rat liver ascites	1·80	−28·0	?	—	—	Ruhenstroth-Bauer et al. 1962
HeLa cells (human)	0·82	−12·7	—	—	—	
Human breast carcinoma	1·40	−21·9	0·083	−26·3	4438	Vassar 1963
,, colon ,,	1·20 sd=0·076	−18·5	0·083	−22·0	3717	,,
,, uterine leiomyoma	1·42	−22·3	0·083	−28·8	4893	,,
Mouse S37 ascites	1·16	−18·2	0·145?	—	—	Cook et al. 1963
,, solid	0·92	−14·4	0·145?	—	—	,,

it appears to be impossible to calculate surface potential with accuracy from measurements in such media, so that we cannot detect specific adsorption of say calcium ions, structural rearrangement of the surface in presence of divalent ions, or calculate electrostatic repulsive forces between cells in a wholly satisfactory way for cells in physiological media.

Confusion between the surface potential and the transmembrane potential has arisen on occasion. It has been usual for workers in this field to suppose that the two potentials are completely independent for cell surfaces. In other words they assume that the cell surface is a completely non-polarisable electrode (reversible electrode) whose surface potential is in no way affected by applied membrane potentials but depends entirely on the chemical nature of the surface and the surrounding medium. In such an electrode ions can be discharged on it with ease. If the surface were a polarisable electrode membrane potentials due to the metabolism of the cell would alter the surface charge and would affect the surface potential, tension, etc. Of which type is the cell surface? Unfortunately we do not know, although it is unlikely that it is purely one type or the other. It is of great importance to find out because it would confirm or destroy our belief in the accuracy and relevance of surface potential measurements (use of the term "polarisation" or depolarisation in neurophysiology does not refer to the same phenomenon).

Electro-osmosis, movement of the medium past a charged surface in a field, may occur in certain circumstances. During the propagation of a nerve impulse flow will be parallel to the surface and slight electro-osmosis of the medium will occur, Ranck (1964) has contributed a theory of memory on the basis of such a phenomenon. Similarly, fluid may induce streaming potentials on cells (see Pidot and Diamond, 1964).

SURFACE MECHANICAL PROPERTIES

Measurements of the mechanical properties of the plasma membrane have been greatly hindered by the difficulty of measuring properties on such small particles as cells and in interpreting the behaviour of a complex structural system. In some cases at least, if not in all, relatively rigid structural elements such as mitochondria, endoplasmic reticulum, just beneath the surface will affect the measurement of surface properties carried out on the intact cell. Thus it might be thought more reasonable to discuss such properties under the heading of cortical properties, but it is preferable to treat them here because the main deductions made from mechanical properties refer to true surface functions such as permeation, cell movement, and perhaps cleavage. Early work on the mechanical properties of the cell surface (summarised by Harvey and Danielli, 1938) was carried out by simply observing cell behaviour and by micro-dissection of cells. It was concluded that the surface was visco-elastic and of low rigidity.

Surface pressure, tension, and free energy

All surfaces have a certain surface pressure due to their molecular packing. Surface or interfacial tension is related to surface pressure thus:

$$73 - \text{surface tension} = \text{surface pressure (dyne/cm) (at } 25°\text{C)}.$$

Surface tension is due to the excess attraction between atoms and molecules in the surface and is ideally numerically equal to the surface free energy in ergs/cm². Low surface tensions indicate close packing of the molecules in the surface. An excellent review of surface tension measurements on cells was compiled by Harvey (1954). Unfortunately all the techniques of measurement are open to considerable objection. When the surface tension is calculated from the shape taken up by the cell under known forces (gravity, centrifugation) on a plate, it is necessary to assume that no other mechanical property of the surface (e.g. rigidity) or interior is responsible for cell shape. Unfortunately, it is usually impossible to make this assumption. Second, the method of measuring the contact angles between a cell and an oil drop is questionable because it is necessary to assume that the cell surface is not penetrated by the oil, and that the contact angles can be used to calculate surface tension. This cannot be done, however, when a surface is deformable as is the surface of the cell (see Lester, 1961; Rose and Heims, 1962). It is probable that inaccuracy in measurement is sufficiently large that we cannot attach any significance to differences of the order of 0·1 to 1 dyne/cm. Harvey (1954) gives value for a variety of marine eggs and other cells; with the exception of rabbit macrophages and mackerel eggs, all cells were found to have tension below 1 dyne/cm. In other words surface pressures of about 72 dynes/cm for most cells suggests close packing of the molecular structure. But because there is a highly ordered lipid layer beneath the outermost part of most, if not all cells, measurements of surface pressure may be misleading. The packing of the lipid will affect surface pressure measurements unless the outermost part of the surface is able to slip over the lipid fairly easily. Further discussion of the possible influence of surface pressure on cell behaviour will be found later in this chapter, and the relevance of surface pressure and energy relationships to cell adhesion will be discussed in Chapter 3.

Surface rigidity and extensibility

Few measurements of surface rigidity and extensibility have been made. The usual technique (Mitchison and Swann, 1954a,b; Rand and Burton, 1964) has been to measure the deformation of a cell surface under a series of given forces. It is then possible by making assumptions (or measurements) about the surface pressure of the cell to calculate the part surface rigidity plays in determining the deformation. By accounting for part of the deformation in this manner the remainder may be treated by the theory of shells (see Timoshenko, 1940), if the thickness of the effective shell (which may be much thicker than the plasma membrane) is known. Using this approach it is possible to calculate isotropic mechanical constants for the cell "surface". Mitchison and Swann (1954a,b, 1955) investigated the mechanical properties of the egg of *Psammechinus miliaris*, both before and after fertilisation.

They measured the deformation produced by applying a micro-pipette to the cell surface when a known suction force was used. If they were measuring the resistance to deformation due to surface tension a plot of deformation against force should be a curve. In fact they obtained a straight line, which showed that the eggs behaved as though they were thick walled spheres. After studying the behaviour of models they concluded that the slope of the plots was proportional to Young's modulus for the shell. For the unfertilised egg they calculated a Young's modulus of $2 \cdot 0 \times 10^4$ dynes/cm^2 for a shell with a wall of $1 \cdot 5\mu$ thick (assuming that the cell had no internal pressure). They suggested that the internal pressure must be low because a 14 per cent reduction in linear dimensions of the egg produced wrinkling of the cell surface. In the second paper (1954b) they gave values of Young's modulus for the unfertilised eggs of five species of echinoderm (see also Wolpert, 1966).

Rand and Burton (1964) used the same technique to study the human erythrocyte membrane, with a view to accounting for the biconcave shape of this cell. They point out that if the membrane is elastic the stresses required to bend the wall will be small compared with those needed to stretch it, but if it is rigid the reverse applies. When the membrane is sucked into a micropipette the resistance to deformation does not increase much at first but above a certain point it increases rapidly with further deformation. These results suggest (if the cell has no, or a small interior pressure) that the membrane is both elastic and rigid. They were able to calculate the membrane resistance to deformation (at small deformations) as $0 \cdot 037$ dynes/cm, whether measured at the edge or the centre of the cell, and were thus unable to explain the shape of the cell. They measured a positive internal pressure of 2 mm of water.

Rosenberg (1963b) examined the effect of compressing cells between coverslips and measured the degree of compression required to produce herniation of their surfaces. It was possible to estimate the increase in surface area required to bring about the onset of surface herniation. This measurement gives a relative idea of the extensibility of the surface but may also include effects due to the extension of inner parts of the cells. This subjective method shows changes in cell extensibility with embryonic age of chick cells and with culture conditions.

One major difficulty is that mechanical constants measured by these methods have only a very relative meaning. Measured values obtained by the methods described here, refer to a surface of some depth which is made of a number of layers and are also a compound of the mechanical constants not only of each layer of the surface but also of each dimension in each layer, since it is very probable that the surface has anisotropic constants. Consequently there is little of the mechanical behaviour of the surface that can be predicted from such values other than movements and changes in shape directly homologous with those obtained in the measuring method.

Surface viscosity

Surfaces have a two-dimensional viscosity. The probable close-packing of the surface of the cell, suggested by surface pressure and other measurements indicates, that it will have an appreciable viscosity, probably of non-Newtonian character. Observations of the movement of cells (see Griffin and Allen, 1960) suggested that the plasmalemma itself is liquid and flows during cell movement, though Bell (1961) doubts whether this occurs in amoebae. This conclusion was derived from the observation of particles adherent to the surface. Bennett (1956) came to a similar conclusion from study of electron micrographs and proposed that the plasmalemma could flow in or out of the inner cytoplasm, being continuous with the endoplasmic reticulum. However, Shaffer (1964), observing the behaviour of slime mould cells came to the conclusion that the plasmalemma was solid. However, his observations might be as well explained by assuming it was a liquid of high viscosity. These observations are confined to non-metazoan cells and all that can be said of the surface of most cells is that it is probable that it is near the liquid–solid transition point.

Few measurements have yet been made of the surface viscosity of cells. Curtis (1961b) applied micro-torsion viscometers to the surfaces of the large cleavage stage cells of *Xenopus laevis*. The viscosity was non-Newtonian, that is it varied with the shear applied. Values for surface viscosity at zero time of shear could be read off the time-viscosity curve, and gave values of $1 \cdot 0$ surface poise. Under the conditions of shear used the surface showed thixotropic behaviour under prolonged shear, i.e. the viscosity dropped, but initially the converse property, rheopexy developed transiently. It is worth pointing out that the exact type of result obtained would depend on the value of shear applied. Of course, because of the layered structure of the plasmalemma it is incorrect to make exact comparison of such measurements with those on true monolayers.

The mechanical organisation of the surface

At this point an attempt will be made to postulate the way in which these mechanical properties will play their part in the control of cell properties.

By analogy with studies on monolayers and other thin structures we can postulate that the surface of a cell can exist in several phase states. Curtis (1962a) made this suggestion in view of Joly's work on monolayers (1954, 1956), and independent work described in Chapter 1 starting from a completely different and observational basis has tended to follow the same trend. If the degree of molecular packing (surface area) is plotted against the molecular cohesion (surface pressure) in a surface a curve will be obtained of the type shown in Figure 8. Each discontinuity corresponds to the transition

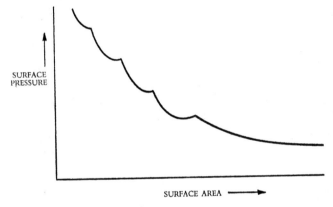

Figure 8. Diagrammatic representation of possible surface area/pressure relationships in a surface such as a plasmalemma. Each point of inflection represents a transition point between one phase state of the surface and another.

to a new phase state. In any one phase state two species of packing will be present (for a surface composed of one chemical substance), and on approaching a discontinuity or transition point one of these species will decline in concentration; past the transition point it will be replaced by a third packing species. Unfortunately we cannot yet say whether a packing species in the plasmalemma lipid would be a single type of lipid molecule or a complex such as Vandenheuvel's sphingomyelin and phosphatidyl/cholesterol complexes (1963) or some larger structure. In Chapter 1 the merits of the bilammelar lipid leaflet structure for the plasmalemma and the hexagonal phase lipid structure for the cell surface were discussed. It was also suggested that the two states might occur in the cell surface, in an interconvertible state. Could these two states be adjacent ones on the phase diagram? In other words is there at least one component packed in the same way in each structure? Obviously it could not be the whole structure, it might however be the cholesterol molecule because this seems to have the same shape in both theories. On the other hand it is equally possible that these two phase states are in fact several transition points apart, there being intermediate phase states between them, and no similar packed structures in either the lipid leaflet or the hexagonal lipid. This would then agree with Kavanau's (1963) proposal that an intermediate state exists. The importance of this concept is twofold. First, a theoretical development of this concept would allow us to calculate surface pressure of a particular packing, and thus to predict surface mechanical properties. By this means it would be possible to arrive at a description of the surface based on our knowledge of its molecular construction, rather than by working backwards from doubtful measurements. Second, if this concept is correct, we can make predictions about the stability of surface structure. When a cell is stretched the surface will be expanded, if there is no recruitment of new surface from synthesis or endoplasmic reticulum, there will ultimately be a breakdown of surface structure into a new state. Cells seem to be capable of displaying the same surface properties (e.g. cell movement and adhesiveness) under various degrees of stretch (see Chapter 5), yet if surface pressure directly affected surface packing in a continuous constant manner, it would be expected that their properties would continuously alter with stretch. Thus there may be an apparent stability of the cell surface properties displayed by any one cell type. This can be explained by the concept of phase states of the surface thus. It is known (Joly, 1956) that an activation energy is needed to pass from one phase state to the other. If this activation energy is fairly large it is improbable that a surface will pass from state A to state B, although the surface pressure is such that state B could be more stable. Thus cell surface properties would be maintained constant over a variety of packing conditions. This concept was also used theoretically by Curtis (1962a) to explain the existence of a small variety of basic cell types (as judged by their behaviour) e.g. fibroblast, amoebocyte etc. It would be very amusing to pursue these concepts to some point of proof of their accuracy or otherwise.

It should be obvious that the phase state of the surface lipid will also be dependent on chemical conditions and two factors in particular will be able to alter it. These are pH and the presence of other lipids or lipid solvents. Work on the effects of pH on cell behaviour is discussed in Chapter 5. Willmer (1961) summarised his work on transformations in cell types of *Naegleria* induced by steroids and work by other authors to conclude that steroids profoundly affected the packing of the cell surface and thus its properties. Glauert *et al.* (1963) demonstrated the devastating lysis of erythrocytes which results from their treatment with Vitamin A, suggesting that the plasmalemma lipid is converted into an unstable phase whose mechanical properties are such that the cell surface cannot be maintained intact.

Such chemical effects as well as physical factors like temperature, pressure, dielectric constant, surface pressure etc. will be able to bring about transformation of surface properties through alteration of the packing of the lipid and perhaps of the monolayers on either side. At high pressure any protein monolayer on the surface will tend to be packed as two-dimensional discs oriented normal to the surface if the molecules are flexible. At lower pressures unfolding will occur. The flexibility of protein molecules at an oil–water interface is pH dependent and very variable, insulin for example being five times as flexible as bovine serum albumin (Crisp, 1958). This folding would affect their enzymatic and antigenic properties. Similar phenomena would probably occur with mucopolysaccharides.

Curtis (1962a) described the surface properties that would result if the surface showed non-Newtonian surface viscosity, and in an earlier paper (Curtis, 1960a) went somewhat rashly far in speculating how this might be related to the various types of cell behaviour and phase states of the surface. Five effects of biological interest were deducible. These were that the changes in surface viscosity under shear, either from other cells or other parts of the same cell surface, would tend to (1) determine the wavelength and amplitude of surface undulations of the cell (see Biot, 1957); (2) affect the density of surface charges on the surface, thus affecting adhesive forces (see Chapter 3); (3) affect cell permeability and (4) affect other mechanical properties of the surface. Unfortunately few studies have been made of the effects of shear on the properties of surfaces with non-Newtonian surface viscosity, but Tachibana and Okuda (1960) demonstrated that the shear set up by rippling monolayers of high polymers at high surface pressures led to changes in surface potential, elasticity and surface pressure. Thus it seems likely that cell movement by setting up shearing forces in the surface, either directly or by the shear of a cell against another surface, will tend to alter surface properties. One interesting point is that as was explained (Curtis, 1962a) these phenomena due to non-Newtonian surface viscosity would tend to lead to a system of negative feedback controlling cell movement. It will be appreciated that any one property such as a mechanical one is affected by and affects a very wide range of other properties so that the solution of what happens when say a change in interfacial pH is made will require sophisticated mathematical handling.

CHEMICAL REACTIONS AND OTHER PHENOMENA

The dielectric properties, surface potential and other physical properties of the surface will affect chemical reaction and properties at the surface. Unfortunately so few experimental studies of chemical reactions at cell surfaces (unambiguously recognised as being at the surface), let alone embryonic cell surfaces, have been carried out that this discussion will have to be largely theoretical, though material discussed in Chapter 1 on the chemical composition of the surface is relevant.

The pH at a surface is partially dependent on the surface potential. The negative charge attracts cations so that an excess of hydrogen ions is present in the double layer, lowering its pH below the bulk pH. On the basis of the Gouy–Chapman theory the magnitude of this excess can be calculated (see Davies and Rideal, 1961). For hydrogen ions:

$$pH_{surface} = pH_{bulk} + \frac{\zeta}{57} \text{ (at 30°C).}$$

In other words, for cell surface potentials of —20 to —30mV the surface pH will be some 0·3–0·4 units below the bulk pH. Danielli (1941) discussed this problem generally for macromolecules and suggested ways of measuring interfacial pH. Since many enzyme and other reactions are sensitive to pH, the interfacial pH may be such that reactions can take place more (or less) readily than in the bulk medium. Davies (1958) points out that hydrolytic reactions may be accelerated in negatively charged systems. However, it does seem that the effects of interfacial pH on cells are insufficiently large to affect reaction kinetics greatly, unless reactions which are very sensitive to pH are found. L. Weiss (1962a) suggested that cell behaviour might depend on this phenomenon, supposing that small pH changes in the bulk medium would affect the interfacial pH differently in varying cells with different surface potentials. If these cells have surface enzyme reactions concerned with adhesion or movement this would provide a method of eliciting different types of behaviour in different cells. L. Weiss (1963a) assayed the supposedly cell bound penicillinase of *Bacillus subtilis* at various bulk pH values and found that the reaction kinetics suggested a surface pH below that of the bulk pH, but still lower surface pH values could be calculated from electrokinetic measurements. This discrepancy could be explained by the penicillinase being deep to the plane of electrophoretic shear or by the charge not being evenly spread on the surface, there being lower charge density at enzyme sites.

Similar phenomena will of course occur with other cations, in particular with divalent ones; and with anions there will be a tendency for their concentration to be reduced near the surface. The interfacial pH will be very sensitive to the presence of other cations, and if considerable amounts of say calcium are attracted into the double layer the pH might rise above its bulk value. The possible effects of this have not been investigated.

A number of other influences on reaction mechanism can be inferred from surface physics. Dipoles will tend to be formed in the electrostatic field, especially at high ionic strengths, and this may directly aid or hinder reaction. Surface pressure and potential will control the orientation of molecules in the surface and by exposing or concealing surface groups may control cell behaviour. The rate of reactions at the surface will be controlled again by the rate of adsorption of reactants from the bulk and the rate of desorption of products into the medium again.

Adsorption, chemisorption and desorption at the cell surface

Whether we adopt the bimolecular lipid leaflet or the hexagonal phase lipid theory of surface structure one generalisation about adsorption applies in either case. This is that at least two classes of adsorption or chemisorption site are present in the plasmalemma, on the lipid and on the non-lipid material respectively. Substances with non-polar groupings at one end and say a carboxyl group at the other will tend to be adsorbed to the lipid with the polar groups projecting into the medium. Other substances may tend to adsorb to the protein or mucopolysaccharide. Chemisorption may occur (for examples see below). The physical chemistry of adsorption and chemisorption on the cell surface will be complex because the various sites may interfere with one another and form a surface of considerable depth. In addition not only may desorption occur to the medium but it may also occur into the interior of the cell, which may allow falsely large adsorption values to be measured.

Although many observations of surface adsorption on cells have been made very few adsorption isotherms have been prepared. Heard and Seaman (1960) prepared isotherms for the adsorption of thio-cyanate and iodide ions on to erythrocytes, and Haydon and Seaman (1962) carried out similar work on the adsorption of methylene blue using changes in surface charge to measure adsorption. The form of their curves suggested that the Temkin adsorption isotherm for ions was obeyed, where $n = b \, log_e c + $ Constant, where n is the number of molecules adsorbed, c the bulk concentration and b a constant. Curtis (1957) attempted to measure the adsorption isotherm for calcium ions on the surface of *Xenopus* embryos and found that the amount adsorbable fell during development from blastula to neurula stages. Steinberg (1962a,b) criticised these results on both theoretical and practical grounds. Using a radioassay method he found that *Triturus pyrrogaster* gastrulae each bound at maximum adsorption some 4×10^{-8} g of calcium compared with the binding of c. 10^{-6} g found for *Xenopus* by Curtis. Steinberg attributes the discrepancy to failure by Curtis to wash the cells free of material which might bind calcium (though great care was taken to do this), whereas it might represent a species differences; at neurula stages values were found similar to Steinberg's. Steinberg objects that the curves published by Curtis do not agree with a possible stoichiometry of calcium binding. This is no objection if it be allowed that the adsorption is of counter-ions rather than calcium bound into the surface for with counterion adsorption no stoichiometry would occur. Singer *et al.* (1962) measured the adsorption isotherms for the adsorption of human gamma globulin to tanned sheep erythrocytes and found that it obeyed the Freundlich (Kuster) isotherm. Straumfjord and Hummel (1959) found that the binding of polyxenylphosphate to sarcoma 180 ascites cells does not obey a simple adsorption isotherm (uptake measured electrophoretically).

Many examples of adsorption on to the cell surface have been or will be mentioned in this book and it is intended here to mention only a few representative examples. Katchalsky (1964) investigated the chemisorption of polylysine, a basic polypeptide to the erythrocyte surface, and Easty and Mutolo (1960) found that a variety of basic polymers were adsorbed to the surface of cells. Haydon and Seaman (1962) observed the chemisorption of aldehydes to the erythrocyte surface. Many workers (see Chapter 1) have detected the adsorption, probably chemisorption, of antibodies to antigens on the cell surface. Glauert *et al.* (1963) suggest that Vitamin A is adsorbed at the erythrocyte surface, and Willmer (1961) has reviewed evidence that lipid is adsorbed by the plasmalemma. Somewhat surprisingly, although it is widely felt that proteins of the medium may become adsorbed on to the surface of cells, this has not yet been clearly demonstrated with the exception of course for antibodies. Devillers (1955) and Hodes *et al.* (1961) have put forward evidence that surfactants such as sodium lauryl sulphate are adsorbed at the surface. Vassar and Culling (1964) detected the adsorption of phytohaemagglutinin to lymphocytes electrophoretically. Further quantitative adsorption studies are badly wanted, for careful investigation of this phenomenon should be a useful tool for elucidating surface chemistry and structure.

A number of interesting relationships between adsorption and other properties of the cell surface can be predicted. Adsorption may cause an increase in surface pressure which in turn may result in the desorption of some surface component not so strongly bound to the surface. Highly solvated molecules may be negatively adsorbed, that is, their concentration will be lower at the surface than in the bulk medium. If the adsorbed molecule is ionised the degree of adsorption will depend on the surface potential, negatively charged surfaces such as cells favouring the adsorption of cations. The electrostatic field may lead to oriented adsorption of molecules: for instance Katchalsky (1964) examined the adsorption of the basic polypeptide on to erythrocytes and found that so much was adsorbed, with adsorption continuing until the surface was positively charged, that it could not be packed as a monolayer of randomly oriented molecules but must be packed with the long axes of the molecules perpendicular to the surface. This effect he attributed to the polarisation induced by the electrical field. Since adsorption is related to and affects surface pressure and potential it will be seen that it is also related to such other factors as bulk chemical conditions, surface mechanical properties, permeation and cell metabolism. A simplified diagram of these inter-relations is given in Figure 9.

It should be realised that the components of the outermost protein or mucopolysaccharide layer of the plasmalemma probably represent the most strongly adsorbable or chemisorbable substances that the cell surface has met during its formation and life. Within the embryo or adult this may provide an admirable means of maintaining a constant and defined surface composition, but in experimental situations new substances may adsorb to the surface causing the usual surface components to desorb. Willmer (1961) suggested that steroid adsorption at the surface alters surface properties, which is probably due to changes in surface composition. During development new surface antigens may be found on cells (see Chapter 1), probably as a result of their adsorption from body fluids. This should serve to remind us that ontogenetic changes in a given cell surface may not be due to differentiation in the surface's own cytoplasm but to the development of synthesis in some other cell of materials which are released into the body fluids.

A further matter of biological interest concerns the possible adsorption of mucopolysaccharides to cell surfaces to form additional

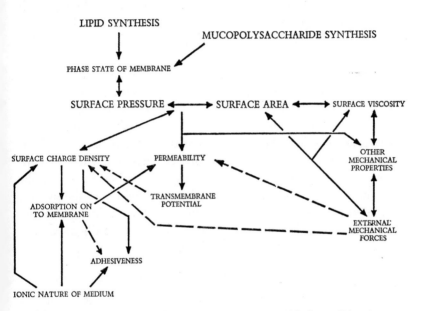

Figure 9. Various relationships between properties and bulk conditions in relation to the functions of the plasmalemma are illustrated. Solid arrows indicate relationships which are either experimentally verified or are theoretically to be expected; broken lines indicate possible relationships which have not been proved yet.

layers outside the plasmalemma. These might (see Chapter 1), be either the so-called "surface coats" or "intercellular cement". Curiously enough the protagonists of the theory of cell adhesion by intercellular cement do not seem to have attempted to demonstrate that cell surfaces will absorb substances such as mucopolysaccharides as additional layers outside the plasmalemma. Gasic and Baydak (1964) showed that mucopolysaccharide adsorption can occur but it is likely that it adsorbed into the outer layer of the plasmalemma replacing or covering some component there. It is in fact improbable for electrostatic reasons, that a predominantly negatively charged material such as a mucopolysaccharide will be adsorbed as a layer to a negatively charged surface such as a cell, although it may be adsorbed by incorporation as part of the surface structure.

The adsorbed or chemisorbed substances may remain part of the surface structure or become involved in reactions leading to their conversion to other compounds which may remain adsorbed or desorb. The enzymatic functions of the surface involve adsorption and are briefly described in Chapter 1. Other almost unexplored reactions of adsorbed substances in the surface have just been described and may lead to changes in surface properties. One particular type of reaction which may be of biological importance is tanning of protein components. Jones (1960), on the basis of staining reactions (evaluated in Chapter 1), suggested that the malignant cell has a tanned surface whereas the normal cell has not. Further work (Jones 1965) suggested that the tanning agent might be a phenolic tannin, which would lead to extensive cross-linking of the surface. Pankhurst (1958) showed that tanning causes a considerable rise in the surface viscosity of monolayers, and as will be discussed later this in turn might affect cell behaviour.

MEMBRANE BREAKDOWN AND CYTOLYSIS

In a number of biological processes the breakdown of the plasma membrane is the main phenomenon. Obviously membrane breakage can be due to mechanical stresses but at other times more specific agents are required. Mechanical rupture of the plasmalemma can result from osmotic swelling of the cells consequent upon the destruction of the normal permeability properties of the plasmalemma by a variety of chemical agents. In other cases these chemicals directly destroy the membrane to a greater or lesser degree. For both of these types of haemo- and cyto-lysis the main interest lies in the effects on membrane structure. Haemolysis provides the best understood examples. Extensive breakdown of the surface of erythrocytes results from the action of lipid solvents, though it is not at present known exactly how they act. Similarly surface active agents such as anionic detergents and saponin produce haemolysis (Ponder, 1955) probably by breaking up the membrane. Ponder (1955) pointed out that at that date many elegant kinetic studies of haemolysis had been made but the molecular happenings were unknown. With the application of electron microscopy more precise understanding of membrane breakdown became possible. Dingle and Lucy (1962) showed that lysolecithin and vitamin A caused rapid haemolysis of red cells, the kinetics of which indicated that the primary site of attack was combination with or adsorption to the surface. Glauert et al. (1963) found that vitamin A alcohol causes a rapid rise in the surface area of the erythrocyte followed by the appearance of many "holes" in the plasmalemma, visible under the electron microscope. Presumably these lipids substitute others in the membrane resulting in a sharp rise in surface pressure which throws the membrane into a violent expansion with consequent folding and rupturing. The appearance of holes in the membrane may be due to the micellisation of the plasmalemma (see Pethica and Anderson, 1953). It is of interest that other fatty acids and lipids which cause haemolysis can be shown to adsorb strongly to the surface, e.g. erucic acid (Ponder and Ponder, 1959).

The third main class of substances which cause membrane breakdown are proteins. Becker (1960) showed that protamine results in a haemolysis whose kinetics indicate surface binding. He also demonstrated that the protamine reacted with a phosphatide which could be extracted from the cells. For these reasons it seems probable that the reaction is directly on membrane structure. The protein complement-antibody complex acts in membrane breakdown in both immune cytolysis and haemolysis. This reaction has been extensively studied so only those results which bear directly on membrane structure shall be presented. Borsos et al. (1961) examined the immune haemolysis of sheep erythrocytes and confirmed by kinetic studies that only a single molecule of complement C_2 was required per cell to initiate haemolysis when antibody was later added. Borsos et al. (1964) examined erythrocyte structure by a negative staining technique after immune lysis, and were able to discover "holes" in the membrane. The number of these holes per cell corresponded closely with the number of complement molecules expected to be attached to each cell. Borsos *et al.* suggest that haemolysis in these cases then results from the osmotic bursting of the cells consequent upon free entry of water through a single hole only 100 Å in diameter. Since water entry would be very slow through such a single hole it is surprising that haemolysis results. Much more extensive surface damage was observed by Goldberg and Green (1960) in cells which had undergone immune cytolysis, the plasmalemma being broken into short sections or plaques which round up into spheres.

Further work on the ultrastructural aspects of membrane breakdown should be of value. In general there are two main mechanisms by which membrane breakdown can be brought about. First, direct substitution in the membrane can occur, either affecting lipid or other components, resulting in the change of membrane structure so that stability and permeability properties may be changed. This substitution probably acts on surface pressure, micellisation etc. Second, substances may remove membrane components without replacing them, but with similar consequences to substances of the first group. It should, however, be remembered that haemo- and cyto-lytic substances may act on components of the cytoplasm inside the plasmalemma rather than on the membrane itself. A type of membrane breakdown takes place during release of certain viruses from a cell. It appears that the virus particles can be budded off the cell enclosed in a covering of cell membrane (Blough, 1963).

MEMBRANE FUSION

Fusion of plasmalemmae takes place during fertilisation (Colwin and Colwin, 1963), and may take place during the formation of zonulae occludentes (see Chapter 3). It has also been described by Hadek (1964) between cortical villi in mammalian eggs. Presumably membrane fusion takes place when two cells fuse, though we have no information on this point. It is possible that one cell is ingested by the other and that the membrane of the ingested cell is then destroyed. At the molecular level two rather different types of fusion might occur. In one, two pieces of plasmalemmae would fuse side to side and in the other, two plasmalemmae would first abut on one another, and then their external cytoplasmic leaflets would fuse, to give a five layered structure. Obviously membrane fusion is an important biological phenomenon but at present there is almost no information about it.

In this chapter we have examined the basic physico-chemical properties of the plasmalemma, which result from its basic structure and chemical composition. The high transmembrane resistance arises from the lipid nature of the surface and allows the existence of a transmembrane potential, which is developed and controlled by the permeability properties of the plasmalemma. Although discussion of this transmembrane potential lies largely outside the scope of this book it is worth noting that embryonic cells display this potential. This suggests that embryonic cells have very similar cell surfaces to those of adult cells. The surface potential arises from the presence of charged groups such as carboxyl ions near the outer surface of the lipid, and as we shall see in the next chapter this potential is probably of great importance in controlling the adhesive behaviour of cells. The various mechanical properties of the surface arise from its structure and probably play a part in controlling cell movement (see Chapter 5), but at present our knowledge of the mechanical properties is unsatisfactory. The surface structure with its regions of high potential gradient, low dielectric constant, short-range specific London forces etc. will control the adsorption and chemisorption of substances to it and of reactions near or in the surface, and these phenomena in turn may affect the surface potential and surface mechanical properties. Finally membrane destruction and membrane fusion are considered as physico-chemical processes.

CHAPTER 3

Cell Adhesion:
(1) Basic Structure and Biophysics

The second level at which we can examine the operation of the physico-chemical properties of the cell surface, deals with the phenomena of the interaction of cells with one another or with non-living substrates. The two main interactions are the adhesiveness and motility of cells. These are cellular properties on which in turn cell behaviour and morphogenetic movements are founded. This chapter will deal with the basic morphology of cell contacts and from this the physical nature of cell adhesion will be developed. In the next chapter the biological and biophysical evidence on cell adhesion will be considered. Cell movement will be discussed in Chapter 5.

THE MORPHOLOGY OF CELL CONTACTS

As is well known, classical histology shows that many cells apparently abut directly on one another, though in other tissues, in particular in adults, large intercellular spaces often filled with secreted extracellular material are found. Nevertheless it is rare for a tissue cell to be isolated from its neighbours over the whole of its surface by a wide intercellular space. With the introduction of electron microscopy it was found that usually a 100–200 Å wide gap appeared between the outermost stained parts of the plasmalemma in those regions where light microscopy showed the cells to be in direct apposition (see Plate III.1). Occasionally these 100–200 Å gaps show considerable folding "membrane knotting" (Waddington et al., 1961), the membranes forming a sort of lock and key; conceivably these structures could tend to hold cells together. Robertson (1960b) reviewed a range of papers which showed that the 100–200 Å gap had been found in many tissues of a variety of species. Unfortunately so little high resolution electron microscopy has been done on embryos that we are not quite sure that this generalisation applies to embryos, though it probably does. In certain parts of certain cells part of the gap region may show structural specialisations or alterations of gap thickness. Most, or all, of these structures are individually of small linear extent: these will be discussed later.

There are at present three attitudes current about the 100–200 Å gap. First, Robertson (1958a, 1960a,b, 1963a) suggested that the gap exists in life but is filled with some colloidal material which does not stain well with conventional electron dense stains. Second, as exemplified by Pethica (1961) and implied by Steinberg (1964a) and Lesseps (1963) it is suggested that the gap is an artefact produced as Pethica and Bangham have put forward (personal communication) by the electrostatic repulsive forces which increase as the tissue is desalted during dehydration. Sjorstrand (1962) also supports the view that the gap is an artefact, though he believes that electron micrographs have been misinterpreted, the plasmalemmae actually abutting on one another. His main evidence for this point of view is that hypertonic and hypotonic saline solutions do not alter the gap width, and that therefore there is no gap for the solutions to enter (but see later). This conclusion is opposite to that obtained by Robertson (1958b) and Curtis (1964). Karlsson and Schultz (1964, 1965) examined electron micrographs of the central nervous system after fixation with aldehydes. They reported extensive apposition of the membranes. This finding has not been wholly confirmed by other workers, but van Harreveld et al. (1965), found that whether or not a gap is seen, depended upon the state of oxygenation of the tissue when fixation took place. Electron micrographs of anoxic brain tissue showed close apposition of the plasmalemmae, but the electron micrographs prepared by freeze substitution of well oxygenated tissue showed extensive 200 Å wide gaps. Physiological measurements of the extracellular space in brain support the idea that there are extensive 200 Å gaps.

The third point of view is that although the gap may contain some macromolecular material in life, it is essentially a fluid-filled gap containing no organised structure (Curtis, 1960a, 1962a). This view has been slightly revised in the light of experimental evidence (1964), it being suggested that 100–200 Å gaps are found only over small areas of contact between living cells, much of the contacting surfaces being further apart. Fixation brings much of the gap to 100–200 Å width.

Beginning with the least telling arguments against the non-existence of the gap, one may remark that if the gap does not exist in life it is at least very odd that it is possible to obtain electron micrographs of cell contacts which show both gaps and close contacts in the same section.

A study of the distance of separation between chick heart fibroblasts and the glass surface to which they were adhering was made using interference reflection microscopy (Curtis, 1964). The advantage of this method was that cells could be examined live. Although the contact was between cell and glass, and not between cell and cell, the results are of interest to the problem. In this method distances of separation between glass and cell are measured from the surface of sudden change in refractive index which lies at the edge of the plasma membrane (probably at the outer side of the outermost monolayer). Live cells were never closer than 100 Å to the glass surface. However it would have been impossible to detect small structures such as zonulae occludentes (see p. 90) which might have approached closer, because of the lack of lateral resolution. A concentration of organic material in the gap such as protein greater than 2% w/v would have been detectable but no evidence for such material between the plasmalemmae was found. Furthermore, hypoionic media enlarged the gap whereas hyperionic media such as 3M NaCl caused the cell surface to close down to less than 25 Å from the glass. If this shrinkage is due to the reaction of a colloidal gap material to the treatment it would have to be capable of a remarkable degree of expansion and contraction and its refractive index change would be detectable. Since the effect of various treatments on the gap thickness correlated well with the known changes in cell to cell adhesion on treatment with the same reagents it seems probable that this evidence is applicable to cell to cell adhesions. At the time of writing no high-resolution electron micrographs of the contacts of cell and non-cellular substrates have been published to my knowledge. The use of interference reflection microscopy on chick heart fibroblasts showed that very frequently much of the region of contact between cell and glass was a gap 200–600 Å thick. Curtis (1964) found no effect of 3M sucrose on the gap width between cell and glass.

Further arguments about the rate of permeation of molecules between cells suggest that the minimum gap is c. 100 Å. The brilliant paper by Farquhar and Palade (1963) describes the fine structure of the kidney tubules in the rat after injection of haemoglobin into the blood some hours before fixation of the tissue. The haemoglobin molecules could be traced in the electron micrographs as regions of dense staining. They found that haemoglobin can diffuse between the plasmalemmae through a 100 Å gap, and through desmosomes (see later); but that special types of junctional complex e.g. zonulae occludentes (see later) prevented the flow of haemoglobin molecules towards the lumen of the tubule. Brightman (1965) perfused brain tissue with ferritin. If the perfusion was carried out before fixation, electron microscopy showed that ferritin molecules (100 Å dia.) penetrated into the intercellular gaps within 12 minutes (the shortest interval used). If perfusion was carried out after fixation no such penetration occurred. This result suggests that fixation either temporarily closes the gap or fills it with viscous material. It can be shown from his results that the rate of permeation is such that the viscosity of the gap material does not exceed three times the value for water. This argues against the action of adhesive cements. These results demonstrate that the gap exists in life and that the contents of the gap are of fairly low viscosity, since the permeation is rapid. This technique can easily be applied to other situations.

The ease and rate of cell dissociation (see Chapter 4 and Curtis, 1964) suggest that a fairly wide gap is present in many cell contacts. It is of interest that Whittaker and Gray (1962) found that salt-free solutions, made isotonic with sucrose, dissociated cells of the central nervous system from one another except in regions where there is electron microscopic evidence for close (less than 100 Å gap) contacts (see also Roots and Johnston, 1964). The nature of the gap is closely connected with the nature of the outermost protein, mucoprotein or mucopolysaccharide components of the cell surface. Finean (1963) pointed out that it is possible that a modification of the outer surface of the plasma membrane takes place as it is built into the myelin structure, where X-ray diffraction results suggest that the outer layer is c. 15 Å thick. If of course this material were highly hydrated in life it might fill the gap between two cells with a thin colloidal medium, representing a sort of tenuous dying away of the surface into the surrounding medium. This view of the cell surface structure has been rather popular recently but I feel that the rather unsatisfactory nature of the evidence in its favour should be demonstrated.

It is of course well known that a slight staining surrounds cells when seen in electron micrographs (Hampton, 1958; L. Weiss, 1962) and it has been fashionable to regard this as being the highly hydrated colloidal material which surrounds cells as part of their surface structure. The electron dense staining outside the plasmalemma, which is remarkably diffuse with most procedures, may be due to materials leached from the cell on fixation, and not to any structure in life. Mercer (1957) used a lead hydroxide reagent as an electron dense stain which produced dense staining in the gap. The mechanism of action of this stain is completely obscure but there is the possibility that it precipitates in gaps in the tissue, and that dense staining represents hollows and not some chemical structure. Farquhar and Palade's results demonstrate that at least in one tissue, any intercellular material as visualised by a slight osmic staining present, is of slight viscosity and exceedingly low concentration. The extensive extracellular colloids such as collagen might of course be regarded as part of the cell surface but this seems to be an over-stretching of terms. Moreover it is quite probable that such extensive extracellular materials are separated from the plasmalemma by a gap. In some protozoa the plasmalemma carried a mucoprotein outer coat (O'Neill and Wolpert, 1961), but similar structures usually have not been seen in metazoan cells. Birks, Katz and Miledi (1959) made electron micrographs of frog muscle after the atrophy which develops after denervation and found a stained layer outside the plasmalemma. This layer was sometimes separated from the plasmalemma by a 100 Å gap (see also Chapter 1 and Plate I.1).

Robertson (1958a) claimed that hypertonic sucrose solutions, e.g. 10M, narrow the gap between the plasmalemmae in axon-Schwann cell contacts from an average value of 121 Å to an average of 90 Å and completely destroy the gap between plasmalemmae in the mesaxon. He argued that these results showed that a mucoid material must be present in the gap; a material which contracts at high sucrose concentrations, for if the gap were filled with a watery fluid sucrose should not affect the width of the gap (for physical explanation see page 86). Before quoting his results in detail it is worthy of remark that after he had treated myelinated and unmyelinated nerve fibres from *Rana temporaria* with hyper- or hypo-tonic solutions, the tissues were fixed for electron microscopy in isotonic media for some hours. It is remarkable that any effects produced by the solutions were not cancelled out during fixation. Cells treated with hypertonic sucrose solutions (9 specimens with tonicity due to sucrose equal to 10 × isotonic Ringer saline, 6 with 6 × isotonic Ringer strength) showed closure of the gap between plasmalemmae from 100 Å or more to less than 15 Å at various points, but these closures were not spread over the whole or even the greater part of the contact surfaces. The greatest lateral extent of gap closure was 7000 Å. Moreover, sometimes the gaps were wider after treatment with hypertonic sucrose solutions than in controls. It may be wondered whether these irregular gap thicknesses are not the result of osmotic shock on the whole tissue compressing some parts of the gap and expanding other parts. If this view is correct, Robertson's results do not indicate the existence of a gap substance. Alternatively it might be supposed that the gap material postulated by Robertson is present only in a few localities in the contact area between two cells. The variability of gap width in Robertson's work and the fact that treatment of a tissue with hypertonic or hypotonic solutions is almost certain to set up mechanical stresses in the tissue during fixation makes his interpretation questionable. He also found that hypertonic saline solutions caused a shrinkage of parts of the gap. But, as will be seen later, this effect of saline solutions can be interpreted in other ways. In the tissue culture system used by the author (1964) it was possible to argue that such mechanical stresses would be unlikely to affect gap thickness. Sjorstrand's (1962) results with hypo- or hypertonic fixation are questionable since he used a small range of tonicities which would not be expected to have any effect on gap dimension.

The possibility that the gap is *sensu stricto* an artefact but one which closely parallels the situation in life has already been mentioned. The experimental evidence for this is at present tenuous but there is a certain amount of theoretical reason for putting this idea forward. During fixation the surface potential may fall due to adsorption of cations, and in consequence the electrostatic forces will diminish so that the gap closes (see later). Further adsorption of cations would cause the surface to acquire a positive potential and reseparate to c. 100–200 Å gap. However Glaeser and Mel (1964) found that the surface potential of OsO_4-fixed erythrocytes was identical with that of live red cells.

The nature of the gap will form a recurrent theme later in this chapter and in the next. There is, as we have seen, much evidence that the gap exists in life. It is probable that the gap frequently contains a certain amount of organic material but, as will subsequently appear, this material appears to be of little connection with aspects of surface behaviour, adhesion, movement or cell behaviour.

Structural specialisations of the plasmalemma

Those structures which are formed at regions of contact between two cells and of which both cells form part shall be dealt with first. With the recent paper by Brightman and Palay (1963) it is hoped that we are at the end of the era in which many different names have been given to structures which are probably homologous. These authors provide a rational classification of contact structures. Nevertheless a variety of names will have to be quoted. Brightman and Palay classify these structures in terms of the degree of closeness of approach of the plasmalemmae, extending from the 100–200 Å gap to those structures in which the two plasmalemmae fuse into a single structure (see Figure 10).

DESMOSOMES AND TERMINAL BARS. THE MACULA ADHERENS

Traditional histology had shown that intercellular bridges appear to connect epithelial cells to one another. Heidenhain (1907) and his contemporaries came to the conclusion that such bridges were of widespread occurrence and that they were responsible for

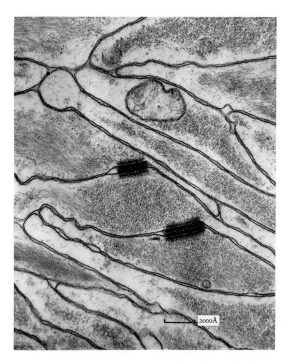

3.1   100–200 Å gaps and desmosomes in muscle tissue.
The plasmalemmae appear as heavy black lines and are
separated by a continuous gap except perhaps at the two
dark desmosomes. By courtesy of Professor D. W.
Fawcett.

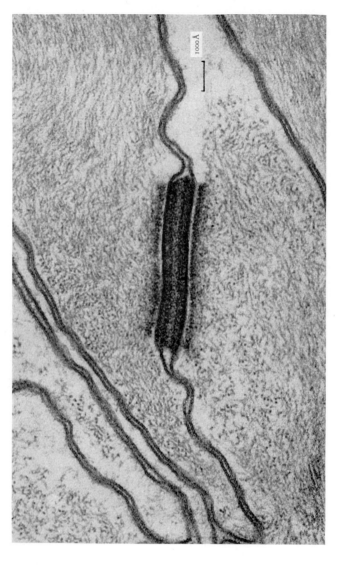

3.2 A desmosome in muscle tissue. 3 and possibly 5 bands of differing stain contrast are visible between the plasmalemmae. Note that the plasmalemmae lie some 100 Å apart. Also note dense staining inside the cyto-plasms which form as part of the desmosome. By courtesy of Professor G. W. Fawcett.

1000Å

cell adhesion amongst other functions. In 1927, Schaffer revived a suggestion by Bizzozero that in fact these bridges represented apposed cell processes: he suggested terming them *desmosomes*. This name has been used more widely by electron microscopists than any

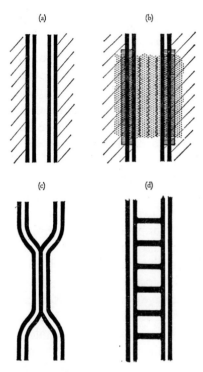

Figure 10.   Cell contacts. Diagrammatic sections of cell contacts. Plasm lemmae are shown as double black lines. (*a*) 100–200 Å gap contact or zonula adherens. (*b*) Desmosome structure. Note the three bands between plasmalemmae, and general staining in cytoplasm. (*c*) Zonula occludens. (*d*) Septate desmosome: here the cross-bars are formed only by the outer lipid leaflet of the cells, but it is possible that both leaflets are involved in cross-bar formation.

other. Porter (1954) and Odland (1958) described desmosomes as being composed of a series of parallel layers of varying density disposed symmetrically on either side of the mid-line of the inter-cellular gap between two cells. The outermost layers lie within the cytoplasm of either cell, and in general the plasma membranes can

D

be traced as faint parallel bands through the desmosome forming part of its layered structure. In certain high resolution micrographs a faint parallel band or bands can be seen within the intercellular gap (see Plate III.2). Most workers do not find that the desmosome extends over any great area of the contact between two cells; Tamarin and Sreebeny (1963) made reconstructions of desmosomes in rat submaxillary gland tissue and found them to be oval shaped patches of axes 4100 Å and 2500 Å. On the other hand some reports of elongated desmosomes stretching right round the cell have been made (Hama, 1960; Mercer and Wolpert, 1962). This immediately raises a question of nomenclature. Terminal bars or *Schlussleisten* were believed by classical histologists to be bands of intercellular cement binding cells together over large areas, in particular in columnar epithelium. Fawcett (1961) showed that in section terminal bars have the same structure as desmosomes although they are more elongate. Some authors, e.g. Palay and Karlin (1959) have tended to term both structures desmosomes. On the other hand Farquhar and Palade (1963) doubt whether the terminal bar is invariably equivalent to the desmosome and suggest that it may represent another type of specialisation of apposed cell surfaces. Farquhar and Palade proposed that the desmosome be renamed the *macula adherens*. Desmosomes have not been reported extensively from early embryonic material (but see Mercer and Wolpert, 1962; Overton, 1962; Overton and Shoup, 1964; for examples).

The faint central laminar bands within the macula adherens have been referred to as the *intercellular contact layer* (Odland, 1958) or as the *median stratum* (Karrer, 1960a). Odland's term carries the implication that cell and cell make molecular contact at this point. Does this band have any real existence in the live animal? Farquhar and Palade found that haemoglobin was able to diffuse into and through the desmosome structure, an observation which suggests most strongly that there is no continuous structure in the macula adherens which binds one cell surface to another. It is possible that this band and perhaps some of the others are electron optical arte-facts of the Fresnel fringe type. Farquhar and Palade (1963) and Brightman (1965) found also that ferritin molecules could pass through this central region. The fine slit between the densely staining outer parts of the macula adherens would act very well to produce diffraction effects (Curtis, unpublished calculations).

THE ZONULA ADHERENS

This structure (see Figure 10), first described under this name by Farquhar and Palade (1963) is almost certainly the same as the 100–200 Å gap and is very similar to the macula adherens in that the plasmalemmae lie some 200 Å apart. (Some authors such as Brightman and Palay (1963) suggest that the gap width in the macula adherens is greater (200–225 Å) than in the surrounding simple gap.) The gap width in the zonula adherens is the same as in the surrounding gap. Farquhar and Palade found that haemoglobin molecules diffuse through the gap, and Brightman (1955) confirmed this with the 100 Å dia. ferritin molecule.

It has been perhaps the rather facile belief that these two structures are specifically organs of cell adhesion, indeed the recent renaming specifically carries this implication. The best circumstantial evidence for their adhesive function arises from early observations of the apparently intercellular contacts of prickle cells in stained material seen under the light microscope. In such material the cells are withdrawn from one another except in the region of the desmosome, but it can be argued that this appearance may either be a fixation artefact, or due to some mechanical property of the cell located near the desmosome which affects its contraction. Overton (1962) points out that the number of desmosomes in the chick embryo increases with developmental age. Since it is widely thought that the cells of most tissues grow more adhesive as they develop, these findings may indicate an adhesive function for desmosomes. Lesseps did not report any desmosomes in embryonic chick reaggregates, where cellular adhesions are developed. Sedar and Forte (1964) found that EDTA treatment of tissues in which desmosomes and zonulae adherens are present led to a considerable widening of the gap and the disappearance of the central part of the desmosome. They interpreted these results as implying that the maculae adherens are the adhesive structures of cells which operate by calcium binding one surface to another (see later). That ferritin molecules can diffuse

into the central band, which Sedar and Forte believe to be the real
adhesive structure, suggests that it has little or no real structure (see
earlier). There is no reason from their findings to believe that other
parts of the cell surface are less active as adhesive structures than the
maculae adherens. It can thus be seen that the evidence for the
adhesive function of desmosomes is exceedingly slender at present.

As Brightman and Palay (1963) point out there are a variety of
possible structures in which the plasmalemmae make contact. The
two plasmalemmae might make contact over a large area or at
repeated small points, or they might merge into a single bimolecular
lipid leaflet. Conceivably the only remaining structure at a point of
contact might be a pair of fused protein monolayers. Some of these
possible structures have been described (see Brightman and Palay,
1963). The results of Petry *et al.* (1961) suggest that desmosomes
may individually last a short time and disappear and appear with
ease.

CLOSE APPOSITION OF PLASMALEMMAE. THE ZONULA OCCLUDENS

Although the exact dimensions of the plasmalemma components
are not yet established with clarity there appears to be a considerable
number of reports of structures whose staining patterns and dimensions suggest that they represent pairs of plasmalemmae, one derived
from each cell, lying in molecular contact with one another, joined
by their outer non-lipid monolayers. Unfortunately different authors
besides using various names, have also reported somewhat different
dimensions for this structure. Yet a knowledge of the dimensions
of the structure is critical for correct evaluation of its structure.

Farquhar and Palade (1963) discovered a structure, which they
termed the *zonula occludens* (Plate III.3) in rat and guinea-pig tissue.
This structure appears to consist of two apposed plasma membranes
in which the outer protein leaflets are fused into a single structure
20–30 Å in thickness (see Figure 10). They found this structure at
regions in the kidney where it would be desirable to prevent permeation of liquid along the 100–200 Å gap. They were also able to
show that haemoglobin would not permeate past such structures in
life, which suggests that they exist in life and that they may girdle
the cell. But no calculations were done to show whether there was

sufficient time for the haemoglobin molecules to diffuse any further so that they passed through the zonulae before fixation. Brightman (1964) found that ferritin molecules did appear to by-pass the zonulae occludentes in brain tissue. There are two alternative interpretations which can be made of this result. Either the zonulae occludentes do not effectively encircle the cell or the zonulae occludentes are arte-facts due to a membrane fusion which pushes ferritin molecules aside. Sedar and Forte found that zonulae occludentes were not apparently affected by EDTA treatment unlike the maculae or zonulae adherens. This suggests that the zonulae occludentes are the main adhesive structures of cells which possess these structures.

It is very probable that the zonula occludens is homologous to a variety of other structures. Sjorstrand and Elfvin (1962) described a similar structure, with a 40 Å wide central fused zone of protein leaflets. They named this the "five-layered attachment zone". The total thickness of the structure was 170 Å. Similar structures have been described by Karrer (1960a) as "quintuple-layered cell inter-connections", by Devis and James (1962, 1964) as "quintilinear membranes", by Muir and Peters (1962) as "quintuple-layered membrane junctions", and by Dewey and Barr (1962) as the "nexus". Most of these authors do not publish precise dimensions for the structures.

The *external compound membrane* described by Robertson (1959) in axon-Schwann contacts appears to be homologous with the zonula occludens; frequently as in retinal cells, it is formed by a cell "throwing" its surface into folds which touch one another.

Internal compound membranes (Robertson, 1959, 1960a,b) have been described from myelin where cell surfaces (of one cell) are fused by their inner leaflets. They are of course also 150 Å wide. Some of Farquhar and Palade's pictures suggest that actual fusion of the lipid leaflet occurs but this structure has not been described or confirmed as yet (see also Robertson, 1963a).

It is possible that in any event some of the instances of zonulae occludentes described in the literature are artefacts. Apposition of the membranes might arise during fixation as had already been suggested. At the time of writing no zonulae occludentes have been described from embryonic material. Since the plasma membranes come into close contact they are probably strongly bonded together (see Sedar and Forte, 1964). There thus seems to be better reason for supposing that these structures are adhesive structures rather than the zonulae adherens or macula adherens. Devis and James' (1964) results suggest that these structures may appear and disappear with ease. The contact relationships of the plasmalemmae in myelin resemble those in zonulae occludentes very closely. Finally it should be mentioned that Loewenstein (1966) has suggested that the zonulae occludens are at regions of the cell surface of low electrical resistance and are concerned in intercellular communication.

CLOSE APPOSITION OF PLASMALEMMAE: SEPTATE DESMOSOMES

In *Hydra* (Wood, 1949) and *Cordylophora* (Overton, 1963) narrow laminae have been described which may join one cell surface to another, see Figure 10. The detailed nature of these junctions has been described by Overton as a paired electron opaque structure c. 70 Å long and of 25 Å width between the plasmalemmae. They have also been described in grasshopper tissue by Tsubo and Brandt (1962). They apparently do not occur in embryonic tissue.

Phagocytosis and fertilisation

It may at first sight seem a little odd to combine these two topics, but both are phenomena in which one particle makes contact with another and is taken in (swallowed) by the other. In doing this the cell forms some kind of adhesion, to the sperm, bacterium etc. which may be specific for a certain type of sperm or particle, and in which finally the particle is included inside the cell. I do not intend to examine the nature of phagocytosis exhaustively in this book, for its role in embryogenesis appears to be slight and little studied, but it forms a definite surface phenomenon although other parts of the cell may play a part in the phenomenon. Unfortunately high resolution electron micrographs of the contact of sperm, bacteria etc. with a cell are difficult to obtain, and few pictures have been published which really reveal what occurs.

Colwin, Colwin and Philpott (1957) examined fertilisation in *Hydroides hexagonus* and showed that the acrosome penetrated the vitelline membrane etc. allowing the sperm to reach the egg surface.

In excellent micrographs Colwin and Colwin (1963a,b) showed that in *Saccoglossus kowalevskii*, an enteropneust, the sperm cell surface fused with that of the egg before the sperm entered the egg. This observation (according to the Colwins) suggests that sperm entry is not exactly akin to phagocytosis, in which Karrer (1960b) described an adhesion in which almost no gap is seen between the particles. Since of course most particles which are phagocytosed do not bear plasmalemmae it could hardly be expected that membrane fusion would occur in phagocytosis. It would be interesting to obtain high resolution pictures of phagocytosis of other cells or cellular particles. Hadek (1963) described the disappearance of the cell membrane on the head of the rabbit sperm as it penetrated the zona pellucida. However it is clear that in both processes the two bodies involved come into "close", i.e. less than 20 Å, contact.

PHYSICO-CHEMICAL CONSIDERATIONS

The morphological studies of cell contacts show that either close approach of cells occurs or a narrow gap is found between them. The occurrence of molecular contact or alternatively the presence of a narrow gap between adhering cells will be determined by the physics and chemistry of cell adhesion. Consequently morphological studies provide a point of departure in considering the mechanisms of cell adhesion, and considerable attention will be paid while considering the physics and chemistry of cell adhesion to the possibilities of accounting for morphological findings.

The physical chemist can view the adhesion of cells as a type of flocculation or coagulation (see La Mer, 1964) and the disaggregation (confusingly known as dissociation) of tissues as a type of redispersion. This conceptual framework is of considerable attraction because the three main theories of cell adhesion represent one or other types of flocculation. Because the cell surface is composed of lipid of low dielectric constant it probably represents a lyophobic colloid system. Three types of interaction can be identified. First, rapid irreversible coagulation of particles may occur so that they come into molecular contact; this type of adhesive mechanism has been proposed for cells by Pethica (1961), Steinberg(1962a, 1964a,b)amongst others. Second, weak coagulation might occur, the particles being held with their surfaces some 100 Å apart by a balance of adhesive and repulsive forces. This type of adhesion would permit easy redispersion of the floc or its repeptisation by removal of ions etc. Curtis (1960a, 1962a, 1964, 1966) proposed and provided evidence that this type of adhesion is found between cells. Third, hydrophilic colloids might adsorb on the surface of the hydrophobic cell surface leading to flocculation due to the chains of the hydrophilic colloid linking (bridging) one surface to the other by ionic or other bonds. This phenomenon is also known as sensitisation and is described by Overbeek (1952) (see also Healy, 1961). This type of flocculation or adhesion of cells is equivalent to the theory that intercellular cements bind cell to cell. Unfortunately our knowledge of these types of flocculation and adhesion is as yet insufficient for us to predict exactly which type occurs in a given living tissue. As will appear, it is almost certain that all three types have been produced at least in experimental situations and that very probably they are also found in various biological situations.

When two bodies are in adhesion to one another any attempt to separate them will be resisted. Generally this will be due to forces of attraction between the surfaces but there are in addition two other forces preventing separation of two bodies. First, if the surfaces of the bodies interdigitate then it may be necessary to deform the bodies in order to separate them. For example, although a key jammed in a lock experiences no attraction to the components of the lock it may be necessary to deform the shape of lock or key in order to separate them. Similarly electron micrographs of cells frequently show considerable interdigitation of their surfaces and L. Weiss (1960a) and Curtis (1962a) have remarked that this may tend to keep cells in adhesion. Second, when two bodies are separated, the medium they are suspended in flows into the gap The rate at which it does this is dependent amongst other things on its viscosity. Since the rate of entry of the medium into the gap is limited, the rate of separation of two bodies is controlled by the viscosity of the medium in part. This effect gives rise to a force opposing the rapid separation of two bodies. This is discussed at length later under the name of the *drainage* effect. Neither of these two effects is an adhesive force but they both tend to keep cells in adhesion.

The other factors maintaining two cells in adhesion are the forces of attraction and the forces of repulsion between them. When two cells have come into adhesion and reached an equilibrium the forces of attraction will exactly balance those of repulsion. On separating the cells mechanically a force of attraction will tend to oppose the force of separation. The distance between the cell surfaces for which repulsive and attractive forces balance may be diagnostic of the mechanism of adhesion, because the various possible forces of attraction and repulsion fall off in several different relations with distance from the surface (see later). For example, if the repulsive forces exceed the attractive forces at all distances from the surface, adhesion will not occur. If the attractive forces exceed the repulsive forces only very close to the surface, adhesion is possible, but a potential energy barrier of repulsive forces will have to be surmounted by the bodies coming into adhesion. They can only surmount this barrier if their kinetic energy (derived either from cell motion or Brownian movement) is greater than that of the energy barrier. There may be conditions in which not only is there a net force of attraction close to the surface but also a region of net attraction further away, separated by a potential energy barrier: in this case adhesion may occur either with the surfaces coming into close contact (attraction in the region close to the surface—known as the primary energy minimum) or with a gap between the surfaces (corresponding to attraction in the region outside the potential energy barrier—known as the secondary minimum). (See Figure 11.)

The possible attractive and repulsive forces in the adhesion of cells were listed by Pethica (1961), from whom with modifications the following list is taken. Attractive forces, arranged roughly in descending order of specificity and strength are:

(1) Covalent bonding forces between surfaces.
(2) Hydrogen bonding.
(3) Ionic (salt) links e.g. $NH_3^+ \cdots {}^-OOC$
$$\text{or } COO^- Ca^{++} \, {}^-OOC.$$
(4) Charge fluctuation forces due to the ionisation and deionisation of individual groups which are at a pH where they are only partially ionised. These fluctuations set up an attractive force in a

nearby surface. Little is known about such forces (see Kirkwood, 1955).

(5) Forces due to charge mosaics on either surface so that areas of opposite sign face each other.

(6) Electrostatic attraction between like charged surfaces of differing surface potential.

(7) Van der Waals forces, both the close range forces acting in the primary minimum and possibly (see Jehle, 1963) of considerable specificity, and the long range forces of little specificity.

(8) Image forces due to the tendency for ions to move away from regions of low dielectric constant (desolvated) which lowers the osmotic pressure in the gap between two bodies, thus leading to their attraction due to the excess of osmotic pressure acting on the rest of their surfaces (see Pethica, 1961; Levine, 1946).

Surface energy or tension effects, representing in the main Van der Waals and electrostatic forces, may be a convenient way of representing adhesive forces when two surfaces come into molecular contact (see Steinberg 1964a,b).

The main repulsive forces which can act are as follows:

(1) Charge repulsion between surfaces of like charge.

(2) Van der Waals forces of repulsion between unlike atoms.

(3) Excess ionic concentration in the double layer leading to osmotic forces of repulsion, compare with image forces (above 8). Solvated layers near the surface may provide barriers to forces of attraction.

The intention is to discuss the various attractive and repulsive forces one by one in the following section.

ELECTROSTATIC FORCES

The electrostatic forces between surfaces are probably better understood than any other type of force concerned with the flocculation of particles. The charged groups on a cell surface, revealed by electrophoresis, give rise to a field which tends to lead to the repulsion of the surface of a cell of like sign and charge density if the second cell approaches close to the first. Surfaces begin to interact appreciably when their double layers just overlap. Since cells are negatively charged under physiological conditions a repulsive force should always exist between them when they are close, though of course this may not be sufficient to overcome some force of attraction. It has been conventional to regard these forces as having a range equivalent to the thickness of the double layer (see Chapter 2), which would imply that cells in say, 0· 10 M 1:1 electrolyte would start to interact at a separation of c. 60Å (twice the double layer thickness). In actual fact some interaction will occur over larger ranges (Overbeek 1952, plotted by Curtis, 1962a). In stronger electrolytes, say 0· 5M monovalent which is quite suitable for some cell types, the double layers will be only 10–20Å thick. The separation of cells at which electrostatic interaction starts may be quite a lot greater than the double layer thickness calculated according to the Gouy–Chapman theory if (a) Stern layers are present or if (b) desolvation occurs at the surface, effectively lowering the ionic strength.

The energies of repulsion V_r per cm^2 are calculated as follows, from Overbeek (1952). For parallel double layers,

(1) $$V_r = f\left(\frac{ze\psi_d}{kT}, \frac{ze\psi_0}{kT}\right) \Big/ z_1^2,$$

where ψ_0 is the surface potential and ψ_d that at distance d between layers, and z_1 is the valency; $e = 2\cdot 71$. For spherical surfaces,

(2) $$V_r = \frac{ea\psi_0^2}{2} h(1 + \exp(-xd)),$$

where a is the particle radius, e is dielectric constant, and x the reciprocal of the thickness of the double layer. The repulsive force or energy is temperature dependent (note error by Moscona, 1961a, who supposed it was not), being linearly proportional to the square of the temperature and similarly related to the dielectric constant of the medium.

The equation for parallel plates is probably best used for the approach of smooth surfaced cells, and figures for the repulsive energy have been calculated for two model systems (see Figure 12) using the treatment of Verwey and Overbeek (1948). It can be seen from Figure 12 that the repulsive energies for surface potentials typical of cells (see Chapter 2) are found in the range where repulsive energies change rapidly for small changes in surface potential. The equation for spherical particles is probably chiefly applicable to particles smaller than cells, but see discussion elsewhere in this chapter.

Examination of equations 1 and 2 shows that the value of the repulsive energy at a given distance from the surface falls if the concentration or valency of the medium electrolyte is raised, an effect equivalent to reduction in double layer thickness. In addition the repulsion is dependent on surface potential, and cation adsorption, in particular of di- and trivalent ions, will reduce the surface potential. The surface thickness potential is determined by the difference in chemical potential of the ions in the bulk phase and in the surface,

$$(3) \qquad\qquad \psi_0 = \frac{\Delta\mu}{z_1 e}$$

Where $\Delta\mu$ is the difference in chemical potential, z_1 the valency and $e = 2\cdot 71$. Consequently an increase in divalent cation concentration is more effective in reducing repulsive forces than a similar increase in monovalent cation concentration. Levine (1946) pointed out that the chemical potential difference between bulk and surface might vary with the separation of the surfaces leading possibly to a considerable discrepancy in the calculated repulsive forces at certain separations of particles, from that given by the Derjaguin–Landau Verwey–Overbeek treatments. If his treatment is correct the equations used in calculating repulsive forces may be inaccurate.

Consequently the presence of divalent ions will act to reduce repulsive forces between two cells in two ways: by reducing the repulsive force at a given distance from a surface for a given surface potential and by reducing the surface potential. Unfortunately most biological media are mixtures of mono- and divalent cations and a wholly acceptable basis for calculation of double layer thickness or repulsive force has not yet been derived. In addition it is obvious that other factors will affect the repulsive forces between cells, such as pH which at values c. pH $4\cdot 0$ leads to a reduction in surface potential. Adsorption of certain substances e.g. proteins, may alter the double layer structure and thus affect the repulsive forces. Similarly changes in the dielectric constant of the medium will affect the repulsive forces. The possibility that surface potentials may be underestimated from electrophoretic measurements has been raised in Chapter 2. In addition it is possible (see Chapter 2) that the cell surface is polarisable so that metabolic processes control surface potential. Sawyer et al. (1964) found some evidence that this might happen in erythrocytes. Loewenstein (1966) has suggested that cells may be very adhesive at the regions of their surfaces where the transmembrane potential is small. This would be expected if the surface potential were affected by the transmembrane potential. It could however be equally likely that when the membrane is in a low resistance state it is also very adhesive.

The relations between changes in the adhesiveness of cells and their surface potential, the medium electrolyte concentration and valency will, if it can be assumed that other forces are not affected by changes in these conditions, provide a test of the action of electrostatic repulsive forces in cell adhesion. Should a good correlation be found between surface potential, ionic conditions and adhesiveness, this will be evidence for the action of electrostatic repulsive forces. If the mechanism of penetration of repulsive energy barriers to close contact (see Pethica, 1961) (see p. 106) were at all common it would be expected to nullify relations between adhesiveness and the magnitude of the repulsive forces.

Electrostatic attractive forces can exist between particles, and Derjaguin (1954) and Bierman (1955) have pointed out that if surfaces have charges unlike in sign or even if they have unequal surface potentials of the same sign, an attractive force may exist between them. A mosaic of positive and negative charges may exist on some cell surfaces (see Chapter 2), each tending to attract an area of opposite sign of a second cell, but very large areas of charge of one sign would be required in order to permit attractive forces to act over any appreciable distance, say 100 Å (Curtis, 1962a). On the other hand it is possible that after two cell surfaces had been brought into contact by other means that "salt" links would develop between the two surfaces, between charges of opposite sign. The question of any specificity of adhesion that might result from this mechanism is discussed in Chapter 6. Pethica (1961) re-examined Bierman's calculations and showed that these forces would be unlikely to act over ranges exceeding 15 Å in range. These various electrostatic forces may act in the close adhesion of cells in particular where specificity is displayed in the adhesion.

COVALENT BONDING

Covalent bond forces will be of very short range and of considerable magnitude. Some estimate of bond strengths can be obtained from heats of formation. On this basis it is possible to calculate theoretical strength for adhesive joints. The experience with systems such as C-C bonds suggests that the actual strengths of joints bonded in this manner would be $3 \cdot 7$ erg/cm^2 for $6 \cdot 23 \times 10^{11}$ bonds/cm^2 (i.e. 6230 bonds/micra^2). Hydrogen bonds and ionic links have somewhat smaller bond strengths of 5–20 kcal/mole. The general experience with adhesives such as epoxy-resins which bind surfaces by chemical links is that they give joints a strength some 10^{-2} of the theoretical values. If we introduce a similar factor of 10^{-2} into consideration of the strength of cell adhesions which are bonded covalently, we are still left with appreciable bonding energies, unless the density of bonds on the adhesive surface is very low. This matter is considered later in this chapter. Covalent or ionic bonds will not form between surfaces unless they are fairly close. If groups on the surfaces react directly with one another, an approach of c. 10 Å between the surfaces is required before reaction and adhesion can occur. If molecules intermediate between the surfaces are introduced, and if these molecules react with components of these surfaces the separation of the surfaces at which bonding will start will be slightly greater than the length of the bonding molecule. This type of adhesion is otherwise known as cement adhesion or one in which flocculation by sensitisation has taken place. The criteria by which cement (sensitisation) adhesions can be recognised are mainly chemical: destruction or promotion of adhesion by specific chemical reactions will help to identify such types of adhesion, but it is essential that any such identification can preclude explanations in terms of other mechanisms.

CHARGE FLUCTUATION FORCES

Charge fluctuation forces due to the presence of mobile protons have been described by Kirkwood and Schumaker (1952) and again by Pethica (1961). The protons flip from one site to another and thus set up an oscillator system. If the two surfaces have in phase oscillation an attractive force is set up. The attractive force is at a maximum when half the groups on the surfaces are ionised and is probably of short range (c. double layer thickness). Pethica (1961) points out that these forces are unlikely to act in cell adhesion under physiological conditions because few chemical groups on the surface of cells have pK values close to 7, and hence are either fully ionised or un-ionised. No quantitative calculation of such forces appears to have been carried out. Some authors (see Jehle et al. 1965) have regarded the London force (see below) as a charge fluctuation force but we will not do this here.

IMAGE FORCES

Image forces due to the desolvation of membrane ions or inter-membrane ions, may act between surfaces. They lower the osmotic pressure in the gap thus driving the surfaces together under the action of osmotic pressure on the rest of the surfaces of the adhering cells, which is now in excess to that in the intermembrane gap. Pethica (1961) attempted to calculate this effect and found that it might give rise to a fairly appreciable force but accurate calculation was impossible. Steinberg (1962a; 1964a) considered whether calcium ions might not lead to cell adhesion by being desolvated with great ease but concluded on qualitative examination of the hydration number for calcium that this was unlikely to occur, since adhesion would appear only at much higher molarities of calcium cations in the medium than are actually required for cell adhesion.

VAN DER WAALS–LONDON FORCE

Van der Waals forces of attraction and repulsion may act between surfaces, but are they sufficient to affect cell adhesion? Of the Van der Waals forces the London dispersion force will probably be the most important, it being the only one with appreciable range. The London force is due to the rapidly fluctuating dipole of the dispersion electron on an atom, which sets up electromagnetic oscillations of c. 5×10^{-5} cm wavelength polarising the dipoles of surrounding atoms so that they are attracted towards the first atom. Atoms of the same element or those of close atomic number will thus have an attractive force between them, but atoms of very dissimilar elements will experience repulsion due to the London forces. It can thus be seen that components of cells are unlikely to give rise to repulsive London forces, since they are mainly of similar atomic weight.

The London dispersion force between two isolated atoms falls off as $1/d^7$ where d is the distance from one atom. However, the force does not fall off so rapidly with distance between two bodies composed of many atoms. Solutions for the London forces between two colloid particles obtained by Hamaker (1937) were used by Derjaguin and Landau (1941) and Verwey and Overbeek (1948, publication delayed) independently and simultaneously in their theories explaining the interaction (flocculation, stabilisation etc.) of lyophobic colloids: incidentally these theories of the stability of colloids are nowadays widely known as the DLVO theory. Unfortunately these early, though still most useful, treatments treated the London force between two particles as though they were separated by a vacuum, though a rough correction to allow for the presence of water in between the particles was known to Verwey and Overbeek (see Hamaker 1937; Overbeek 1952). Vold (1961) examined the action of these forces when surfaces carried adsorbed layers on them. For two parallel plates the attraction energy V_A between them (conventionally expressed as negative, repulsive forces being positive) is given by

$$(4) \qquad V_A = -\frac{A}{48d^2\pi}$$

where A is the Hamaker London constant for the material of the plates, and d half the distance between the plates. This equation is only valid when d is very much less than the thickness of the plates, and is not valid for very small values of d, i.e. 5 Å. For two spherical particles,

$$(5) \qquad V_a = -\frac{A\,a}{12H},$$

where a is the particle radius and H is the closest separation of the particles.

At appreciable separations between particles, for example those approaching a quarter of the London wavelength the forces are retarded owing to the time it takes for a wave to travel from one particle to the other (see Casimir and Polder, 1946).

Lifshitz (1956) and Dzyaloshinskii et al. (1959) approached the problem of the action of the London force from a macroscopic point of view and were able to solve for the case of two particles separated by a third medium such as water. The calculation of the London constant for the Verwey–Overbeek approach is based on the optical constants of the particles involved (see Moelwyn-Hughes, 1964) and similarly the approach of Lifshitz and Dzyaloshinskii is based on the electrostatic values of the dielectric constants derived from optical constants. Although the particular form of the dielectric constants used in this theory cannot as yet be accurately evaluated, the theory does predict that an attractive force exists between two surfaces if there is a difference between the dielectric constants of surfaces and the intervening medium. The larger the difference, the greater the force. The particular interest of this relation is that it is highly probable that the right conditions for attraction exist with cells separated by an aqueous medium, because the dielectric constants of cell and medium differ so largely.

A certain degree of doubt as to the action of London forces in colloid systems still exists, probably because the theory was first put forward on theoretical grounds, it being some twenty years before convincing experimental measurements were carried out. Experimental evidence for the existence of a long range force has been produced by Derjaguin et al. (1954) (see Kitchener and Prosser, 1957 who repeated their experiments) by Overbeek and Sparnaay (1954) and Schenkel and Kitchener (1961) by Black et al. (1960) and by Hachisu and Furusawa (1963) amongst others. The first two groups of these authors identified the force of attraction with considerable certainty as the London force. Scheludko (1962) and Voropaeva (1963) obtained a most elegant confirmation of the action of retarded London forces in liquid films. Schenkel and Kitchener investigated a system which resembles the biological one more closely than any other since they used 10 micra diameter spheres. Srivastava and

Haydon (1964) measured the London-Hamaker constant for $2 \cdot 0$–$4 \cdot 0\mu$ diameter spheres of paraffin wax in $0 \cdot 02$M KCl solutions and found that these particles adhered in the secondary minimum with a London-Hamaker constant of $1 \cdot 78 \times 10^{-13}$ erg.

Values for the London constant can be derived from optical constants of the bodies involved (see Overbeek, 1952; Lifshitz, 1956 and Moelwyn-Hughes 1964). These calculations give values of between 1×10^{-13} erg and 6×10^{-12} erg. Attempts to make experimental measurements of the London constant in non-living systems, (Albers and Overbeek, 1960) have given values in this range. In cells we might expect the lipid layers of the plasmalemma and endoplasmic reticulum to have London constants of c. 10^{-12} erg but estimates of the average constant for the whole cell (required for calculation of attractive force) have at present to be guessed at. Pethica using the simple treatment of equations 2 and 5 (1961) suggested a value of c. 10^{-14} ergs for the whole cell; but the actual value will probably depend on the lipid content of the cell in question.

The identification of the action of London forces in cell adhesion is fraught with difficulty. Younger and Noll (1958) suggested that they acted in the adhesion of virus particles to cells, but only produced qualitative evidence. Wilkins *et al.* (1962a,b) attempted to calculate the London constant for sheep polymorphonucleocytes by studying their rate of flocculation in the presence of various cations. Rosenberg (1962) found that the proportion of human conjunctiva cells adhering to a substrate of barium-stearate multilayers in a given time was reduced with increasing number of multilayers. Adhesion also depended on the nature of the substrate in which the multilayers were formed. Since the multilayers are up to 1200 Å thick he concluded that long range forces, probably London forces, might be acting in this situation. Curtis (1960a, 1962a) suggested on a mixture of theoretical and qualitative grounds that these forces act in cell adhesion, but the evidence which he had published previously was somewhat insufficient. A fuller study of the possible action of London forces in cell adhesion will shortly be made. Jehle *et al.* (1965) suggested that specific London forces might act between cells to produce specific adhesion.

The effect of Brownian motion energy in cell adhesion

The magnitude of the adhesive energies resulting from these forces should be compared with the energy of Brownian motion (thermal random energy) which will continually act to randomise any existing disposition of the cells under forces of attraction or repulsion, and which may break adhesions if they are of very low energy.

It will be appreciated from the beginning of this section that the Brownian motion of cells and particles may be most important in controlling the adhesion of cells in the primary minimum. In addition it will affect the adhesion of cells in the secondary minimum, determining the stability of such adhesions. Consequently calculation of the Brownian motion energy of a particle will provide a useful value against which to measure the stability of adhesions. A suspended particle in the medium is struck by molecules of the medium with an average energy $3/2kT$ where k is Boltzmann's constant and T the absolute temperature. Although a large particle receives many molecular hits per second it is exceedingly improbable that it will receive more than a few more consecutive strikes in any one direction than in another. Consequently the larger the particle is the smaller and smaller mean free path it will have due to Brownian motion. Conversely the smaller a particle is the fewer hits it will receive a second and the larger the mean free path it will have. Thus, the actual average kinetic energy of a particle due to Brownian motion will be the same whatever its size, though the variance (absolute) will be greater for larger particles than small. Consider this effect at the cellular level, small particles of negative charge, say viruses or large macromolecules, will occasionally have kinetic energies up to $5kT$ for a radius of say 500 Å, which will be sufficient considering the small mass of such particles to carry them through quite large energy barriers (say 1000 kT for the whole cell surface), due to electrostatic repulsive forces, because the cross-section of interaction between cell and virus is only a small fraction of the total area of the cell. Large particles, say cells, will receive an identical average kinetic energy from Brownian motion but having larger mass and circa (radius c. 50,000 Å) will experience proportionately greater repulsion from any repulsive forces because of the large cross-section of interaction. Hence they will be less likely to penetrate any potential energy barriers to close approach. But of course very infrequently cells will have larger kinetic energies then $5kT$, but if they have an energy of $5kT$ 0·1 of the time, an energy of $50kT$ will be found only 0·001 of the time. The Brownian energy $\frac{3}{2}kT = 6 \times 10^{-14}$ erg $(25°C)$.

On the above basis it will be seen that a fairly small repulsive energy c. 1 erg/cm² in the potential energy barrier will be sufficient to prevent the close contact of two cells of say adhesion area 1 micra² (10^{-8}cm^2) being brought about by Brownian motion. Protagonists of the theory that cells adhere with their surfaces in molecular contact (close contact) (Pethica, 1961; Steinberg, 1964a,b) have realised this difficulty and have offered an alternative explanation to avoid it. Bangham and Pethica (1960) and Pethica (1961) suggested that cells adhere at least initially by fine projections on their surfaces, perhaps equivalent to microvilli. They pointed out that the projections of the cell surface might be treated as spheres of say radius 0·1 micra. If this is reasonable, then as they suggest other conditions being suitable, a fine projection might receive sufficient Brownian (thermal) energy to overcome the small repulsion energy that it would experience, whereas a spherical body of large size would not. For spherical cells of 3 micra radius in 0·145M NaCl with a surface potential of —15 millivolts and a London constant of $A = 1 \times 10^{-14}$ erg they calculated that the potential energy barrier would be some 95kT at maximum, probably too large for penetration by the cells. But spheres of 0·13 micra radius would experience a repulsion of only $12kT$ in approaching each other, probably insufficient to invariably prevent their molecular contact. Thus they suggest that microvilli come into close contact because they behave like small spheres and in turn the cells become zippered together from these initial points of contact. Lesseps (1963) used this hypothesis to explain the adhesion of cells. There appear however to be three reasons for questioning this concept. First on biological grounds, it is unlikely that two microvilli will be frequently projected forward one another along the same axis, which is the requirement to apply Bangham and Pethica's treatment. More serious however are two physical objections: first the projection of a microvillus cannot be treated as the projection of a freely mobile particle because it is connected to the cell and any displacement of it towards another cell under thermal energy will be resisted by the cell surface etc. and this effect will lessen the energy of the microvillus. As a result the microvilli will possess less thermal energy per unit area to overcome potential energy banners than would small spheres, whereas water molecules etc. in the gap between the cells hitting the surfaces will impart all their kinetic energy($3/2kT$ each) to acting each surface to prevent the surfaces from approaching. Second the formulae for the repulsion of approaching cylinders will not be the same as for spheres. Finally it should be mentioned that Lesseps' electron micrographs did not clearly show the contacts he proposed.

The interaction of London and electrostatic forces. The theory of the stability of lyophobic colloids.

There is good reason (Chapter 2) to believe that cells carry charged groups under physiological conditions on their surfaces. These charges are capable of giving rise to repulsive forces. Later in this chapter it will be shown that there is evidence that the adhesiveness of cells is directly inversely related to their surface charge density. These conditions suggest that electrostatic repulsive forces act in cell adhesion. But in order that adhesion shall occur an attractive force must exist. As we have seen several types of attractive force might exist, but I feel that it is most profitable to consider the action of the London dispersion force in detail. The combination of the dispersion force with the electrostatic repulsive force forms the basis of the theories of colloid stability of Derjaguin and Landau and of Verwey and Overbeek. Both these groups of workers calculated the sum of net adhesive and repulsive energies at various distances between two particles for known conditions of surface potential, London constant, particle shape and ionic strength and valency for the medium between the particles. Both groups found that the energies of attraction and those of repulsion frequently balanced (equalled one another) at two distinct distances of particle separation (see Figure 11). The particles will be most strongly attracted together when the adhesive forces are maximal as in the primary minimum. If a very large repulsive energy barrier is present between the two minima many types of particle will be unable to move from one minimum to another. Since the secondary minimum region has no appreciable barrier (if any) to prevent approach of distant particles to this position of attraction there will be conditions where cells

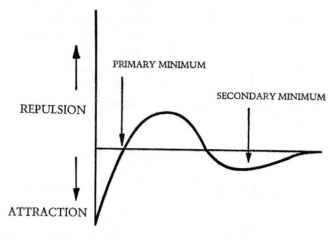

Figure 11. Diagram of energy of interaction between particles in terms of separation of particles (abscissa).

adhere in the secondary minimum. Schenkel and Kitchener (1961) provide good evidence that this occurs with their polystyrene particles (see also Van den Tempel, 1960).

The immediate biological importance of these treatments is that they explain the existence of the 100–200 Å gap found between cells at their closest approach (see earlier): this being a common value for the separation of particles when they are held in adhesion at the secondary minimum. However, this is not sufficient evidence for the action of such a system of adhesion. Pethica (1961) calculated the position and depth (i.e. the value of the maximum energy of attraction in the two minima) of the primary and secondary minima for cells of 1 micron radius in an 0·145M monovalent electrolyte with surface potential of −17 mV and the London constant=1 × 10^{-14} erg. He found a secondary minimum at c. 35 Å separation of the particles which was of energy 1·6 × 10^{-13} erg, which being equal to

$4kT$ implied negligible stabilisation against Brownian movement. He thus concluded that the secondary minimum was unimportant in cell adhesion. However this treatment of Pethica's appears to be somewhat irrelevant, because it is very well known that cells do not adhere under the conditions he chose. Consequently his dismissal of the secondary minimum and even more remarkably his suggestion of a system of cell adhesion through the contact of cell protrusions (see earlier) appear to be not well founded. It would be far more reasonable to choose conditions under which cells are known to adhere and to calculate the sum of attraction and repulsion energies for these conditions. The presence of divalent cations appears to be essential for adhesion (in normal physiological conditions), and not only do they affect double layer thickness and the way in which forces of repulsion fall off from the surface but also they reduce surface charge density and potential. Unfortunately it would be difficult to calculate the energies of repulsion and attraction in a mixed mono- and di-valent electrolyte such as cells live in under physiological conditions, but by making use of a biological observation we can circumvent this difficulty. Cells will adhere well in a medium 0·025M in calcium ions, pH 7·0, provided that the osmotic pressure is maintained with some substance such as sorbitol and that traces of Na and K are provided (not required for adhesion but to maintain cell viability). The double layer thickness in such a system is not far different from that for a monovalent saline of c. 0·1M. The energy of interaction of particles such as cells has been calculated using equations (1), (2), (4) and (5) for a divalent cation medium 0·025M (the assumption has been made that a 2:2 electrolyte is present, for ease of calculation) for plane cell surfaces of potential — 10mV and — 20mV and various Hamaker constants.

A further consideration, probably of considerable importance, is whether we should treat adhering cells as being rigid spherical particles or parallel plates. Pethica (1961) only considered cells as spheres, but on biological grounds this appears to be rather unjustifiable, partly because cells with rigidly curved surfaces such as erythrocytes are not particularly adhesive (though there may be explanations for this) but chiefly because adherent cells line up with their surfaces parallel over long distances and the regions of adhesion are surfaces of very large radii of curvature (see earlier). It will be of interest to compare results for both treatments.

Interactions for plane parallel surfaces are shown in Table II and Figure 12. For cells with a Hamaker London constant of 6×10^{-14}

TABLE II. *Values of potential energy of interaction in erg/cm² for parallel plates in the secondary minimum.*

Separation				
	A (in erg)			
−20 mV	6×10^{-14}	1×10^{-14}	6×10^{-15}	1×10^{-15}
60 Å	$+3 \cdot 56 \times 10^{-3}$	$+6 \cdot 35 \times 10^{-3}$	$+7 \cdot 05 \times 10^{-3}$	—
80 ,,	$-1 \cdot 89 \times 10^{-3}$	$+1 \cdot 76 \times 10^{-4}$	$+3 \cdot 41 \times 10^{-4}$	—
100 ,,	$-1 \cdot 56 \times 10^{-3}$	$-2 \cdot 2 \times 10^{-4}$	$-1 \cdot 21 \times 10^{-4}$	—
120 ,,	$-1 \cdot 12 \times 10^{-3}$	$-1 \cdot 84 \times 10^{-4}$	$-1 \cdot 10 \times 10^{-4}$	—
−15 mV	6×10^{-14}	1×10^{-14}	6×10^{-15}	1×10^{-15}
60 Å	$+1 \cdot 6 \times 10^{-4}$	$+2 \cdot 95 \times 10^{-3}$	$+3 \cdot 6 \times 10^{-3}$	—
80 ,,	$-2 \cdot 14 \times 10^{-3}$	$-7 \cdot 3 \times 10^{-5}$	$+9 \cdot 2 \times 10^{-5}$	—
100 ,,	$-1 \cdot 57 \times 10^{-3}$	$-2 \cdot 43 \times 10^{-4}$	$-1 \cdot 37 \times 10^{-4}$	—
120 ,,	$-1 \cdot 12 \times 10^{-3}$	$-1 \cdot 86 \times 10^{-4}$	$-1 \cdot 11 \times 10^{-4}$	—
− 10 mV		1×10^{-14}	6×10^{-15}	1×10^{-15}
60 Å	—	$+1 \cdot 24 \times 10^{-3}$	$+1 \cdot 24 \times 10^{-3}$	$+2 \cdot 50 \times 10^{-3}$
80 ,,	—	$-3 \cdot 5 \times 10^{-4}$	$-3 \cdot 5 \times 10^{-5}$	$+1 \cdot 53 \times 10^{-4}$
100 ,,	—	$-2 \cdot 52 \times 10^{-4}$	$-1 \cdot 46 \times 10^{-4}$	$-1 \cdot 27 \times 10^{-5}$
120 ,,	—	$-1 \cdot 87 \times 10^{-4}$	$-1 \cdot 12 \times 10^{-4}$	$-1 \cdot 76 \times 10^{-5}$

Surface potential at −10mV, −15 mV and −20 mV. London–Hamaker constant (A) values between 1×10^{-15} and 6×10^{-14} erg. Electrolyte, $0 \cdot 025$ M, 2:2. The interaction energies are given at four different separations of the plates. Positive values, repulsion; negative values, adhesion.

erg and surface potential − 10 mV a secondary minimum is found at c. 80 Å separation of the surfaces and this has a stabilisation energy of $2 \cdot 5 \times 10^{-3}$ erg/cm² and for − 20 mV the minimum is found at c. 85 Å separation with a stabilisation energy of 2×10^{-3} erg/cm² in $0 \cdot 025$M electrolyte ($CaCl_2$). These may seem to be very large stabilisation energies but it should be remembered that the adhesive regions of cells rarely exceed 4×10^{-7} cm² and that consequently the stabilisation energies are unlikely to exceed 4×10^{-10} erg ($10^4 kT$) and may often be much smaller. Reducing the Hamaker constant to 6×10^{-15} would increase the separation at -10 mV surface potential to c. 112 Å and at -20 mV to c. 115 Å and the stabilisation energy at the latter value to c. 15×10^{-5} erg/cm². It can thus be appreciated that surface potential and the value of the

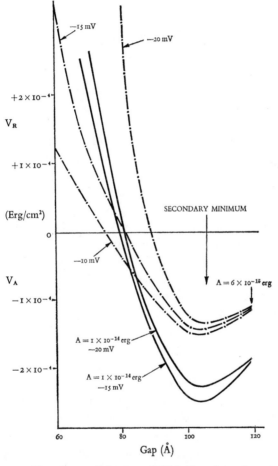

Figure 12. Plot of part of the data of Table II to show that appreciable stabilisation in the secondary minimum will occur between particles (parallel plate) of surface potential $\leqslant 20$ mV and London constant $\leqslant 1 \times 10^{-14}$ erg in 0·025 M 2:2 electrolytes. Ordinate, potential energy of interaction in erg/cm², repulsion positive; abscissa, distance between surfaces in Å.

Hamaker constant have a profound effect on the separation and stability of cell adhesions. For example, for a surface potential of -25 mV and $A = 10^{-15}$ erg an adhesive area of $10\,\mu^2$ will only have a stability of c. $10kT$ which will be insufficient to prevent dispersion of the cells even in still media. Moderate shearing forces will be able to reseparate cells of rather larger Hamaker constant and smaller surface potential. These calculations show that there are no theoretical objections to London dispersion forces holding cells in the secondary minimum and that in fact the observed separations between cells are well predicted by the theory (see earlier for data on gap dimensions), that effects of surface potential predicted are similar to observed correlations between surface potential and adhesiveness (see later), and that small values of the Hamaker constant are required. This last point was raised by Pethica (1961) who supposed cell adhesion would only occur with large values of $A = 10^{-12}$ erg. In fact if such large values were found adhesion would occur with only some 10Å separation of the surface and with improbably large energies. Wilkins *et al.* (1962a,b) found a value of c. 5×10^{-15} erg for the Hamaker constant for leucocytes and such a value would as we can now see explain observed adhesive behaviour far better than higher values.

Consider now the interaction of spherical particles of 10μ radius. For a Hamaker constant of 6×10^{-14} erg and $\psi_0 = -10$ mV a secondary minimum is found at c. 33 Å separation with a stabilisation energy of 1×10^{-11} erg and for $\psi = -20$ mV it would lie at c. 38 Å separation with an energy of c. 5×10^{-12} erg. It can immediately be appreciated that these stabilisation energies are less than are found even for $1\,\mu^2$ contact areas on the parallel plate solution. With $A = 6 \times 10^{-15}$ erg stabilisation would not occur except with very low surface potentials. These results suggest that if cells can only interact as spheres then they cannot adhere effectively or only very weakly. If the cell is able to convert such an adhesion to a parallel plate type adhesion by spreading its area of contact a stable adhesion in the secondary minimum would result. These predictions accord well with observations that curved undeformable cells such as erythrocytes do not adhere in normal media (though this may also be explained by supposing that they have lower values of A than other cells), and point contacts between deformable cells are very impermanent becoming either extensive contacts or breaking up.

When cells approach each other they will first begin to interact in the region where the retarded London force acts. Consequently at separations close to the London wavelengths of the materials concerned the attractive force will be diminished. However, when a gap of $<$ 100 Å lies between the surfaces, interaction will be in the direct (non-retarded) region.

It can also be seen that the effective value of A might vary from one cell type to another and that this could overshadow effects due to differences in surface potential. Cells with appreciable amounts of lipid in their interiors would have higher values for the London dispersion forces than those with little lipid. Cell deformability would also affect adhesion. But in the main these calculations show that there are good theoretical reasons for supposing that adhesion between cells can occur in the secondary minimum with London dispersion forces, since the gap thickness, strength of adhesion and dependence on surface potential amongst other factors are well accounted for. Under conditions of low surface potential the London forces will lead to adhesion in the primary minimum.

Types of adhesion

We can now predict on physical grounds that there will be at least three main types of adhesion between particles. For all three systems interactions of the type treated by the theory of the stability of lyophobic colloids will determine conditions under which surfaces come close enough for adhesions to form, but once the surfaces have approached close enough additional factors may act in the case of adhesion in the primary minimum, and will do so for adhesion by bridging. For these reasons the question of the reversibility of adhesions by changing from non-adhesive conditions to adhesive conditions and back becomes of considerable importance in the diagnosis of the adhesive mechanisms of cells. It is likely that only adhesion in the secondary minimum will be fully reversible. Thus surface potential may determine whether adhesions form in all three types of mechanism but may not produce reversibility except when adhesions take place in the secondary minimum. It should be realised that adhesion in the primary minimum is under certain conditions a prediction of the lyophobic colloid stability (DLVO) theory.

(1) CLOSE OR MOLECULAR CONTACT

This will only occur if the particles possess either little repulsive force because of their low surface potential or small size, or possess considerable kinetic energy for their size so that they are able to penetrate potential energy barriers. Very small particles will acquire sufficient kinetic energy from Brownian movement to penetrate large potential energy barriers because of their very small area of interaction. Cell movement e.g. that of sperm may act in some cases to provide this energy (this is discussed later). Bond forces, various electrostatic interactions or London forces acting in the primary minimum will bind the particles together with very considerable energy ranging from 10^2 to 10^5 erg/cm^2 unless bond sites are very rare, in which case lower energies will be found. Consequently redispersion of adhering particles by shearing forces will be difficult and require large energy input. Morphologically such adhesion will be characterised by a very small, say 5Å, gap or none at all between the original surfaces. Image and charge fluctuation forces may also bring about this type of binding. Considerable specificity in adhesion may be observed if it is dependent on either the presence or stereochemistry of certain groups on the surface.

(2) ADHESION IN THE SECONDARY MINIMUM

This type of adhesion can occur if the particles possess little kinetic energy relative to the energies of repulsion they encounter in adhesive interactions. In practice this requires that the particles have appreciable surface potentials and interaction-distance curves of the type shown in Figures 11 and 12. A London constant of between 5×10^{-15} and 10^{-12} erg may produce this minimum. Characteristic of this type of adhesion will be (i) it will only occur between fairly large bodies (c. > 500 Å radius) and under many conditions only > 2000 Å, (ii) a gap will be found between adhering particles at their closest approach, (iii) the strength of adhesion will be low compared with that for close adhesion, and unlike that may vary from almost negligible adhesiveness to strengths of c. 10^{-4} erg/cm², (iv) the degree of adhesiveness will be sensitive to the surface potential and ionic strength and valency of ions in the medium, (v) redispersion of the particles will be easy because of their low adhesive energies, (vi) little or no specificity will be observed in such adhesions provided the particles have appreciable electrostatic repulsive forces to prevent close interaction, but graded degrees of adhesiveness may be found between differing types of particles depending upon their surface potentials, charge reduction by cation binding, ionic strength and London constants, and (vii) formation and destruction of adhesions will be completely reversible. For convenience this type of adhesion will often be referred to as type 2 adhesion hereafter.

(3) ADHESION BY SENSITISATION (BRIDGING)

This type of adhesion has been only briefly mentioned previously, partly because much less is known about its physical chemistry than for the other two types. That molecules from the medium should bind two particles together requires that strong forces of attraction or repulsion be absent, except of course for those bond forces between cell surface and binding molecule. If this is not so, attraction may draw the surfaces together so close that binding cannot act, or repulsion drive particles so far apart that no bridging can occur. Bridging might conceivably occur when two cells have been brought into a weak secondary minimum or type 2 adhesion and act to strengthen the adhesion. We know from the work of Healy (1961), Healy and La Mer (1964) and Kragh and Langston (1962) amongst others that sensitisation is dependent upon ionic strength, pH, and the specific chemical nature and amount of the bridging agent, which presumably reacts with surface groupings on the cell. Nevertheless in some cases the chemical specificity may be more general, being brought about by a range of different macromolecules all say possessing phosphate groups on their chains. Adhesion produced by antibodies or agglutination of cells may be due to a form of bridging action. It is harder to list typical features for this type of adhesion but it seems probable that the following are tolerably diagnostic, (i) a gap will be found between the particles at closest approach but the width of this gap is somewhat unpredictable, (ii) adhesion could be highly specific so that one bridging agent would only work for one cell type, though it might also be relatively unspecific, (iii) the range of adhesive strengths will be very large but if bonds are dense per unit surface area, say 1 per 2000 $Å^2$, adhesive strengths will be very large and approach the maximum strengths of type 1 adhesions, (iv) the bonding will usually be very sensitive to chemical conditions, for example if the bridging agent binds ionically pH changes may lead to deionisation of one or other of the groups involved and de-adhesion will result; but unfortunately this cannot be regarded as an invariable rule, (v) according to Healy and La Mer (1964) in the presence of sufficient bridging agent to combine with more than half the binding groups on each surface, adhesions will weaken and (vi) complete reversibility of adhesions is very unlikely. (In biological terms the sensitisation theory is a cement theory.)

The measurement of adhesiveness

In the preceding section the question of the quantitative aspects of adhesiveness has arisen; in addition the biologist wishes to know whether one cell type is more adhesive than another in order to test interpretations of phenomena of cell behaviour such as sorting out in reaggregation. Absolute values are required to provide some test of mechanisms of adhesion, but relative values may be sufficient for attempts to explain a given biological situation. Any measurement of adhesiveness of say a cell refers of course to its adhesion with one specific type of surface. Unfortunately techniques for the measure-ment of adhesiveness are, even in ideal situations, extremely unsatis-factory, particularly those methods which measure the force required to break an adhesion. Two surfaces may be detached from one another by a large force acting in a very short time so that the surfaces are parallel over their whole extents as detachment occurs. Or peeling may occur so that separation of the surface spreads from one region. This latter type of de-adhesion requires much smaller forces which however must act for longer times. Bikerman (1957) and Jouwersma (1960) examined the peeling of adhesive joints and pointed out that one important factor is the deformation of at least one surface, which requires of course additional energy to that required to break adhesive bonds. Steinberg (1964b) appears to have ignored this point, though he does stress the important fact that a major consideration in assessing adhesiveness is the energy involved and not the magnitude of the force required to break an adhesion, which varies at various times during detachment. In addition there is some evidence (L. Weiss, 1961a, 2a) that the rheological properties of cell surfaces change as forces which break adhesions (distractive forces) are applied. Finally some part of a force applied to separate a cell from another surface may in practice be involved only in deforming parts of the cell far from the region of adhesion.

It is consequently hardly surprising that very few measurements of the absolute energy of adhesion of cells have been made although a number of systems of measurement have been set up which are potentially capable of being analysed to make this calculation. L. Weiss (1961a) introduced the technique of shearing cells off a surface to which they were adhering by applying a shearing force through the medium above them. Unfortunately the shear in his machine varied from place to place over a sheet of cells, since he used a parallel plate system for applying the shear and not the cone-plate system of viscometry which ensures that shear is equal over the whole of the measuring area. He found that shear rates of c. 10 dyne/cm^2 led to cell detachment. We can use this figure to make some assessment of the strength of cell adhesions to surfaces such as glass, plastics etc. which Weiss used. For a circular cell of radius 10 micra we can calculate, assuming that the cell is detached by being separated 200 Å without any peeling, a binding energy of 60×10^{-11} erg. However this may be an underestimate because peeling occurs. Or more likely an overestimate because most of the shearing force goes into deforming the cell and because it is assumed that the attractive forces remain constant over 200 Å. However it seems probable that we can accept value in the range 10^{-5} to 10^{-12} erg/cm^2 as possible binding energies for cells. This may at first sight seem an unfortunately large range of values, but since adhesion by direct chemical bonding could have energies up to 10^3 erg/cm^2 it appears to give a result of some use. L. Weiss (1961a,b) calculated adhesive forces of 10^5 dyne/cm^2 from surface tension considerations on the basis that the surfaces came into molecular contact. In order to equate this result with his measurements of the force required for de-adhesion, he suggested that either perhaps only 1 per cent of all the surface was involved in adhesion (this second point was based on a misunderstanding of the operation of the surface contact microscope) or that de-adhesion took place by rupture in the cytoplasm. Neither of these assumptions seem likely; the question of whether de-adhesion occurs by cytoplasmic rupture is discussed later. Using results on the rate of de-adhesion of cells from glass some estimate can be made of the probable forces of attraction at c. 200 Å separation. With interference reflection microscopy Curtis (1964) found that cells in a hanging drop culture de-adhered under (presumably) their own weight from glass surfaces when the ionic strength was very low. It was possible to observe that a peeling process took place starting at the perimeter of contact of the cell, so that greatest resistance to the force of gravity occurred at the start of de-adhesion. Such a cell of contact area $300^2 \mu$ might exert a force of $4 \cdot 1 \times 10^{-6}$ dyne under gravity and start de-adhering at c. $12 \cdot 0 \mu^2$ per sec. In this case the adhesive energy might be equivalent to $3 \cdot 2 \times 10^{-5}$ erg/cm² or $3 \cdot 2 \times 10^{-13}$ erg/μ^2 (9×10^{-11} erg for whole cell) at c. 100 Å separation of cell and glass, on the assumption that an attractive force of this value acts over 100 Å and then drops away to nothing. These examples show that accurate calculation is very difficult unless the way in which forces of attraction decline with distance is known, so that accurate integration to obtain the resultant adhesive energy can be carried out.

Curtis (1967) introduced the use of flocculation kinetics to measure cell adhesiveness. In this technique cell suspensions are placed in a known shear gradient. Under these conditions the frequency of collision between the cells can be calculated. By measuring the actual rate of aggregation the probability that a collision results in an adhesion can be calculated. This probability is known as the stability ratio and can be used to calculate the energy of adhesion. Early results show that many cell adhesions have an energy of c. 10^{-4} erg/cm² (or 100–500 kT per cell to cell interaction).

It can now be appreciated that the adhesiveness of cells is probably of low energy compared with many conventional adhesive systems such as epoxy-resins, and that none of the three main types

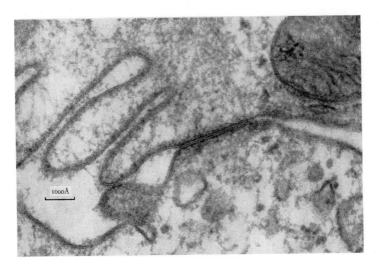

1000Å

3.3   Zonulae occludentes in kidney tissue. By courtesy of Dr. M. Farquhar.

of adhesive mechanism is ruled out by reason of this low value. Although adhesion in the primary minimum could give far stronger adhesions than are found in cellular systems they need not necessarily do so, for direct chemical bonding between two surfaces may be so sparse that the resultant adhesion is weak; for instance if bonds with an energy of 20 kcal/mole exist between two surfaces, it is only necessary to have 6 bonds per $0.1 \mu^2$ to provide a binding energy of 8×10^{-5} erg/cm^2. One conclusion from this is that if one visualises staining between cells in electron micrographs as a dense mass of bonds between the surfaces one is misinterpreting reality. Either cells bind chemically by direct contact or through cements, with very sparse bonding or by adhesion in the secondary minimum. These observations on the strength of cell adhesions are of course very incomplete at present and it may turn out that some contacts are very strongly bound in life. The finding of strong binding will be a proof that adhesion in the secondary minimum is not taking place, but the present finding of weak adhesion is only negative evidence which allows the existence of any mechanism.

Another way of stating these observations is to say that cells are easily redispersed by shearing forces. This is in fact a common observation using live cells and is often made use of in the disaggregation of tissues which is assisted by pipetting the tissue through a narrow orifice or by pressing tissue through bolting silk (Wilson, 1907). Colloid chemists tend to regard the easy redispersion (or peptisation) of particles as evidence that adhesion in the secondary minimum is taking place (Overbeek, 1952; Wilkins, et al. 1962a,b). Suppose two particles are unable to bond chemically and possess very small London dispersion forces, $(A < 10^{-17}$ erg) then if they bear appreciable surface charge they will be unable to adhere, but if they lack charges weak adhesion might result in the primary minimum, yet with the possibility of redispersion. This example appears to contradict the generalisation just mentioned but it is probably a very rare event. Wilkins et al. claimed that the fact that flocculated sheep leucocytes could not be redispersed in their experiments was evidence against cell adhesion in the secondary minimum but there are two objections to this interpretation. First the cells were at their charge reversal point so that they would possess no

E

electrostatic repulsive forces to establish a secondary minimum, secondly it is well known that cells can often be redispersed, particularly in near physiological conditions. Obviously once two cells come into contact (molecular) in the primary minimum the potential energy barrier will tend to prevent their being reseparated, provided their surfaces possess a reasonable surface potential, even if the actual forces of attraction are no longer appreciable. The evidence from the ease of redispersion of cells should be cautiously taken as suggestive evidence for the existence of adhesion in secondary minimum, though an alternative explanation is available.

Compared with these few estimates of the absolute strength of adhesions many measurements of the relative strength of cell adhesions have been made. Early methods depended on allowing cells to adhere to glass and then detaching them under gravity or with a shake (Dan, 1936, 1947b; L. Weiss, 1959a), a slightly more advanced system of centrifuging cells off a surface to which they were adherent was tried by L. Weiss (1961a,b), Easty et al. (1960), Berwick and Coman (1962). Garvin (1961) estimated cellular adhesiveness by passing cell suspensions down columns of glass beads and counting the percentage not adhering to the glass. In all these methods the percentage of the population leaving or being added to the surface was used as a measure of adhesiveness. Cell to cell adhesiveness has been harder to measure but Coman (1944), Coman (1961) and Berwick and Coman (1962) have tried to compare the cell to cell adhesiveness of different types. They used a micromanipulation method to pull cells apart and compared the bending of a microneedle, on which one of the cells was impaled, at the point of separation of the cells for various cell types. All these techniques have provided comparisons of qualitative interest, which will be discussed later, but it is very sad that few measurements of quantitative nature have been made, for such would be highly diagnostic of particular adhesion mechanisms if carried out over a range of conditions. Malenkov et al. (1963) measured the force required to separate mouse cells by micromanipulation and found that forces of c. 0·05 to 0·2 dyne were needed. They attempted to calculate the adhesive interaction from these results but since there was no assessment of how much energy went into deforming the cells their results, like all others in which micro-manipulation has been used so far, are in effect only qualitative.

The effect of contact area on cell adhesions

This problem has been chiefly raised by L. Weiss (1961a) who made the important point that in measurements of the strength of cellular adhesiveness the actual area involved in the adhesion should be known in order to calculate the actual strength of adhesion of cells. Thus this knowledge is required to assess possible adhesive mechanisms from adhesive strengths. Weiss made no attempt to measure the contact area, but believing that adhesive strengths should be due to forces of values of the order 10^5 dynes/cm^2, concluded from the low shear values required to detach cells from glass that only a very small part of cell-surface contact area was involved in adhesion. Alternative solutions to Weiss' problem exist. Curtis (1964) found that although the area of closest approach (probably of strongest adhesion) of cell to glass of such adhesions was only c. 0·1 to 0·05 of the total area of cell facing substrate, over most of the whole contact area the gap between cell and substrate was sufficiently narrow for adhesive forces to act. Electron micrographs frequently show regions of 100–200 Å gap over extensive parts of cell to cell adhesion, though as we have seen these may be partially artefactual. Although it seems improbable that only some 10^{-3} to 10^{-4} of the cell surface is involved in adhesion, as L. Weiss would suggest, or that so small a portion of the cell surface carries all the adhesive functions of the cell (see electron micrographs), it is quite possible that only one tenth of the potential adhesive area is actually involved in adhesion, in which case values for the strength of adhesion from measurements should be raised tenfold, but this would not greatly alter our conclusions from such measurements.

The kinetic energy of cell movement and adhesion

Earlier it was mentioned that some cells might possess sufficient kinetic energy to overcome potential energy barriers of repulsion. The kinetic energy of cells derived from their movement ranges from very small values to in extreme cases c. 1×10^{-1} dyne locomotive force for *Chaos chaos* (L. Weiss, 1964b). In the main cells have

comparatively large interaction areas so that the electrostatic repulsive forces will be up to c. 10^{-5} dyne in value (using the parallel
plate solution) and able to prevent the force of movement of many
tissue cells from driving the two cells into close contact. But in the
example given by Weiss the locomotive force of the cell is much
larger and moreover he considers that the cells behave as spherical
bodies so that they would experience even smaller electrostatic
repulsive forces. Thus in the simple treatment given by Weiss it
would appear that actively moving cells might be able to overcome
the potential energy barrier to close adhesion. Two factors act to
prevent this occurring. First cells carry a thin layer of medium with
them as they move, hydrodynamically bound to them, so that before
two cells make contact they will begin to exert force to prevent
their continued motion towards one another; moreover as these
water layers touch the cells may undergo deformation from a spherical shape to one better treated by the parallel plate solution. Second,
the drainage phenomenon may provide an apparent force preventing
contact. Gillespie and Rideal (1956) and Elton (1948) considered the
action of this phenomenon in non-living systems, and Curtis
(1962a) suggested that such phenomena might occur in cellular
systems. When two surfaces approach the medium between them
must be drained away into the surroundings, and the rate at which
it does so will be controlled by its viscosity, the dimensions of the
gap between the cells and the forces under which the gap is being
closed. Similarly when two surfaces are being separated the flow of
medium into the gap between them will tend to limit the rate of
separation, for example separation under a given force will be slower
in a medium of high viscosity than in one of low viscosity. In other
words there will be an apparent force trying to prevent change in the
thickness of the gap between cells. This force may nearly equal but
can never exceed in magnitude the actual force causing the surfaces
to approach or separate. Curtis (1962a) pointed out that the approach
of cells to one another under gravitational forces would be very
delayed and cells might take 100 seconds to settle down on to a
surface, a prediction partially confirmed by Taylor's (1961) measurements on the time of settling of tissue cells on to glass. Weiss (1964b)
examined the case of drainage between two approaching *Chaos*

chaos cells, which have a locomotive force of 0·15 dyne, and considering one of the cells to be a sphere and the other flat surfaced he found that this drainage force would not appreciably retard the approach of the two cells. If however the cells were both plane parallel surfaced in the possible contact region the drainage phenomenon would reduce the effective locomotive force of the cells to 7×10^{-5} dyne at 100 Å separation and it would fall to lower values as they approached closer. Consequently the electrostatic repulsive forces ($< 10^{-4}$ dyne) would be able to prevent close contact except for small areas of interaction. Furthermore in these calculations the viscosity of medium close to the cells has been assumed to be that of the bulk medium whereas there may be reason to think (see Chapter 2) that it is considerably higher (thus further increasing the apparent repulsive force due to drainage). Thus the drainage effect, the probable flattening of curved cell surfaces (except when they are rigid) as cells approach close to one another and the hydrodynamically bound layer of medium tend to prevent cell movement bringing about close contact. Unfortunately we cannot as yet provide the necessary values of viscosity, deformability etc. to present a quantitative account of all these effects. Weiss (1964b) does not consider the probable flattening of approaching cell surfaces and the resulting drainage effect on interaction and thus comes to the conclusion that cells must come into molecular contact under the action of their locomotive forces. He thus raises the question of why is it that cells such as those of *Chaos* which are not adherent fail to adhere. He suggests that there must be no system to keep them in adhesion once molecular contact is made. This is of course very unlikely because London dispersion forces acting in the primary minimum if nothing else would serve to keep the cells adherent. Moreover, if Weiss' contention were correct one could not expect to find any correlation between surface potential and adhesiveness which of course (see later) does exist. There is however one type of interaction of cells in which the kinetic energy is almost certainly sufficient to bring about close contact. This is sperm-egg interaction. Using data given by Rothschild (1961) on the kinetics of sperm movement one can calculate that bull sperm have a locomotive force of c. 1×10^{-5} dyne, and their interaction area with a cell surface is unlikely to exceed 10^{-7}cm^2 for which the potential energy barrier would have a repulsive force of c. 10^{-9} dyne. The shape of sperm and the acrosome filament would particularly prevent drainage effects from being of appreciable action, because a fine filament approaching a surface has a very small cross-sectional area in interaction with the egg surface.

CHAPTER 4

Cell Adhesion:
(II) The Biological Evidence

Three main systems of adhesion have been suggested for cells on theoretical grounds and in this section I intend to discover to what extent actual biological evidence supports one or other mechanism. As will appear, it is very probable that any one of the three mechanisms will be found to occur in at least some situations. But on biological grounds we can immediately separate off the adhesion of unlike cell types, e.g. sperm and egg, and adhesion under non-physiological conditions. Unless it is specifically stated to the contrary, cell adhesion will refer to the adhesion of tissue cells under normal or near normal physiological conditions. Unfortunately owing to our difficulty in separating phenomena of cell adhesion from those of cell movement a certain degree of ambiguity subsists on some of the evidence presented. For example, Deuchar (1961) has shown that adenosine triphosphate accelerates the aggregation of amphibian mesoderm into the somite blocks. But this might be brought about either by accelerated motility of the cells so that they move into blocks more rapidly or to increased adhesiveness of the cells. Then again effects on cell locomotion may be due, though they need not, to alterations in the adhesiveness of the cells concerned because cells must adhere in order to move. This ambiguity affects our interpretation of evidence from complex cellular systems such as reaggregates or whole animals. On the other hand, evidence derived from studies of the dispersion, disaggregation (unfortunately sometimes known as dissociation) or separation of tissues into single cells is unlikely to be much affected by this consideration because it is usually carried out too rapidly for cell movement to affect the result. These two processes, disaggregation or de-adhesion and their converse, adhesion with its associated phenomena of reaggregation and the sorting out of cells in reaggregates, and in more complex tissue systems, provide much evidence on adhesive mechanisms. In addition, biophysical evidence will be considered.

THE DISAGGREGATION OF TISSUES

Although it is sometimes supposed that the separation of cells must be due only to the destruction of their adhesive mechanisms there are two alternative processes of disaggregation. The first is that mechanical forces applied to shear two cells apart may be stronger than the bonds between the cells so that the cells will be separated, although there will be no reason to suppose that the ability of the cells to re-adhere has been destroyed. Indeed, Whitefield (1964) has suggested that the chelating agent, ethylene diamine tetra-acetate, (some trade names for this chemical are Sequestric acid, Sequestrone, Complexone I or Versene) which separates cells of the slime mould *Dictyostelium discoideum*, does so by stimulating pseudopodal activity with the result that the cells tear apart from one another. The second method may occur in tissues in which there is a large amount of extra-cellular matter. The destruction or solution of this material by enzyme action would cause the cells to separate even though the adhesive mechanism by which the cells were bound to the extracellular material may not be affected by the enzyme at all.

In the last chapter the ease with which cells may be mechanically dispersed was remarked upon and it was suggested that if cells adhere by direct chemical bonding between surfaces then there must be remarkably few bonds between adhering cells. For this reason the mechanical dispersion of cells will be considered in greater detail here. Squeezing tissues through fine mesh (e.g. bolting silk as used for plankton nets) has proved to be an effective method of disaggregating sponge cells, see Wilson (1907) and Sanyal and Mookerjee (1960). Indeed it is usual to have recourse to mechanical methods of dispersion to complete cell separation even though the tissues have already been treated with a disaggregating chemical. Unless very powerful disaggregating agents are used it is necessary to complete cell dispersal by shearing the tissue through a pipette orifice, by agitating the medium around the cells or by using other shearing methods. The reason for the recourse to mechanical dispersion to complete chemical dispersion is that if a certain treatment weakens cell adhesion the cells will not separate unless some force acts to pull them apart.When a very considerable weakening of the adhesive forces is brought about by a certain chemical treatment, Brownian movement energy may be sufficient to break the adhesions, but usually this is insufficient and recourse has to be had to mechanical dispersion. The disaggregation of, for example, amphibian embryonic tissues after chemical treatment will be aided by the action of gravity on the cells so that those at the edge of the tissue fall free of it. Many other tissues require the use of a pipette. These considerations indicate that in order to show the disaggregating effect of a certain reagent it is necessary to carry out control mechanical agitation of the untreated tissue. This has very rarely been done except by L. Weiss (see papers individually described hereafter).

Temperatures in the range of 0°C–40°C do not bring about disaggregation of tissue (see, for example, Steinberg (1962a) who investigated this point in chick embryonic tissue reaggregates). But L. Weiss (1964a) found that foetal rat skin cells were detached from glass by shearing forces more easily at 37°C than 20°C.

Treatment of tissues with alkaline media generally aids disaggregation. Holtfreter (1943a) observed that amphibian embryonic tissue disaggregates above pH 9·5, though Steinberg (1962b) found that a pH 10·5 was necessary to disaggregate *Triturus pyrrhogaster* embryos. Essner *et al.* (1954) observed that rat hepatoma cells disaggregate spontaneously at pH values $> 7·0$. However, when a non-cellular material is definitely involved, relations may be different, for example, Chambers (1940) noticed that the hyaline layer around sea-urchin eggs dissolves on lowering the pH and Chambers and Zweifach (1947) consider that the cement material around capillaries breaks down at low pH values. Lowering the pH of media below pH 6·0 does not bring about disaggregation, or separation of cells from surfaces such as glass.

Little is known about the effects of other monovalent cations on cell disaggregation, although it is well appreciated in an empirical way that tissues do not disaggregate in hypertonic media although they may be very damaged of course. In unpublished work I found that *Xenopus laevis* neurulae disaggregate more easily in calcium-free

media if the NaCl concentration of the medium has been reduced to
0·005M, the tonicity being maintained with sorbitol. Curtis (1964)
found that lowering the ionic strength of tissue culture medium to
0·001 resulted in the de-adhesion of cells from glass surfaces. More
is known about the effects of divalent cations, and in particular about
the action of calcium. Rappaport (1966a,b) and Rappaport and
Howze (1966a,b) have reported that the chelating agent sodium
tetraphenylboron disaggregates adult mouse liver and other
tissues. They attribute its action to its ability to chelate potassium
and suggest that potassium plays a major role in cell adhesion.
However, sodium tetraphenylboron also chelates calcium and they
omitted calcium or sufficient calcium for adhesion from their disaggregation media. When 1×10^{-3}M sodium tetraphenylboron is
used with $1·5 \times 10^{-3}$M $CaCl_2$ no disaggregation occurs.

Roux (1894) showed that calcium-free media aided the disaggregation of *Rana fusca* embryos and Herbst (1900) found that the
blastomeres of echinoderm embryos separated in calcium-free seawater. In the latter case it is clear that the disaggregation was certainly mainly due to the breakdown of the hyaline capsule around
the egg. In recent years it has become more common to use chelating
agents to effect disaggregation. These reagents remove other di- and
tri-valent cations and consequently, as L. Weiss (1960a) points out,
effective disaggregation by EDTA is not a proof that calcium is
involved in cell adhesion. But when the evidence from disaggregation and aggregation are combined, the importance of calcium
and magnesium appears. The most widely used chelating agent has
been ethylene diamine tetra-acetate (EDTA) but imino-diacetic
acid, *o*–cresolphthalein complexone (Curtis unpublished) and
citrate (Feldman 1955) and Laws and Strickland (1961) have been
tried with success. Anderson (1953) and Zwilling (1954) introduced
EDTA as a disaggregating agent for mammalian and chick material
respectively, and Curtis (unpublished) applied it to the separation of
amphibian embryonic cells. Coman (1954) perfused adult rat livers
with EDTA and observed tissue disaggregation and cell separation
using electron microscopy. Easty and Mutolo (1960) re-examined
this system and were unable to obtain cell dispersion. Since they
used EDTA at *p*H 6·0 under which conditions it has little binding

power for calcium it is not surprising that disaggregation failed. Leeson and Kalant (1961) obtained extensive disaggregation of rat liver when EDTA was used at pH 7·4, at which pH the stability constant of the calcium complex is appreciable. It was found (Curtis, unpublished) that EDTA would not disaggregate amphibian embryonic tissue unless used above pH 7·8 where the stability constant of the calcium complex is c. 10^6. Similar conditions are required to remove tissue culture cells from glass surfaces. An apparently rather confused story applies to the disaggregation of slime mould cells. De Haan (1959) found that EDTA disaggregated all stages in the life cycle, but Gerisch (1961), who used this reagent at pH 6·0 was unable to obtain extensive disaggregation of fruiting stages. This result is apparently to be explained by use of the reagent at pH 6·0, but Whitefield (1964) was unable to obtain disaggregation of fruiting stages at pH values up to 8·0.

It is of considerable interest that citrate is an effective disaggregating agent for amphibian tissues at pH 7·0 where its calcium complex has a stability constant of c. 1700, whereas EDTA has to be used at pH 8·0, and is ineffective when used under similar conditions to citrate. This difference is difficult to explain, though it may be that EDTA is partially adsorbed to the cell surface at lower pH values so that it cannot act as a chelating agent.

Despite the remarkable efficiency of chelating agents in effecting disaggregation (if final mechanical dispersion is resorted to) with little damage to the cells, it has been more usual to use enzymatic treatments for this purpose. Where extracellular material is massive (and not necessarily of any adhesive function) its solution by enzyme action will release cells from the meshwork it formed, without necessarily affecting the adhesive functions of the cells. The tendency to use enzymes for disaggregation can be attributed partly perhaps to reasons of tradition and partly to the efficacy of enzymes in lysing extracellular matter. The earliest reference to the use of enzymes for disaggregation I know is that by Schiefferdecker (1886) who treated epithelial cells with pancreatin. The use of pancreatin (trypsin plus deoxyribonuclease etc.) and trypsin has been revived by Rous and Jones (1916), Billingham and Medawar (1951) and Moscona and Moscona (1952) (see also Harber, 1964). L. Weiss

(1958) found that incubation of cells with trypsin led to loss of material from them and Steinberg (1961) showed that a jelly-like substance (probably of deoxyribonucleic acid nature) might result from the lysis of cells during trypsinisation. This jelly-like material, thought by Moscona (1962) to be the cementing material of cell adhesions, was considered by Steinberg to be derived from the nuclei of lysed cells. Steinberg also showed that the use of deoxyribonuclease with trypsin led to the preparation of cell suspensions free of jelly, as Boyse (1960) had done, a result empirically noticed also by Rinaldini (1958), who recommended the use of pancreatin or impure trypsin samples. Madden and Burk (1962) also suggested the use of a mixture of DNAase and trypsin. Hebb and Chu (1960) found that tissue culture L strain fibroblasts detached from glass substrates with trypsin did not re-adhere as easily to glass as ones which had been mechanically detached; from this they suggested that trypsin damages the cells but alternative interpretations should be easily made. Edwards and Fogh (1959) found that microvilli in cells disappear after trypsinisation. The jelly-like material often found between the cells after disaggregation is also observed when high pH or chelating agents are used. Its nature is discussed at length in connection with reaggregation since there has been controversy as to whether it represents an adhesive cement or an unfortunate artefact of disaggregation. When trypsin is used for disaggregation it is normal to use it at 37°C and near pH 7·5 and to incubate the tissues in a trypsin solution made up in a calcium and magnesium-free saline. Mechanical disaggregation usually has to be resorted to to complete disaggregation. Although there is some ambiguity it appears that the action of the enzyme probably lies in its proteolytic action. Moscona (1963b) found that trypsin inhibitors such as the soya bean inhibitor prevented disaggregation of embryonic tissue by the enzyme, but Easty and Mutolo (1960) discovered that trypsin inactivated with di–isopropyl–fluorophosphate separated cells agglutinated with protamine, though it did not disaggregate normal tissue. L. Weiss (1963b) observed that trypsinisation aided the mechanical dispersion of cells from glass surfaces in which they had been in adhesion: he used a range of mammalian cells in culture. Very many other authors have routinely used trypsinisation to obtain cell suspensions. It has been usual to disaggregate with impure trypsin so that it is possible that other substances actually bring about disaggregation. Even the purest trypsin commercially available contains some six proteins. It is interesting to note that highly purified ficin is inactive in disaggregation as is the "chemical peptidase" N–bromosuccinimide.

Papain has been occasionally used for disaggregation, Mateyko and Kopac (1963) treating various mammalian tissues with it in the course of their most interesting work reviewing a variety of methods of disaggregating cells. Whitefield (1964) applied it with success to the separation of slime mould "bodies" into cell suspensions. Easty and Mutolo (1960) found that papain disaggregated adult rat liver with ease though Essner et al. (1954) and Terayama (1962) were unable to disperse the closely related tissue, rat ascites hepatoma, with this enzyme. Pronase, an enzyme of bacterial origin, which can cleave a wide range of peptide bondings, appears to have been first used with success for the disaggregation of mammalian embryos by Mintz (1962); and Gwatkin and Thomson (1964). It seems probable that it will act with success on other tissues. Hyaluronidase (sometimes in combination with trypsin) has been used for the disaggregation of grasshopper embryos (St. Amand and Tipton 1954), but Easty and Mutolo (1960) and Laws and Strickland (1961) were unable to disperse rat liver with this enzyme. Hyaluronidase is ineffective on amphibian embryos, and L. Weiss (1963b) could not weaken adhesion of tissue culture cells to glass with it. Rinaldini (1958) used elastase to disaggregate embryonic tissues with success. Cavanaugh et al. (1963) observed that embryonic heart tissue (avian?) was disaggregated by collagenase with a higher proportion of viable cells than could be obtained after trypsinisation, and Laws and Strickland (1961) used this enzyme to disaggregate liver. Easty and Mutolo discovered that lipase, alkaline and acid phosphatases were ineffective in disaggregating rat liver. L. Weiss (1963b) found that trypsin and neuraminidase treatments aided detachment of mammalian tissue culture cells from their adhesion to glass. It is unfortunate that the purity and activity of the enzyme preparations used has been very rarely reported on.

A variety of miscellaneous agents have been reported as disaggregating tissues. Mateyko and Kopac (1963) found that caffeine disaggregated certain tumour tissues but not others. Could the effect be due to the high pH that would result from adding the base to a medium? N–acetylcysteine appears to be a fairly effective disaggregating agent for human tumour tissue (Vassar, personal communication), but it is inactive on amphibian embryonic tissue. Since this chemical is powerfully mucolytic it may act only when there is appreciable extracellular material to lyse. Coman (1960) observed that the carcinogen, 20–methylcholanthrene, caused some disaggregation *in vivo* when injected into the ear of the rabbit. Child (1953) used sodium azide to obtain cell disaggregation in the echinoderm *Patira*, but this may have been due to the high pH produced by the decomposition of azide. Yamada (1962) found that the surface active agents of the polyoxyethylene sorbitate series (tweens) disaggregated rat ascites hepatoma cells with ease, increasing the surface charge/density of the cells as measured by colloid titration (for critique see Chapter 2) some sixfold. Devillers (1955) also noticed that other surface active agents disaggregate embryonic tissues. But, as Hodes *et al.* (1961) found, there is always the danger of cell lysis with such agents.

RE-ADHESION AND REAGGREGATION OF CELLS

These studies on the separation of cells from one another or from non-living substrates can be compared with those on the formation of adhesions between cells or of cells with other substrates. This comparison poses the question of whether mechanisms of adhesion are the exact corollary of those of de-adhesion and whether cell to cell adhesion depends on the same mechanism as the adhesion of cells to other substrates. Three main types of experimental technique have been used in these studies.

(1) Cells are allowed to settle in a still medium on to a substrate which may of course be cellular. The cells may move around and on making contact build up aggregates (for discussion of mechanism see Chapter 6), or they may adhere to the substrate with little or no movement or with movement which fails to build up aggregates. In the first case measurement of the number or size of aggregates may provide some subjective assessment of the adhesiveness of the

cells, and in the second the proportion of cells that remain attached to the surface on some small shearing force being applied may give some measure of adhesiveness. Neither of these techniques is wholly satisfactory because aggregation under these conditions may reflect the influences of the rate of cell movement, its direction and frequency of turning, aspects of cell to substrate adhesion and possibly complex cell behaviour (Chapter 5 and 6). Detaching those cells which have not become adherent after a certain time is, strictly speaking, a measure of de-adhesion and may additionally be influenced by such factors as the rate of settling of the cells (Taylor, 1961). But this method has been used by Dan (1936), Buschke and White (1949), Ambrose and Easty (1960) and L. Weiss (1959a) to give a rough measure of cell substrate adhesion.

(2) Cells are centrifuged together or on to their substrate. If cells are adherent they may form tissues, if very non-adherent the pellets will redisperse very easily. Trinkaus and Lentz (1964) have used this technique for preparing aggregates but it does not appear to have been used as a test of adhesiveness (see also Hayes, 1965). A number of obvious objections attach themselves to this technique.

(3) Cell to cell adhesion and probably cell to substrate adhesion can be brought about by shaking cell suspensions. This technique was introduced by Gerisch (1960) and used later by Moscona (1961b) for the aggregation of chick embryonic tissues, and it has proved successful with cells of other groups; (see Moscona (1962) for mammalian embryonic cells, Giudice (1962) for echinoderm cells). Although Moscona used a rotary shaker which imparted a swirling motion to the medium of the suspension cultures it is possible to get equally good reaggregates with shakers which give a reciprocating movement. This technique has the attraction that it is possible to avoid the complication of competition between cell and glass for the adhesion of another cell (provided cells do not adhere to the walls of the vessel). But unfortunately reaggregation takes an appreciable time to complete or even to obtain small reaggregates so that if cell adhesiveness is judged by aggregate size one may in fact be taking a measure of cell adhesiveness averaged over a variety of conditions. Subjectively, of course, we tend to relate aggregate size to the degree of adhesion of the cells but this approach can be a little naïve.

Moscona (1961a,b) suggested that at a given temperature, in a given medium under specified conditions of shaking, the larger the aggregates the more adhesive the cells. An apparently simple justification of this view is that the larger the aggregate the more adhesive must its elements be in order to prevent the aggregate breaking up under the influence of the shearing forces in the moving medium. Curtis (1964) studied the kinetics of aggregation in this system and the main features can be deduced from Figure 13. First adhesions are mainly of cells in pairs, but in due course the common-est aggregate type becomes the 3 to 4 cell aggregate, and then in turn these are replaced by classes of larger aggregates till macroscopically visible aggregates form. A certain proportion of cells do not come into adhesion and these are not contaminating erythrocytes. Two reasons may be advanced for their failure to adhere. Firstly, except in very dense suspensions, the chance of a single cell hitting another cell or group of cells will fall off rapidly as the number of bodies in the suspension is reduced by aggregation, so that such cells may merely represent those that were never in effective collision with others. But in fact, the number of such single cells left in the sus-pension, say after 24 hours, is under many conditions too large to be accounted for by failure of collision, and they may represent damaged cells. These measurements provide us with a very suitable method for assessing cell adhesiveness as we shall shortly see. Moscona's criterion for the measurement of adhesiveness is reag-gregate diameter, in consequence of which very small reaggregates are considered to be a sign of slight cellular adhesiveness. This treat-ment obscures an important feature. This is that a cell type may only form small aggregates but a very high proportion of the total cell population may be found in the aggregates, yet another cell type forming larger aggregates may have a rather smaller proportion of the total population aggregated. Which type is the more adhesive? Obviously the most important distinction is between non-adhesive (*i.e.* remaining as single cells) and adhesive (forming aggregates). Yet this distinction is obscured by Moscona's method in which cell types forming very small aggregates are classed as non-adhesive. If a population is made up of ten thousand cells in individual aggre-gates of ten, is it composed of cells which are more, or less, adhesive than a population composed of two aggregates of five thousand cells each? There is in fact some reason for considering the cell type with the larger aggregates the more adhesive but on Moscona's criterion the aggregates composed of ten cells a piece would appear

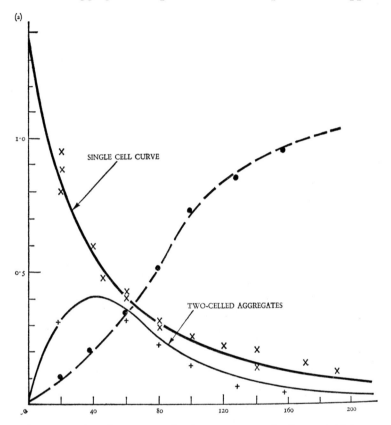

Figure 13. Flocculation curves for the aggregation of 5-day chick limb-bud embryonic cells in Hank's medium (50 per cent): Medium 199 (50 per cent). Ordinate, cell density in millions of cells per ml.; abscissa, time in minutes. Shear gradient c. 2 recip. sec. Cells were brought into collision by stirring in a shaking machine. The curves for the number of single cells and for the number of cells in two cell aggregates are the ideal curves if flocculation kinetics (slow) are obeyed. The points are experimental points and it should be noted how well they follow theoretical expectations. The ideal curves are calculated from the number of single cells/ml. at zero time and the time taken for the number of single cells to fall to a quarter of their initial density. (From Curtis and Greaves, 1965.)

to be of non-adhesive cells, which is obviously incorrect. Curtis
(1963), Curtis and Greaves (1965) and Curtis (1967) introduced the
following system of measurement in order to meet these points:
the aggregation is treated as a flocculation process. Flocculation is
the adhesion of particles due to their collision brought about by
Brownian motion, sedimentation or by flow of the medium (*i.e.*
shaking). Flocculation due to agitation of the medium is termed
orthokinetic flocculation and in this process there is a certain
probability that a collision produces an adhesion. This probability
can be used to calculate the absolute adhesiveness of the cell (see
Chapter 3). A very simple form of this new method was used
(Curtis, 1963) in which the number of cells remaining unaggregated
after, say, 16 hours was compared with the number of cells placed
in the shaker flask. A more precise method (Curtis, 1967) is derived
as follows: The total number of cells and the aggregates formed by
flocculation $N_{\infty t}$ at time t is compared to the number at zero time
$N_{\infty 0}$. The following equation expresses the conditions affecting
flocculation.

$$\ln \frac{N_{\infty t}}{N_{\infty 0}} = \frac{4G\phi\alpha t}{\pi}$$

where α is the probability that a collision results in an adhesion, ϕ
the fraction of the medium occupied by cells and G the mean shear
rate. The stability ratio α can be used to evaluate the absolute
adhesiveness not, however, using the treatment of Fuchs (1934)
but a new one developed by Curtis (in press).

The vagaries of methods of assessing adhesiveness from reaggregation phenomena have led to very differing interpretations
which it is desirable to consider immediately before proceeding to
a general survey of studies of cell adhesion. Moscona (1961a,b)
studied the reaggregation of embryonic chick cells after they had
been dispersed with trypsin. A medium containing fresh serum
(10 per cent—usually horse serum, as has been customary) was used
for the suspension medium. At 37°C Moscona found that reaggregation was extensive as judged by the formation of large visible
aggregates, whereas at, or below, 15°C reaggregation did not
apparently occur. From this he concluded that the reaggregation of

cells is dependent on the metabolic formation of a substance responsible for maintaining cell adhesion, this occurring only above 15°C. In 1960 he had described the deposition of some material on the flask wall by actively aggregating cells and observed strands of it between the cells. This material, termed ECM, short for extracellular material, was proposed by Moscona (1962) as the agent of cell adhesion which presumably acted by bridging between cells. This concept appears to have been put forward earlier by De Laubenfels (1934). A further experiment was to prepare cell free supernatants from actively metabolising cell suspensions and to add these fractions to cells about to be aggregated at 25° C (at which temperature aggregates are normally small). This treatment resulted in the formation of large aggregates, Moscona (1962). He suggested that the supernatant aided adhesion because it contained an adhesive substance which cells at 25°C were unable to make in sufficient quantity for extensive adhesion. Moscona and Moscona (1963) found that puromycin, an inhibitor of RNA action in protein synthesis, prevented aggregation of chick embryonic tissue. They also found that actinomycin D at a concentration of 0·5 μgm/ml produced a certain degree of inhibition of aggregation of these cells. Aggregation ceased after about 4 hours. They interpreted this result as indicating that a certain supply of m-RNA was already available for synthesis of the adhesive cement when actinomycin was applied. However if actinomycin D simply lowers adhesiveness somewhat by some other mechanism it would be expected from aggregation kinetics that the aggregates would be smaller. Similarly, Garber (1963) observed that glucosamine tended to prevent aggregation of chick embryonic material, its action being more marked at lower temperatures. Nakanishi et al. (1963) found that chloramphenicol partially inhibited the aggregation of dissociated embryonic chick cells. In all these experiments aggregation was judged by the presence or otherwise of aggregates which would be macroscopically visible and by the diameter of these aggregates. The results were used to support the hypothesis that a metabolically formed substance was essential for the re-adhesion of cells.

It can be easily seen that these experiments are most interesting evidence for the existence of some metabolic action on adhesion in the aggregation of chick embryonic cells in the media used. But it is a large logical step to conclude that there is the formation of some adhesive substance. There is now considerable evidence that the metabolic activity is in fact the destruction of an inhibitor of cell adhesion present in various sera. The evidence for this was derived in the following way: consideration of the quantitative methods used by Moscona and by Steinberg (1962a) suggested that aggregation was in fact taking place at 6°C but that it was somewhat suppressed. Steinberg had noticed that embryonic chick cell suspensions aggregate at 6·5°C if unagitated. By reaggregating embryonic chick liver, heart or limb-bud tissue in serum-free media, large aggregates formed rapidly whether it was carried out at 1°C, 6°C or 37°C (Curtis, 1963). Examination of the quantitative kinetics of reaggregation in serum-free media showed that very extensive adhesion takes place at 1°C, up to 80 per cent of the cells reaggregating from an initial density of c. 1×10^6/ml, but in medium containing horse serum only some 20 per cent of the cells aggregate and then only in reaggregates smaller than 50 cells. This result indicates that serum contains some factor which inhibits aggregation, but which is presumably destroyed by cells which can metabolise it, when the temperature is high enough.

Further evidence in support of this was derived by the isolation of the factor from horse serum and its identification as a protein, molecular weight, 150,000 (Curtis and Greaves, 1965; Curtis, 1966b). The factor was also found in human, chick and calf sera. The purified protein is a very effective inhibitor of reaggregation at low temperatures. These results, together with the fact that dead cells (Steinberg, personal communication) aggregate in serum-free media, support the idea that the metabolic evidence points to the destruction of an inhibitor of reaggregation. Further evidence on this point was obtained by examining the kinetics of aggregation at 37°C in the presence and absence of serum, from which it was discovered that reaggregation in the presence of inhibitor was delayed for some hours, while aggregation in the absence of serum started immediately. In addition the adsorption of the protein to cells could be detected. Glucosamine, puromycin and actinomycin D do not inhibit aggregation when the inhibitor is absent, though they do so in its presence. The action of this and related factors will be discussed later. It can thus be concluded that there is no dependence of aggregation on temperature (within the range 1°C to 37°C) for chick embryonic cells, but that in the media which it has been conventional to use, an inhibitor of adhesion was present. This inhibitor could be metabolically destroyed by the cells. Some further support for this conclusion can be derived from Steinberg's finding (1962a) that pre-incubation of disaggregated cells as a suspension culture at 30°C does not influence the subsequent effect of temperature in a serum containing medium, which argues against Moscona's suggestion and would be explained by the existence of adhesion inhibitors in serum. Steinberg (1963a) had also demonstrated that the ECM visualised by Moscona could be lysed with deoxyribonuclease and that after this had been done, reaggregation still took place: this result could, however, be answered by supposing that either very small amounts of the substance are required or that it is synthesised very rapidly. Moscona A. and Moscona M. (1966) reported that aggregation of trypsinised cells did not take place at 2°C in serum-free media and that puromycin also prevented the aggregation of trypsinised cells in serum-free media (Moscona M. and Moscona A. 1966). Curtis (in press) confirmed this, provided that heavy trypsinisation is carried out: in other work EDTA was used for tissue dispersal. Trypsinised cells do not reaggregate in the cold because trypsin contains the aggregation-inhibiting protein. The conclusion, however, from all this work is that there is no evidence at all for the existence of binding by extracellular material in chick embryonic aggregates.

However, Moscona and his school attempted to get evidence for a similar mechanism in the reaggregation of sponge cells. When sponge cell suspensions of different species are mixed any one of the resulting aggregate bodies tends to be of cells of one species type, so that there appears to be some very specific mechanism in the aggregation (see Chapter 6 for further discussion). It would obviously be of great interest to show that this specificity is dependent on some specific type of adhesive mechanism such as a series of adhesive substances. Humphreys (1963) and Moscona (1962) both published papers on what apparently is the same work. They found that cells of the sponges *Microciona prolifera* and *Halicona occulata* alone amongst other species tried (Humphreys, personal communication) would not aggregate at 5°C when they had been disaggregated by agitating sponge fragments in calcium- and magnesium-free artificial seawater, though reaggregation at 24°C was good. If, however, the sponges were disaggregated mechanically by pressing the tissues through a fine net reaggregation took place at both 5°C and 24°C. Reaggregation was carried out in a shaker system and aggregation (and thus adhesivity) assessed by the size of aggregates. There is thus an apparent parallel with Moscona's (1961a,b) experiments with chick embryonic cells. The work was carried further, however, by preparing fractions from the supernatants of chemically disaggregated cells. These supernatants were found to assist the aggregation of chemically disaggregated cells at 5°C if the supernatants came from like species. Whereas Moscona only reports testing the supernatants in systems containing cells of both species, Humphreys does this and also describes testing the action of supernatants on cells of a single species: this is of some importance since cell colour was used to identify species type and this test being unreliable it could be that Moscona's mixed aggregates merely showed good aggregation of some cells and not of others without any true specificity. If supernatant factors from both *Haliclona* and *Microciona* cells were added to disaggregated cells of both types, separate aggregates of each cell type formed at 5°C. It is unfortunate, as will be shortly seen, that neither author reported the initial and final number of cells in their systems nor did they try the action of the supernatants at 25°C. Both authors concluded that their results indicated the existence of an intercellular material responsible for the specific adhesion of cells. Margoliash et al., (1965) and Humphreys (1965b) claim to have isolated the material active in the promotion of aggregation of *Haliclona* as a large macromolecule containing amino-acids and amino-sugars, but the two reports differ as to the size of the molecules. Humphreys suggests that the molecules are of the right size to account for the gap dimensions. Humphreys (1965a) found that the aggregation of sponge cells was not affected by the presence of puromycin.

It can of course, easily be seen that the same logical mistake may have been made in the interpretation of these experiments as in the interpretation of the experiments with embryonic chick material. It could be that what has been demonstrated is the existence of some inhibitor of aggregation at 5°C, destroyed by adding a metabolically active substance, in which the supernatant factor does not act as a cement but as perhaps an enzyme.

Curtis (1962b) found that chemically disaggregated cells (using EDTA) of the sponges *Microciona sanguinea* and *Halichondria panicea* could be aggregated at 3°C provided that aggregation was carried out in still medium. Cells from *Suberites suberites* and *Hymeniacidon perleve* behaved in a similar manner and it was possible to prepare disc-shaped aggregates of up to 10 cm in diameter at 3°C. The supernatant from the disaggregation medium contained much organic material adsorbing at 2600 Å, but even after such prolonged washing that no detectable organic material was left in the suspension medium cells aggregated well at 3°C. It was estimated that any organic substance present must have been diluted at least 10^5 times Consequently, the effects Humphreys and Moscona found may in fact be due to some reaction of the cells to placing them in a rotary shaker. It is of course well known that only a slight increase in shaking rate prevents the adhesion of chick cells, though the reason for this is not fully understood at present, and possibly lower shaking rates are required at 5°C than 25°C to obtain aggregation. Supernatant materials could possibly modify the reaction of cells to the shaking rate. Similar phenomena in connection with amphibian reaggregation will be discussed shortly. Galtsoff (1929) showed that many sponge species contain substances which are lethal for other species. He prepared simple extracts by squeezing the sponge bodies in sea water, much as is done in disaggregation. These extracts invariably caused cell movement to cease and frequently resulted in the cytolysis of species other than that from which the extract was derived. Galtsoff likened this reaction to a heteroagglutination phenomenon. The phenomenon was studied in various subtropical sponges including *Reniera cinerea* and *Pachychalina* spp. The extract of *Reniera* agglutinated or lysed all other species of sponge cell tried. It can now be seen that this phenomenon might have occurred in Humphreys and Moscona's experiments when supernatants of different species origin from the cells being reaggregated were being used. If this were so their evidence on specificity would be nullified and the assistance of aggregation of cells of a given species by a supernatant of the same origin might be not more than an unspecific influence on the resistance of cells to high shear rates. It is for these reasons that it would have been very desirable to have data on cell number in their systems at the start and the end of aggregation. It might be argued that this same phenomenon could have occurred in experiments carried out by Curtis (1962b) when mixed species were used, but the washing of the cells was very thorough and the identification of cell types by cytological detail and staining is a more reliable method than Moscona or Humphrey used. Thus it will be appreciated that the work on sponge aggregation urgently needs further research to clear up these contradictory results and it would be unwise to draw any conclusions at present. We can now return to a general survey of results of investigations of cell adhesion in aggregates etc.

Temperature does not appear to affect cell to cell adhesion in aggregates (see earlier), aggregation not being temperature dependent over the range $1°C$ to $37°C$. L. Weiss (1961b) suggested that the adhesion of foetal rat skin cells to glass was temperature dependent because the percentage of cells spreading down on to glass increased with temperature. He also found that the ease with which cells were detached from glass by shear became more marked with a rise in temperature. He thus concluded that adhesion and de-adhesion were not opposites of each other. However, it is improbable that the spreading of cells on a glass surface is only a phenomenon of adhesion because it requires the formation of pseudopods to spread the cell. Moreover, since medium viscosity decreases with temperature, the rate of de-adhesion as well as rate of adhesion would be expected to be accelerated by temperature, and Weiss' techniques may partially measure rate rather than value of adhesiveness.

pH effects on reaggregation are a little unclear. Steinberg (1962b) found that amphibian embryonic cells would not reaggregate at pH 4·0 in still medium whereas reaggregation started and became more marked as the pH was raised above 4·5 to 7·0; these experiments were carried out in a medium containing calcium ions.

Curtis (1963) claimed that he could obtain reaggregation of *Xenopus* gastrula cells at pH 4·0 and slightly below (cytolysis set in at lower pH values). However, the aggregates could be obtained only by shaking the cells. He explained Steinberg's inability to obtain aggregation as being due to the inhibition of cell movement at pH 4·0, movement being required for aggregation in still media. However, Curtis' aggregates were composed of only up to ten cells: possibly the reason for this is that the cells become so rigid at pH 4·0 that they cannot undergo the changes of shape required to extend the area of contact betwen them, so that a chain of cells is built up with very small individual areas of contact between each pair of cells, which is easily broken up by shearing forces. The adhesion of two cells is very strong in these conditions. Curtis also found that aggregation of chick embryonic cells at pH 4·0 took place very rapidly in shaking suspensions and the cells were very adherent. There is some reason to think that at pH 4·0 the mechanism of adhesion is different from that at pH 7·0 and this will be discussed later. Dan (1947b) measured the adhesiveness of the eggs of the sea-urchin *Anthocidaris crassipina* to glass and found that it decreased from pH 2·7 to 8·0. Measurements of the absolute adhesiveness of chick embryo limb-bud cells (Curtis 1966c) show that it increases 10 times for every 0·8 pH unit drop in the range pH 8·0 to 5·0. But Fenn (1922) found that the adhesiveness of leucocytes for glass increased over the range pH 6·2 to 8·0. Of course aggregation and adhesion will not take place at pH values high enough to cause disaggregation (see earlier).

Unfortunately little is known about the action of other monovalent ions on adhesion processes of cells. Lithium salts appear to increase the adhesiveness of embryonic chick cells to glass as judged by features of complex cell behaviour (Weston, personal communication). Dan (1936,1947b) observed that the adhesiveness of sea-urchin eggs to glass was strongly inversely correlated with their surface potentials and that decreasing the concentration of monovalent cations raised the surface potential. However, Dan did not compensate either ionic strength or tonicity when the concentration of any one ion was changed. Rappaport *et al.* (1960) reported that the ease with which a variety of tissue culture cells adhered to glass surfaces was dependent on treatment of the glass with Na_2CO_3 solutions before culture. This result was attributed to the presence of additional sodium ions in the glass and to increased surface negative charge as a result of the treatment. However, the data they give show that the changes in charge (measured by absorption of crystal violet) and sodium binding produced by the treatment are far too large to be confined to the surface. They interpret the results to mean that a rise in surface charge density aids adhesion as does a rise in exchangeable sodium. If the changes occurred only at the surface of the glass some 3×10^2 crystal violet molecules would be adsorbed per $Å^2$. In other words adsorption in depth of crystal violet must occur and in consequence they cannot safely assume that the same phenomena occur at the surface of the glass as in its layer open to sodium exchange. Other explanations of their results will be made later. Nordling (1965a) described the effects of various methods of cleaning glass on the subsequent adhesion of cells.

Divalent cations have been shown to be essential for the aggregation of sponge cells (Spiegel 1954). Roux (1894), Herbst (1900) and others have noticed the importance of calcium for the maintenance of adhesion in embryonic tissue. Curtis (1957) and Steinberg (1958, 1962b) have shown that calcium, magnesium or strontium are essential for cell adhesion in amphibian aggregation, but Armstrong (1966) found that chick embryonic cells would not aggregate when Ba or Sr were the only divalent cations present. De Haan (1959) came to a similar conclusion as Curtis about the adhesion of slime mould cells. L. Weiss (1960b) found that calcium was essential for the attachment of tissue culture cells to glass as did Rappaport et al. (1960). Garen and Puck (1951) noted the importance of divalent cations for the attachment of T1 virus to Escherichia coli. Embryonic cells adhere well to cation exchange resin beads in the presence of divalent cations (Puck and Sagik, 1953): this reaction can be interpreted either as evidence for specific calcium binding of one cell to another or as evidence for the lyophobic colloid theory of cell adhesion, but this phenomenon will be discussed later. Garvin (1961) found that polymorph neutrophil cells require calcium and magnesium to adhere to glass. Taylor (1961) observed no effect of the absence of these cations on the detachment of embryonic mouse cells in serum free media from glass, but he used EDTA to bring about removal of Ca and Mg in unspecified conditions. Curtis (1964) found that high concentrations of calcium c. $0 \cdot 1M$ lessen the gap distance between cell and glass when the cell is in adhesion with the glass. Collins (1966a,b) measured the effect of calcium on the surface charge density of embryonic chick heart, liver and neural retina cells and found that at c. $1 \cdot 8 \times 10^{-3}M$ calcium reduced the charge density by about 10 per cent. Wilkins et al. (1961a,b) were able to flocculate leucocytes with high concentrations of calcium, c. $0 \cdot 1M$ and provided evidence that the cells were adhering with no gap between them. These authors also tried the effects of tri- and tetra-valent cations on leucocytes and by using sufficient concentrations were able to bring the cells into adhesion, probably in the primary minimum. Dan (1947) found that ceric ions greatly increased the adhesiveness of sea-urchin eggs for glass.

More complex ions have a variety of effects; organic polycations such as polyvinylamine hydrobromide or protamine sulphate agglutinate cells. In such cases their action can almost certainly be attributed to a bridging type action between cell surfaces the cationic charges combining with the negative charges on the cell surfaces. The effect of polyanions on the adhesion of cells was tried by Nordling et al. (1965b), who found that they tended to prevent cell adhesion to glass when serum was present, but a rather inexplicable clumping of cells resulted. The action of macro-ions such as the majority of proteins, in which the charge groups are of both signs, is harder to interpret, unless they are antibody molecules in which case they presumably act as highly specific bonding agents. Katchalsky et al. (1959) flocculated erythrocytes with polylysine, and Katchalsky (1964) showed that the polylysine was adsorbed as a dense oriented layer on the erythrocyte surface. Easty and Mercer (1962) examined erythrocytes which had been agglutin-ated by the macrocations polyethylene-imine, protamine sulphate and polyvinylamine hydrobromide and found gaps of c. 250 Å width between the cells, a faint electron dense staining was visible between the cells. They suggest that these reagents produce ionically bound bridges between the cells. It is, however, just possible that at least on occasion these reagents act by reducing surface charge so that adhesion in the secondary or primary minimum occurs. It should be mentioned that such treatments invariably kill the cells. Easty and Mutolo (1960) had already made the interesting observation that trypsin inactivated with DFP disaggregated such artificial tissue. These artificial tissues cannot be redispersed with ease by mechanical means. Little is as yet known about the effect of peptides on cell adhesion. Bradykinin, vasopressin and oxytocin do not affect the adhesion of chick embryonic cells in aggregates.

The effect of proteins in aiding cell adhesion has been studied mainly with reference to adhesion to glass and plastic surfaces. Very few studies appear to have been made in aggregation systems. Lieberman and Ove (1957,8) obtained an α–globulin fraction from serum which promoted adhesion of cells to glass. Fisher *et al.* (1958,9) found that HeLa cells grew in compact or perhaps aggre-gated colonies on glass when human albumin serum and an α–globulin fraction from calf serum (fetuin) were added to the culture medium. Unfortunately, adhesion was judged by the number of cell colonies which grew from an inoculum and this took several days so that not only might the results be rather a test of mitotic promotion than adhesion but they might also be due to changes in the medium depending only secondarily on the presence of the factor. Lieberman and Ove (1958) produced evidence that their fraction was not identical with fetuin. Neither of these groups of workers offered evidence that their protein fractions were single proteins or even free from peptides etc. Taylor (1961) separated human conjunctiva cells from glass with trypsin or EDTA and chick or mouse embryonic cells with trypsin. He then measured the time taken for these cells to attach to glass and other surfaces (attach-ment was measured as the resistance to a de-adhesion on mechanical shock in various media). Cells adhered to glass more slowly in the presence of fresh horse serum than in its absence, they also adhered (though still more slowly) to glass coated with adsorbed serum (though the film may have desorbed under the conditions of culture). If cells had been violently detached from a glass surface, then this surface deterred cell adhesion, contrary to the results of L. Weiss (1961b) but Weiss incubated the cells in contact with the surface overnight before testing the adhesiveness of the cells, whereas

Taylor made immediate tests. Formalin-killed cells behaved in a similar fashion to live cells. Ovalbumin, pepsin and histone also delayed cell adhesion whereas salmine accelerated it. Cells adhered and spread well on hydrophobic surfaces such as cellulose acetate. Taylor's experiments of course appear to be in part a measure of the rate at which cells approach a surface and spread out laterally as well as of adhesion and it is hard to interpret their exact relevance to the mechanism of adhesion. Easty, Easty and Ambrose (1960) allowed mouse Landschutz ascites carcinoma cells to settle on glass in various media; serum (nature unstated) prevented adhesion over short periods of time. Heat-killed cells (50°C treatment) behaved in the same way as live cells. L. Weiss (1959) allowed HeLa cells to settle on glass overnight and then measured their resistance to de-adhesion. High concentrations of human serum (heat treated) promoted adhesion by comparison with low ones. Fresh serum appeared to promote adhesion slightly more strongly than heat-treated serum, but after overnight culture extensive destruction of any component of fresh serum which might impede or assist cell adhesion would occur. Curtis (1965a) found that this destruction does indeed occur in the presence of cells. Weiss also observed that the lipoprotein-rich fraction of serum, termed G2, from human serum, aided cell adhesion. He also found (Weiss 1959b) that serum was necessary for the adhesion of HeLa cells to agar and silica gel surfaces, but not to collagen and fibrin surfaces. Moskowitz (1963) examined the aggregation of mammalian cells of cell lines in shaker flasks using a variety of sera of mammalian origin and found that different sera produced either large smooth aggregates or small rough clumps; one cell type would not aggregate in horse serum. Curtis (1964) found that horse serum contained at least two components affecting cell adhesion, one promoting it and one inhibiting adhesion. The inhibitor was isolated as a pure protein (Curtis and Greaves, 1965). The molecular weight of this aggregation inhibiting protein was 150,000. Since it is soon destroyed by the metabolism of cells and is somewhat heat labile, it is possible to explain some of the discrepancies found between the results of various workers who studied the effects of sera either before or after heat treatment, or who examined sera either before or after they had been incubated for long periods with the cells. In addition, the fact that serum contains an inhibitor of adhesion means that those experiments in which some serum fraction was recognised as a promoter of adhesion, whole serum being used as a control, are misleading. Furthermore, it is obvious that the isolation of pure proteins is essential to clear up the confusion in work with serum factors. Curtis (1966b) also isolated an adhesion-promoting protein from horse serum; in most samples examined this protein was present in lower activity than the aggregation-inhibiting protein.

Pentinnen *et al.* (1958) and Saxen and Pentinnen (1961) found that fresh human serum contained a factor which caused cell clumping on glass (whereas heat inactivated serum allowed migratory growth). Hayry *et al* (1966) and Myllyla *et al* (1966) continued this work and found that a lipoprotein material affected cell adhesion, perhaps by forming a steric barrier (*sic*) to cell adhesion. No extensive evidence was produced to show that such a barrier existed. Adhesiveness was measured by following the attachment rate which represents a variety of phenomena as well as adhesiveness *e.g.* medium viscosity. Unfortunately cell clumping is a complex phenomenon and may either result from effects on adhesion or cell movement. Moskowitz and Amborski (1964) obtained some slight evidence that β and γ globulin fractions from equine serum produced the same results, but cultures were scored only after 5 days treatment. Michl (1961) isolated a fraction from serum which aided cell spreading on glass and perhaps cell adhesion as well, later Michl (1965) claimed that this factor was carbamyl phosphate.

A further line of evidence has arisen from observation of macromolecular components derived from cells which may aid adhesion *in vitro*. Merchant and Kahn (1958) and Kuchler *et al.* (1960) observed that fibrous proteins were formed in the medium of suspension cultures; these proteins bound the cells into clumps. However, there is considerable evidence in these papers that the cells were moribund or dead.

Taylor (1961) mentions that adsorbed films build up on a glass surface on which cells have been cultured. In order to make this observation cells were removed from the glass with a jet of fluid and the dried slide was then examined by ellipsometry. Taylor discovered that the spreading and attachment of cells was delayed on such surfaces. Rosenberg (1960) extended this work by measuring the deposition of a material on the glass by cells grown on them. Ellipsometry of the dried surface after the cells had been violently removed revealed deposition of a material on the glass. Rosenberg suggested that this material was an adhesive substance synthesized by the cells and L. Weiss (1961b) obtained evidence which provides some support for this idea. Weiss found that a surface from which rat fibroblasts had been violently sheared provided a more adhesive surface after overnight culture for trypsinised cells of the same type than a surface on which cells had not been previously cultured. But control and experimental cells were cultured in different media when differences were observed which suggests alternative explanations. In view of Taylor's results Weiss' result seems a little surprising. Four criticisms can be made of Rosenberg's work: first, the material may be formed as the cells are torn off the glass; second, the measured thickness of the material refers to the dry state so that it might be thicker but much more tenuous in life; third, since the cells had been recently trypsinised, a procedure known to make them leaky, the substance might represent an abnormal macromolecular component derived from the cells, and fourth, the measuring method averages film thickness over a 1 mm diameter disc so that one cannot say whether the thickness of the material between a cell and the glass might be 0 Å or 1,000 Å. The method of shearing cells off glass with violent shock runs great risks of leaving parts of the cells behind. L. Weiss (1961b) used a violent shearing method to remove cells from glass and found that fresh cells adhered more strongly to this surface than to freshly cleaned glass. Weiss suggests that this material is equivalent to that described by Rosenberg, though Weiss would prefer to consider this material part of the cell surface. Weiss and Coombs (1963) demonstrated that some substance was left behind on the glass after cells had been sheared off the surface: immunological methods were used. There are, however, reasons (see later) for treating Weiss and Coombs' result with reservations. Aub et al. (1963) discovered that crude lipase contains a co-factor which clumps Ehrlich ascites tumour

cells. Since the cells cannot be redispersed by standard disaggregation procedures afterwards, it is probable that the type of adhesion
produced is not similar to that in life.

Antibodies have, of course, been frequently used to produce cell
adhesion both in situations where complement is required (immune
adherence test) and in those in which it is not (mixed agglutination,
haemagglutination reactions). Although immunologists have rarely
studied such reactions with reference to adhesive mechanisms,
enough is known to make some generalisations. The following is
taken from Nelson (1963) and Kabat and Mayer (1958). The reaction
is generally irreversible and results in very strong adhesion of the
cells involved. The kinetics of such reactions suggest that rapid
flocculation is taking place. It can also be shown that whether adherence occurs or not is unrelated to the surface potentials of the
cells involved (Sachtleben and Ruhenstroth-Bauer, 1962). These
criteria (combined with the specificity of such reactions) strongly
suggest that a bridging type of adhesion is involved. Easty and
Mercer (1962) examined erythrocytes which had been agglutinated
with an anti-erythrocyte gamma-globulin using electron microscopy. They found that 250 Å gaps were present between the cells
with a faint staining in this gap. L. Weiss (1965) found that various
antisera tended to produce cell detachment from glass—since most
antisera contain the aggregation inhibiting protein which lowers the
adhesiveness of cells this result is not surprising. Oda and Puck
(1961) observed that antibody treated HeLa and hamster ovarian
cells adhered to glass in the absence of complement, but that further
adhesion of cells ceased after complement was added. However, cell
to cell adhesion can be promoted if both antibody and complement
are added (Nelson, 1963) in other systems. These observations can be
reconciled if antibodies produce adhesion by bridging.

Such simple experiments do not necessarily provide any information on the normal mechanism of cell adhesion because although
antibodies may either assist or inhibit adhesion they may do this by
creating new mechanisms of adhesion or de-adhesion. (This difficulty
can be appreciated from the work on aggregation mechanisms using
antibodies.) Antibody mediated adhesion occurs as the normal
method of adhesion, in the immune adherence of macrophages to

tissue cells (Granger and Weiser, 1964), a process which results in the destruction of the target cells and most of the macrophages. Such a process may occur in graft rejection. Spiegel (1954, 1955) found that sponge cells would not aggregate in still media after treatment with specific antisera. Since the cells were probably dead or immobilised this is hardly surprising; it would be interesting to try aggregation after such treatment in a shaker. However, even if aggregation was then still blocked this would not necessarily indicate that the antigens reacting with the antibody are normally used for adhesion, because new mechanisms to prevent adhesion might arise from the presence of antibodies. Surprisingly when Spiegel aggregated cells of two species of sponge, simultaneously in the presence of both appropriate antisera aggregation took place. Similar studies of slime mould adhesion by Gregg (1956, 1960), and Gregg and Trygstad (1958) can be criticised in the same manner. Work on the specificity shown in fertilisation has shown that antigen–antibody like reactions are involved (Tyler, 1948) but it is less likely that these are directly concerned in life with adhesion of sperm to egg surfaces, rather than the adhesion of sperm to jelly coats.

However, many of the reactions which can be produced with extracts of eggs and sperm lead to adhesive results *in vitro*: for example, the fertilisins (agglutins) derived from jelly coats lead to the agglutination of sperm of the same species. This agglutination is often reversible, probably because the fertilisin molecules are split by sperm action leaving only univalent fragments combined with the sperm surface (Tyler, 1948). Bacteria and other particles coated with a suitable antibody are more easily phagocytosed by leucocytes than uncoated particles (see Nelson, 1963 for review and Nossal, 1959). It is unfortunately unclear to what extent adhesive phenomena play a part in phagocytosis. Do phagocytosed particles form close adhesions with the surface before they are incorporated into the cell? Electron micrographs e.g. Karrer (1960b) of phagocytosis are at present unsatisfactory and the molecular details of the process are unclear. However there is a slight evidence (Mudd et al., 1934) that phagocytosis cannot be explained as being related to the surface potential of either cell or particle and this finding suggests

F

that the lyophobic colloid theory of adhesion cannot explain phagocytosis.

It can be seen that a large variety of substances affect the adhesion of cells to one another and to other surfaces. The majority of these substances are charged and their actions can be explained in terms of this property or in their effects on the charges of other molecules (see later in this chapter). The effects of non-polar substances and some compounds of molecular weight c. 200–1000 have been much less studied and in general such work as has been carried out only impinges tangentially on the problem of cell adhesion. For example, Willmer (1961) showed that various steroids affected cell form and morphology in the amoeba *Naegleria*, effects which may conceivably in part at least be on cellular adhesion. Polet (1966) found that hydrocortisone appeared to increase the adhesiveness of human amnion cells to glass. In the more complex systems such as the whole animal it is very hard to elucidate the mechanism by which a given agent acts on cell adhesion. Gasic and Galanti (1966) found that the formation of S–S bonds appeared to be required for the adhesion of glutaraldehyde-fixed sponge cells, but the relevance of this finding to the adhesion of living cells is unclear. Yamada and Saton (1964) found that nitrogen mustards affect cellular adhesiveness. B. Jones (1966) and P. Jones (1966) found that ATP prevented aggregation of chick embryonic fibroblasts and that it also disaggregated aggregates of cells. ADP appeared to aid cell adhesion. The complex interactions involving serum proteins which affect mammalian cell adhesion have already been described and it seems likely that similar considerations apply in other animals; for example, Rizki (1961) found that glucosamine-HCl fed to Drosophila tu^{10} mutant strain tends to prevent lamellocyte adhesion; from this he suggested that the cells are cemented by a mucopolysaccharide substance; but comparison with the conditions affecting chick cell aggregation in sera suggests that this is a somewhat uncertain conclusion.

BIOPHYSICAL EVIDENCE: SURFACE CHARGE AND ADHESIVENESS

The measurement of the surface charge density of cells by electrophoresis and the calculation of surface potential has been described in Chapter 2 (see also Table I for details of such measurements). In the last chapter it was pointed out that surface charges give rise to an electrostatic repulsive force which would be expected to affect cell adhesion. Consequently the evidence for and against such an effect will be discussed. As early as 1936, Dan (see also Dan, 1947) measured the adhesiveness of various echinoderm eggs to glass and the zeta potential of the eggs for a variety of ionic conditions. Adhesiveness was measured by observing the resistance of the cells to detachment under gravity from a glass surface. He found that as surface potential increased, adhesiveness decreased. These pioneer studies have not been repeated in so elegant a manner as yet for other cell types.

Although surface potential measurements have been made for a variety of cell types other attempts to correlate surface potential with adhesiveness have been rather unsatisfactory because the adhesiveness of the cells has not been directly measured but only judged from aspects of cell behaviour. Most correlations have been made for malignant cells as compared with their normal progenitors. Coman (1944, 1953) proposed that the malignant state was characterised by the cells being of lower adhesiveness than in the normal condition and was able to provide measurements on a few cell types to support this claim. The discovery that malignant cells did not possess the property of contact inhibition (Abercrombie and Heaysman, 1954, see Chapter 5) has been interpreted, in particular by Ambrose (1961) (see also Abercrombie and Ambrose, 1962) as being due to decreased adhesiveness of malignant cells. In consequence it was of considerable interest to discover whether malignancy is associated with increased surface potential of the cells. However, it should be pointed out that no measurements of adhesiveness have been carried out either on those cell types on which electrophoretic measurements have been made (other than by Dan) or on cells which do or do not possess the property of contact inhibition, so that there is an unpleasantly large gap in the experimental data. Ambrose et al. (1956) measured the electrophoretic mobility of hamster kidney cells and their stilbocstrol induced tumorous form and found that the surface potential was higher in the tumorous form than in the normal cell type. Similarly Lowick et al. (1961) reported that a rat hepatoma gave cells with a mobility greater than the normal liver cell type. In both papers disaggregation with a chelating agent was used to prepare the cell suspensions. Lowick *et al.* also showed that the electrophoretic properties of tumour cells, at least in the case of mouse Ehrlich and Landschutz ascites tumours, remained fairly constant over a number of years of transplant from animal to animal. Forrester *et al.* (1962, 1964) continued this work by examining the malignant transformants of the hamster kidney cell line produced by viral infection. They showed that increased surface potential was associated with the development of malignancy, as judged by behaviour in tissue culture. Similarly Purdom, Ambrose, and Klein (1958) showed that the surface potential of mouse MC1M tumour sublines cells increased as the sublines went from solid to ascites forms: of course this change is not necessarily associated with malignancy.

But Vassar (1963) measured the electrophoretic mobilities of a number of freshly obtained human normal and tumour cells and was unable to discover any correlation between mobility and tumorous properties let alone malignancy. Simon-Reuss *et al.* (1964) found no correlation between electrophoretic mobility and the "malignancy" of a small range of mouse normal and tumour cells. Vassar (personal communication) found that there was no correlation with malignancy as judged pathologically. It is obvious that insufficient work has been done to associate increased surface charge with malignancy (or to disprove it) let alone to demonstrate that it is correlated with decreased adhesiveness. Nevertheless, the work of Ambrose's school suggests that such a relation may be demonstrated. One difficulty in the correlation of electrophoretic data with evidence from aspects of cell behaviour lies in the fact that electrophoretic measurements are usually made in monovalent cation electrolytes, whereas cell behaviour takes place in media containing calcium ions etc. It may be that the surface potential of various cell types is reduced to differing degrees by calcium, so that a better correlation between adhesiveness (as judged by cell behaviour) and surface potential will be obtained for a range of cell types when surface potential measurements are made in the presence of calcium ions. Nordling and Mayhew (1966) suggest that there is no relationship between cellular adhesiveness and surface potential: their results are discussed on page 176.

Wilkins *et al.* (1962a,b) found that sheep polymorphonuclear lemocytes could be flocculated (more correctly "coagulated") at their zero charge point. Since these cells do not adhere strongly in life but were very adherent after coagulation it seems probable that they were adherent in the secondary minimum in life but in the primary minimum after coagulation. Despite this dissimilarity from conditions in life these results are best explained as being due to the reduction in electrostatic forces of repulsion consequent on lowering of the surface potential by binding of the various cations used; there is no other reason why adhesion should be strongest at zero charge point. Similarly Curtis (1964) observed that embryonic chick fibroblasts in culture tended to de-adhere from their adhesions with glass when conditions were such that electrostatic repulsive forces would be of greater magnitude than in normal tissue culture conditions. When cells were in normal physiological salines they adhered to the glass with a separation corresponding to a secondary minimum adhesion, but on changing the media so that the cells would be at or near their zero points of charge, the separations changed to <30 Å, corresponding to adhesion in the primary minimum. Both the works of Wilkins *et al.* and Curtis suggest most strongly that electrostatic forces of repulsion act to control cell adhesion, whether it occurs in the primary or secondary minimum.

Similarly the large volume of work on the effects of ions on cell adhesion discussed earlier supports the view that the electrostatic forces of repulsion control adhesion to a considerable extent. Decrease of ionic strength, rise in pH and the absence of divalent cations would all be expected to increase the repulsive forces and thus to reduce adhesion, and this is found of course in practice. The opposite changes in ionic conditions would be expected, if they act through the surface potential and electrostatic repulsive forces, to increase cellular adhesion and they do so. In particular the fact that chelating agents disaggregate tissues can be well explained on this hypothesis. Calcium ions reduce surface potential as Forrester *et al.* (1962) have shown; and consequently their removal will increase surface potential and the repulsive forces will increase resulting in de-adhesion.

Little is known about the effect of the surface potential of non-cellular substrates or the adhesion of cells to them. Cells of varying kinds will adhere to glasses of various kinds (see Nordling *et al.* 1965a), silica, polyethylene, polypropylene, polystyrene, polyamides (nylons), cellulose acetate or nitrate, collagen, various adsorbed protein surfaces (Taylor, 1961), plasma clots (fibrin; P. Weiss, 1929), silk and spider silk, (Harrison, 1914), keratin, agar gels, (L. Weiss, 1959b) metallised surfaces, ion exchange resin particles, paraffin wax and even polytetrafluorethylene. Of course many cells are grown in media, components of which, such as proteins, may adsorb to the surface to make it suitable for cell adhesion, but many cells of marine and freshwater species may adhere in simple salines. Rosenberg (1960) suggested that cells secrete substances which may aid their adhesion, but as we have seen in Chapters 3 and 4, the evidence for this is somewhat unsatisfactory. However Rappaport *et al.* (1960) were able to obtain adhesion of tissue culture cells to glass in media free from macromolecules, and many embryonic cells will adhere to glass in simple saline media. Thus, although evidence may be discovered to the contrary, it seems probable that cells will adhere directly to most, if not all, of the surfaces mentioned above, without the interposition of an adsorbed layer of protein etc. Can we discover any effect of surface potential on adhesion of cells to these surfaces? Unfortunately electrophoretic measurements have not been carried out on such surfaces. Rappaport *et al.* (1960) attempted to measure the surface charge density of various glass surfaces (for method and criticism see earlier) and concluded that increase in the number of surface sites which can and have bound sodium was correlated with an enhanced suitability of the glass for cell adhesion. This conclusion led them to suggest an ion bridging mechanism (Type 3) of cell adhesion to glass. However, the demon-stration of increased sodium binding in glass does not necessarily mean that more negatively charged groups are present. It may mean that negative charges are more easily suppressed by a given concentration of sodium ions so that the surface potential of glasses to which cells will adhere is lower than that of glasses to which cells will not adhere. Thus, the Rappaport's results may in fact support a type 2 mechanism of adhesion (adhesion in the secondary minimum). Rubin (personal communication) has found that tissue culture cells will not adhere to polystyrene surfaces if too large a surface charge density is induced on the plastic by acid treatment (a result compatible with type 2 mechanisms).

The wide range of non-living substrates to which cells will adhere suggests that no precise chemical specificity controls cell adhesion. The substrates, with the exception of paraffin wax, polyethylene, polypropylene and polytetrafluorethylene, are all negatively charged at physiological pH values, and some such as cellulose nitrate may have high surface potentials. In practice surface oxidation and contamination of the ideally uncharged surfaces may provide a certain surface potential. However, these observations are suggestive that cells do not require either specific chemical groupings nor high surface charge densities for adhesion.

Similarity of adhesion of cells to one another and to other substrates

Although the evidence which has just been discussed suggests that cell adhesions have no specificity it has been postulated that the adhesion of cells to non-living substrates, in particular glass, is by a mechanism different from that which acts in cell-to-cell adhesion. Coman (1961) and Berwick and Coman (1962) have strongly supported this view. These workers measured the adhesiveness of one cell to another by a micromanipulation method and the adhesiveness (which they term stickiness) of cells to glass by counting the percentage of cells detached from a glass surface on shaking. Coman found that the rabbit V_2 carcinoma cell showed greater adhesiveness to glass than the rabbit normal epiderm cell, but that it was less adhesive to other V_2 cells than the normal cell was to other normal cells. Similarly Berwick and Coman found that EDTA treatment reduced cell-to-cell adhesion but did not affect cell-to-glass adhesion (in contradiction to much evidence to the contrary), although many cells were detached from the glass during the test. Neuraminidase was found to affect cell-to-glass adhesion and not cell-to-cell adhesion. They thus concluded that cell-to-cell adhesion and cell-to-glass adhesion depended on different mechanisms.

Some criticism of Berwick and Coman's conclusion can be directed at the techniques for measuring adhesiveness. The micro-manipulation method may measure internal rheological properties rather than adhesiveness and the chemical treatments used may affect these internal properties. Further the extent of the actual areas of adhesion may vary in each type of contact and treatment, so producing this apparent division of cell adhesion into two types, because the area of contact will affect the measured adhesiveness (see p. 121). However, these criticisms are speculation. Kojima and Sakai (1964) attempted to demonstrate that adhesiveness (judged subjectively) and stickiness are due to the same mechanisms, but their work contains several technical mistakes.

But there are much stronger reasons for believing that cell-to-cell, and cell-to-glass (and other such substrates) adhesion are alike in mechanism. In Chapter 3 evidence has been put forward that there is a c. 100–200 Å gap in the adhesions of both cell to cell and cell to glass. The pH dependence of adhesion is, as seen earlier in this chapter, similar for both forms of adhesion. Likewise the effects of the presence or absence of divalent cations appear to be largely (if not wholly) similar in cell-to-cell adhesion and cell-to-glass adhesions. Enzymatic disaggregation of cells either from one another or from glass or other surfaces appears to be similar. Both forms of adhesion do not appear to be temperature dependent (see earlier for discussion of evidence). Such measurements of adhesive strength as have been made suggest that there is no reason on this ground for suspecting that cell-to-cell adhesion differs from cell-to-glass adhesion. The inadequately investigated relationship of surface potential to adhesiveness appears to be similar in both systems. Both types of adhesion under physiological conditions are fully reversible. Thus there appears to be much evidence in favour of the view that the two types of adhesion are really the same.

Are adhesion and de-adhesion converse processes?

In discussing the evidence on mechanisms of cell adhesion, material is derived from studies on both cell adhesion and processes of de-adhesion. Is it justifiable to consider the two processes to be exact reverse of each other? If it is not, evidence from these two sources cannot be easily united to support a theory of adhesion. L. Weiss (1961b, 1962a,b, 1963b, 1964) had suggested that though cells come into adhesion by means of their plasmalemmae, de-adhesion occurs by much rupture inside the cytoplasm, with only occasional breaks occurring between plasmalemmae or plasmalemma and a non-living substrate. If his view is correct, evidence derived from studies of de-adhesion may in fact deal with properties of the inner cytoplasm and not with mechanisms of adhesion. L. Weiss and Coombs (1963) showed that after cells are sheared off a glass surface, a material with immunological properties of the cells is left behind on the glass, which may support this idea. Of course cells in culture move around on the culture dish surface prior to the experiment and it is well known that small portions of the cell, in particular pseudopods, become detached laterally from the main cell body when another pseudopod succeeds in hauling the main mass of the cell body off in another direction. Such small fragments might because of their small size and relatively greater adhesiveness tend to be left in adhesion when whole cells are removed from the glass by shearing forces. Weiss and Coombs did not consider this pos-sibility but it would in fact give rise to considerable query about Weiss' hypothesis that cell detachment processes are not the converse of those of adhesion.

The evidence Weiss puts forward in support of this hypothesis is that (i) cell fragments can be detected on a surface after the cells have been sheared off a surface with which they had been in adhesion; however, if these fragments could have arisen by lateral breakages of the cells before detachment his interpretations would be very questionable. In addition (ii) Weiss (1964a) points out that the opposite temperature dependence of adhesion and of de-adhesion suggest that the phenomena are different, but we have already seen that this would in fact be expected (on considering drainage rela-tions) if the processes were basically the same. As theoretical evi-dence Weiss (1961a) pointed out that many adhesive joints break through the surrounding adherents rather than through the joint itself, but this does not occur in all cases and whether this occurs or not depends upon the relative strengths of joint and adherent as well as upon such factors as the rate of peeling etc. He also (Weiss, 1961a) pointed out that if molecular contact of the surfaces occurs, the strength of adhesions would be such that far greater forces would be required for cell detachment through breakage of the joint than are in fact found, so that the break must occur in some weaker structure. The alternative explanation of his paradox is of course that adhesions do not involve molecular contact and are in fact very weak. The evidence as we have seen supports this alternative.

However, there is much stronger evidence that the processes of adhesion and de-adhesion are fundamentally the same. The dependence of both processes on the presence or absence of divalent cations, the effects of pH, the absence of a temperature effect on adhesion or de-adhesion unless drainage effects are important, the similar rates of adhesion or de-adhesion under a given force and the extensive evidence that surface potential is closely related to adhesion or de-adhesion, all suggest that both processes depend on a common mechanism.

SYNTHESIS OF THE EVIDENCE ON THE MECHANISM OF CELL ADHESION

At the start of this discussion it will be necessary to introduce a certain amount of material discussed at length later in Chapters 6 and 7. This material is concerned with the specificity shown by cells in their adhesions. It is of course obvious from data from tissue culture experiments that a large variety of cells will adhere to a wide range of non-living substrates. Similarly experiments and observations on tissue cell behaviour in embryos and reaggregates (see Chapters 6 and 7) suggest that although there are differences in the adhesiveness of one cell type or another (in a given situation) there is no marked specificity in the adhesions such that one cell type will adhere or largely adhere only to its own type. The only adhesive system in which a marked specificity probably occurs is the interaction of the unlike cells, egg and sperm of one species. In previous reviews it has been frequently supposed that there is a specific adhesiveness such that cells of like type adhere to their own type. This supposition has been made to explain phenomena such as sorting out in reaggregation. The specificity proposed has been either one confined to cells of like tissue type (histospecificity:

Moscona, 1960, 1962, 1963a; Steinberg, 1958) or a more generalised one confined to cells from one species of animal; (in view of experiments on sorting out of sponge aggregates of mixed species, see Spiegel, 1954a, 1955; P. Weiss, 1953a.) Spiegel (1955) and others viewed either type of specificity in terms of a unique surface chemistry of the chemical groups involved in adhesion for each cell type. This type of theory, of course, runs into the difficulty that cells of different types will usually adhere to one another so that there can be no absolute specificity. There may however be chemical groupings on the surface allowing some cross reaction and bonding between cells of different type, but which lead to the preferential adhesion of cells of like type seen in the sorting out process. However the similarity in the chemistry of adhesion of cells (see earlier) argues against this hypothesis, though it should be admitted that more information on the chemistry of adhesion of a wider variety of cell types is required to settle this point. A contrary idea (Curtis, 1960a; Steinberg, 1958, 1962b, 1963b, 1964a,b) is that the surface of most cells is of similar chemistry (at least in those elements concerned with adhesion) and that most (Curtis) or all (Steinberg) adhesion is carried out by the same basic mechanism. Both Steinberg and Curtis consider that there are quantitative differences in the degree of adhesiveness from one cell type to another, which may account at least in part for features of sorting out of cell types in reaggregates (see Chapter 6). This hypothesis can of course appear in a variety of forms. Originally Steinberg (1958) suggested that although all cells adhered by calcium bridging of carboxyl groups, a specificity of adhesion was superimposed on this by the dimensions and nature of the pattern of the carboxyl groups on the surface. In this theory cells with like surface patterns of carboxyl groups would adhere more strongly because like patterns would match up. Curtis (1962a) criticised this on the theoretical ground that the geometry of matching-up would not bring about effective sorting out of cells. The reason for this is that two like patterns can be put together in every orientation giving a great range in the number of carboxyl groups on one surface directly facing carboxyl groups on the other; similarly unlike patterns could be faced to give a wide range of groups opposite one another. In other words the degree of adhesion between two like or unlike cells would be very variable and each class would overlap preventing effective specific adhesion.

Steinberg (1962b, 1963b) dropped his 1958 hypothesis, and advocated no more specificity than quantitative differences in adhesiveness. However, Steinberg's 1958 hypothesis has wider possibilities which will be briefly examined. An effective specificity of adhesion cannot be based on the idea of patterns of surface groups which have one continuous preferred orientation over the whole cell, because of the improbability of always bringing two cells of the same type into that orientation which leads to strongest adhesion. But if only small portions of the cell's surface carry a pattern of one constant orientation there are situations in which effective specificity might result. Similarly, if surfaces carried widely spaced bull's-eye type patterns an effective specificity of adhesion might result. But all these theories presuppose that two surfaces can come close enough for a rigid matching up of surface patterns, and even 100 Å seems to be too large a gap for this (see Curtis 1960a). As will be seen in Chapter 6, explanations of sorting out in reaggregation in terms of specific mechanisms of adhesion do not meet the facts, whereas explanations in terms of one generalised adhesive mechanism which is quantitatively variable are more satisfactory.

In summary it appears that although there is no evidence for a specificity of adhesion of tissue cells which depends upon a chemical uniqueness or stereochemical individuality of a cell type this idea cannot as yet be completely dismissed.

The adhesion of sperm to egg appears to be remarkably specific (Tyler 1946, 1948), but on closer examination the specificity probably mainly resides in whether or not fertilisation reactions will take place allowing the sperm to penetrate the egg membranes. The phagocytosis of bacteria by leucocytes in or from animals which have been "sensitised" with a suitable antibody appears to be remarkably specific in some instances but not in others (Nelson 1963; Suter 1956), but we are not certain to what extent this is an adhesive phenomenon.

We can now summarise the biological types of adhesion, and interpret this evidence in the light of the biophysical mechanisms of adhesion discussed in Chapter 3. A morphological classification will appear to be most useful.

1. The adhesion with a 100–200 Å separation between plasma-lemmae. This type appears to be of low specificity. The adhesion is relatively weak, redispersion being easy. It tends to form between bodies both of which are too large or too non-motile to possess appreciable kinetic energy. These adhesions are probably controlled by the surface potential of the bodies and change into other types of adhesion when zero point of charge is approached. Ionic conditions, in particular the presence or absence of divalent cations, have a marked effect on adhesion. Enzymatic treatment will dissociate tissues adhering by this means. Evidence for a cementing material is unsatisfactory at present, except in artificial tissues made with basic polypeptides etc. Whether or not the enzymatic treatment raises the surface potential is uncertain at present. These adhesions appear to be fully reversible.

2. The close adhesion, characterised by a separation of surfaces of less than 20 Å. In some cases these adhesions appear to be highly specific but in others (e.g. the coagulation of cells at low pH) are probably unspecific. Bodies of relatively high kinetic energy appear to become involved in such adhesions. The bodies bound by these adhesions will not redisperse and the adhesions appear to be very strong. Such small evidence as exists suggests that the bodies come into adhesion when surface potential is low or zero. Evidence on the effects of enzymes etc. is lacking.

The category in which to class desmosomes appears (see Chapter 3) to be the first type, but structures such as zonulae occludentes are harder to place with certainty, though they may be close adhesions.

I will now examine which particular theory, or theories of adhesion (see p. 113), will best account for these types of adhesion, remembering that more than one type of mechanism may act in the adhesion of a single cell. Before doing this it is as well to remember that at present much of the evidence which may lead to a particular view is inadequate in quantity and that some theories (such as those that favour the cementing of cells) will appear rather unsatisfactory not because they are necessarily so but because existing evidence in their support is inadequate.

(a) EXPLANATION OF ADHESION BY THE BRIDGING THEORY

The cementing or bridging (also described as type 3) theory of cell adhesion has probably been in existence longer than the other theories, if only because if our macroscopic experience of such processes as building. The morphological evidence for this theory is, as we have seen earlier, unsatisfactory, although this may of course be in part a reflection of the inadequacy of our techniques of electron microscopy. Similarly, although optical microscopy shows much intercellular material between cells, the important criterion for any theory of adhesion is whether the intercellular material actually abuts on the cell surface directly or is separated from it by a gap visible only by electron microscopy. Even if a gap is present and free from appreciable amounts of organic matter, we can still adopt the cementing theory, because only a very few molecules are required per adhesion.

It has been assumed that the action of enzymes in disaggregating tissues is evidence for a cementing material acting as an adhesive agent, but as has been seen such a result might arise from effects of the enzyme in increasing surface potential or from the presence of other factors in the enzymes. Of course many tissues, which contain appreciable extracellular matter are disaggregated with greater ease after enzyme treatment. This probably results from no effect other than that the removal of extracellular material allows greater shearing forces to act on the cells to disperse them mechanically and permits the escape of cells which though of little adhesiveness would otherwise be trapped in a mesh of extracellular material. However, further work is required to elucidate these problems. It has also been supposed that effects of changes in ionic conditions on cellular adhesiveness are evidence for the existence of cementing materials, cements lysing at high pH and low ionic strengths. However, as Curtis (1964) pointed out, this evidence is better explained as an effect on repulsive forces due to surface potential, for otherwise cementing materials have to be of improbable properties. This argument does not of course dispose of the possibility that adhesion is carried out by such cementing materials with electrostatic repulsive forces playing an additional part in controlling adhesion. Various subtheories of the cementing hypothesis have or can be put forward. If covalent bonds attach cells to cements and then in turn to other cells it is surprising that adhesion is so sensitive to small pH changes as well as other ionic changes. However, this can be explained by a subtheory proposed by L. Weiss (1962a). He suggested that cell adhesion is mediated by a cementing material which is always liable to be lysed by enzymes secreted by the cells, so that slight changes in the ionic conditions of the environment will lead to changes in enzyme activity controlling the adhesion. By such a means adhesion by covalent bonds might be very sensitive to ionic conditions. There is however no evidence for this subtheory at present and the fact that adhesion of cells is maximal at low pH values corresponding to the zero point of charge argues against this idea. The difficulties of proof of theories of adhesion by cements which are antibodies to antigens on the surface have already been discussed, though of course such adhesive mechanisms occur in experimental situations. The theory that there is a calcium-sensitive cement will be discussed shortly under the calcium bridging theory. However, this topic requires further investigation.

We have already seen (p. 139) that evidence derived from studies on aggregates which is believed to point to the production of a cementing material is unfortunately unresolved at present. From what is known about the chemistry of cell adhesion there appears to be no reason for predicting the existence of some cementing material, specific or otherwise, in the adhesions of normal or tumorous tissue cells. For those cells which have been artificially bridged together, as judged by probable behaviour of the reagents used, with such agents as antibodies, several differences appear when they are compared with normal adhesions. Although we cannot be certain that bridging has occurred in these cases (such electron micrographs as have been made being rather ambiguous) these artificial adhesions differ from the adhesion of normal cells in having different pH sensitivities (polylysine), in being undispersable (antibody agglutinated cells), in being independent of the presence or absence of divalent cations, and (possibly of less importance) in being between cells which become killed by the treatment.

Although it cannot wholly count against the cementing theory, the comparative weakness of cell adhesions and their low specificity is evidence which inclines one against such a theory. As mentioned earlier comparison with known bridging mechanisms would suggest that if they occurred in cells then adhesion would be very specific (either in terms of the cells involved or in terms of the adhesive cement) and very dependent on the ratio of numbers of bridging molecules available for sites on the surfaces to the number of these sites. No such evidence has yet been found. The macromolecular requirements for cell adhesion are unclear at present, and the recent discovery of the serum factor which inhibits cell adhesion until it is metabolically destroyed suggests that some of the evidence for the various macromolecules identified as being required for adhesion should be questioned. For these reasons there is no good basis for considering such macromolecules as possible natural cementing materials. The cementing or bridging theory can of course explain the existence of a gap of rather undefined dimensions between the plasmalemmae of cells in adhesion but does not account for the existence of close adhesions. This latter type of adhesion is better accounted for type 1 mechanisms (see p. 114).

(b) EXPLANATION OF ADHESION IN TERMS OF MOLECULAR CONTACT OF CELLS

Type 1 mechanisms are of course those in which molecular contact of the plasmalemmae occurs because there are insufficient repulsive forces to prevent it. Adhesion results because bonding occurs between the surfaces or because London forces in the primary minimum are able to hold the surfaces together. Although it might seem rather obvious that such mechanisms would best explain adhesions in which close contact occurs, these mechanisms have also been used to explain all types of cell adhesion (Steinberg, 1958, 1962a, 1964a,b; Pethica, 1961) and in consequence the applicability of such theories to other types of adhesion will be examined. The main evidence that adhesion occurs by direct molecular contact of the plasmalemmae arises from the work of Wilkins et al. (1962a,b) who brought leucocytes into adhesion by coagulating them at zero point of charge. The kinetics of adhesion and the. fact that the cells could not be redispersed suggested adhesion in the primary minimum. They suggested that this occurred in the normal adhesion of other cells, but there is no evidence that tissue cells in life are at or very near their zero charge points (see earlier). Part of their argument was based on the fact that they believed cells coagulated with cupric ions to be alive, but there are reasons for doubting this (see Curtis, 1964). Further evidence against their theory of course is derived from the comparative ease with which the cells of tissues can be redispersed, and the existence of a gap between plasmalemmae.

Several other papers have also interpreted the adhesion of tissue cells as dependent on a mechanism bringing the plasmalemmae into close contact. It has been found that treating the adhesion of two molecularly contacting liquids or a liquid and a solid in terms of their surface free energy or tension relationships gives a fairly good estimate of the strength of the bonding involved. Steinberg (1963b, 1964a,b) and L. Weiss (1961a,b) suggested that cell adhesion can be treated as resulting from the surface energy (or surface tension) relations of contacting bodies, as earlier workers had also done (Mudd, McCutcheon and Lucke, 1934). This type of theory requires that surfaces come into molecular contact. Steinberg (1962b, 1963b, 1964a,b) appears to have suggested that cell adhesion can best be treated in this fashion on the basis of his results on sorting out in reaggregates. He found that the cells he used could be arranged in a simple linear series of the character of their position in sorted out aggregates (see Chapter 6 for details) and proposed that the property involved was adhesiveness. He went on to state it could best be represented by surface free energy relations. It seems that Steinberg does not intend to state that cell adhesion is carried out by this mechanism but rather he would prefer to regard it as a suitable formal representation of adhesive mechanisms without any precise implication as to the actual mechanism. One unexplained problem with this treatment, if followed strictly, is as follows: if we consider two liquid bodies of dissimilar surface free energy, the body of greater free energy will tend to spread over the other; and as Mudd *et al.* pointed out, phagocytose it. But for a population of cells of two types of dissimilar surface free energy. Steinberg suggests, not that phagocytosis will occur, but that sorting out instead will take place. Lester (1961) and Rose and Heims (1962) have pointed out that the type of treatment used by Steinberg is physically incorrect, and cannot be applied to adhesive joints (see also Tait, 1918). Moreover, the estimated adhesive strength of such bonds for bodies with surface tension values similar to cells is far greater than actual measured values for cells (see earlier and L. Weiss, 1961a,b).

A somewhat different type of adhesive joint which can be evaluated in terms of surface free energy relationships should be mentioned here. If two facing surfaces are some small distance apart and are joined by a meniscus of liquid, an adhesive pressure is set up. The reason for this is that the medium in the joint is at lower pressure than the air or any second medium surrounding parts of the body outside the joint. This type of joint acts only if the medium inside the joint is different from that outside. In cell adhesions there is little reason to suppose that the medium outside the cell has a different surface tension on different parts of the cell surface, though this is not impossible. A special form of the theory that cells adhere in molecular contact has been developed by B. Jones (1966) and P. Jones (1966). They suggest that cells adhere in molecular contact by coulombic forces and that when cell adhesions break the surface contracts under the influence of ATP, thus concealing positive charges on the surface. As a result the surfaces repel each other by their negative charged groups. There is no evidence that surfaces can contract in this remarkable way, though it is clear that ATP affects cell adhesiveness, perhaps due to surface adsorption.

(c) EXPLANATION OF CELL ADHESION BY THE CALCIUM BRIDGE MECHANISM

A specific mechanism of adhesion favoured by Pethica (1961) and Steinberg (1958, 1962b) amongst other authors is the so-called "calcium bridge" theory. Although other workers, including Rappaport et al. (1960), have supported a calcium bridging mechanism in general terms Steinberg and Pethica have specifically proposed that the mechanism brings cell surfaces into direct contact. In general the theory proposes that monovalent anionic groups, probably carboxyl groups, on the plasmalemma surface are bound ionically to calcium atoms whose other valency is taken up by combination with a second anionic group. Either this second group is supposed to be on a second plasmalemma so that the two cells are bound by calcium atoms directly or this second group is part of an intervening cement molecule. In the latter case the cement molecules are supposed to be bound by calcium links at either end to the plasmalemmae (see Figure 14). Both forms of the calcium bridge theory find support

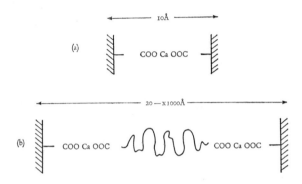

Figure 14. Two possible mechanisms are illustrated whereby calcium might act in cell adhesion. Both are types of "calcium bridging". Dimensions of gap between cell surfaces (hatched) are given in both cases. (a) Direct bridging between surface; (b) bridging with an intermediate cement, shown by irregular line.

from the well known dependence of cell adhesion on the presence of calcium and several authors have proposed the theory on these grounds alone. The pH dependence of adhesion is harder to reconcile with this theory.

Steinberg (1962b), see also Collins (1966a,b), measured calcium binding of amphibian embryonic cells at various pH values and found that binding fell off below pH 6·0. He also observed that apparently the reaggregability of cells declined below pH 6·0. He thus concluded that this was evidence for the action of calcium bridging in cell adhesion. However, Curtis (1963) showed that the adhesiveness of cells did not decline below pH 6·0; rather it increased to pH 4·0, and that Steinberg's results were probably due to a failure to take account of the fact that cell movement ceases below c. pH 6·0. It may of course be that the mechanism of adhesion at pH 4·0 (probably at the zero point of charge) is not the same as that at pH 6·0 and above, so that Curtis's results do not completely disprove the possibility of calcium bridging, but it is correct to say that at present this type of observation provides no evidence for the action of calcium bridging. Since cells adherent at pH 4·0 cannot be re-dispersed, unlike those at pH 7·0, it seems probable that adhesion at these low pH values occurs in the primary minimum. This difference might give heart to the supporters of the calcium bridge theory until it is pointed out that the difference implies that adhesion at pH 7·0 occurs in the secondary minimum which means that direct binding of the plasmalemmae cannot occur and that a calcium bridged cement is improbable. The gap dimensions of c. 100–200 Å again is evidence against the type of calcium bridging favoured by Steinberg and Pethica. Wilkins *et al.* (1962a,b) coagulated leucocytes with calcium and the kinetics suggested adhesion in the primary minimum, but it is of importance that the concentrations of calcium required for this were some ten times that in which normal cell adhesion takes place, and the cells were at their zero charge point. These experiments and arguments of course suggest the extreme importance that consideration of surface charge effects should play in assessing the situation. If we accept the calcium bridge theory it should be remembered that surface charge effects may give rise to over-shadowing phenomena and may, for example when electro-static repulsive forces are minimal, produce a different type of adhesion from that which occurs in life. Supporters of the bridge theory have usually failed to pay attention to these effects. The evidence pointing to the requirement of divalent cations for main-tenance of cell adhesion in life can of course be well explained in terms of the lyophobic colloid theory of adhesion whether adhesion occurs in the primary or secondary minimum. Even in the case of the experiments of Wilkins *et al.* there is no evidence that the final adhesion was by calcium bridging and not by London forces acting in the primary minimum. For these reasons it can be stated that there is at present no good evidence for a calcium bridging mechan-ism of cell adhesion.

Rappaport *et al.* (1960) conjectured that high surface charge densities on glass might be correlated with increased cell adhesion to the glass, the surface charges providing binding sites for cell adhesion. Similarly high surface charge densities such as erythrocytes have, ought, according to the calcium bridge theory, to be correlated with increased adhesion. In fact the evidence is to the contrary. Moreover even if only a small proportion of the charges were involved in calcium bridges and the remainder in providing electrostatic repulsive forces to oppose the bridges, adhesion would still occur by bridging when intervening cement molecules were present because the bond energies of the bridges would be some 10^4 times greater than the repulsive energy. In other words if such bridging occurred adhesion would take place even at very high surface potentials. The evidence is against this.

(d) DISCUSSION AND SUMMARY

We can thus see that attempts to explain the adhesion of tissue cells in life, where electron micrographs show a 100–200 Å gap, by hypotheses proposing the direct molecular contact of plasmalemmae run into a number of difficulties. First, evidence on the dimensions of the gap between cells is against such hypotheses, although structures such as *zonulae occludentes* may represent the required contact for these theories. Second, there is no clear evidence from the effects of *p*H on cell adhesion that direct contact occurs at physiological *p*H. The effects of divalent cations in promoting adhesion can be better explained as effects on electrostatic repulsive forces rather than effects on calcium bridging. The fact that tissues can normally be easily redispersed mechanically argues against the theory and is further strengthened by the observation that cells artificially adherent with their plasmalemmae probably in direct contact, cannot be redispersed, and are therefore different from normal cells. The comparative weakness of cell adhesions is a further reason for disbelieving these hypotheses. The correlation of surface potential with decreased adhesiveness is hard to explain if the surfaces come into molecular contact with ionic bridging though it would be expected if adhesion can occur in the primary or secondary minimum with London forces. Enzymatic effects on cell disaggregation can be accounted for on such theories but can of course be explained on other hypotheses.

On the other hand it is now clear that other types of cell adhesion can be well explained by the hypothesis that the surfaces come into direct molecular contact, and as we shall shortly see this can be accounted for by the lyophobic colloid theory of adhesion. First, there are those adhesions formed in experiments at or near the charge reversal point (for example leucocytes, Wilkins *et al.*). Then there are those adhesions which are morphologically characterised by surfaces in molecular contact. Sperm–egg contacts are probably of this type, because of the close contact between surfaces (Colwin and Colwin, 1963) and the degree of specificity shown (though it is still unclear whether the specificity lies only or partially in the adhesion between sperm and egg membranes) (Colwin and Colwin, 1963; Tyler, 1946). *Zonulae occludentes* are also probably of this type. It is a little unclear whether they are in fact adhesive structures. As yet there appears to be no report of the disaggregation properties of a tissue containing *zonulae occludentes* or similar structures, other than those of Whittaker and Gray (1962) and Sedar and Forte (1964), though we would expect such tissues to be non-redispersable if the zonulae have any appreciable adhesive function, as Whittaker and Gray and Sedar and Forte found. Incidentally, the finding that tissues containing desmosomes have adhesive properties such that there is no reason to suppose that the plasmalemmae come into direct contact, is further reason (see also Chapter 3) for believing that these structures have no particular adhesive function.

(e) EXPLANATION OF CELL ADHESION BY THE LYOPHOBIC COLLOID MECHANISM.

Since we have found the hypotheses that all cells are bound together by cements or by direct molecular contact so unsatisfactory, it will be considered whether the lyophobic colloid mechanism can be applied to explain cell adhesion. This theory predicts the existence of two types of adhesions: those with the surfaces in molecular contact (primary minimum) and those adhesions with a gap between plasmalemmae (secondary minimum) (types 1 and 2). First this theory alone of all others predicts the existence and even dimensions of this gap, and fairly good agreement is found between prediction and actual measurements. Its second prediction is that cell adhesions with a gap will be comparatively weak and easily redispersed, a prediction verified for the adhesions of many tissue cells, though this is possibly not a completely diagnostic feature since other adhesive mechanisms operating on very small areas of adhesion might produce the same result. The pH and ionic relations of the adhesions of tissue cells are better explained on this theory than on others. The pH and ionic concentrations and their effects on surface potential, which together in turn affect the electrostatic repulsive forces, are expected on the lyophobic colloid theory to prevent adhesion at high surface potential, high pH or low ionic strength and to aid adhesion under the converse conditions. Divalent cations will be particularly effective in promoting adhesion in the secondary minimum and of course all these conditions do, so far as we know, have the predicted effects on adhesion. It is of interest that Collins (1966a,b) has found that c. 2×10^{-3}M calcium reduces the surface potential of chick embryonic cells by about 10 per cent which is of the correct order to produce adhesion in the secondary minimum (see Chapter 3). As we have seen there is no reason to expect such behaviour on a cementing theory and though a calcium bridging mechanism might share some of the features of a generalised lyophobic colloid mechanism, adhesion in the absence of divalent cations would not be expected on the calcium bridge theory, whereas Curtis (1964) was able to obtain such adhesion in monovalent electrolytes. Again, the fact that cells will adhere to a very wide range of surfaces is accounted for by the lyophobic colloid theory in which the precise chemistry of the surface groupings is unimportant, whereas it is difficult to explain this on other theories. The lack of a marked temperature effect in adhesion would also be expected on the lyophobic colloid theory, as might an absence of specific adhesion. It should be realised that the lyophobic colloid theory of adhesion actually predicts two types of adhesion; first, adhesion in the secondary minimum, evidence in favour of which for tissue cells has just been given, and second, adhesion in the primary minimum. Adhesion in the primary minimum will take place if insufficient electrostatic repulsive forces are present to prevent close approach of the plasmalemmae, and this latter adhesion will of course be characterised by great strength and plasmalemmae in or almost in molecular contact. Sedar and Forte and Whittaker and

Gray found that close adhesions are not dispersable and thus very strong. There is no particular reason for supposing that calcium bridging would necessarily act in these primary minimum adhesions for which the London forces would be sufficient.

On the other hand, one main criticism may be held against this theory. The action of enzymes in aiding disaggregation is at present difficult to explain on the lyophobic colloid theory. It is of course possible that enzymes act by raising surface potential and thus bringing about disaggregation but there is no evidence on this point at present. Should it be established that enzymatic treatment of tissue cells does not affect surface potential this would become a serious objection to the lyophobic colloid theory. Heard et al. (1961) found that trypsin was apparently without effect on the surface charge density of mouse fibroblasts but the control cells had been mechanically disaggregated, a procedure which might conceivably alter surface charge. On the other hand, Malenkov et al. (1963) found that trypsinisation raised the surface potential of mouse tissue cells. Yet again Hayry et al. (1965) found that trypsinisation of HeLa cells lowered their surface potential. Somewhat similarly it is difficult to understand why iodoacetamide should affect cell adhesion to glass in the presence of serum, as Garvin (1961) found, though conceivably it acts to affect the metabolic destruction of serum factors which affect adhesion. Again we are still unsure as to how the action of serum factors and similar macromolecules should be interpreted, though explanation on the lyophobic colloid theory is possible. It should be remembered that if the lyophobic colloid theory is true for the adhesion of tissue cells it will not necessarily apply to all experimental adhesions. Thus, although it can explain the adhesion of cells in the primary minimum after the surface potential has been brought to zero, it does not explain the adhesions produced by treating cells with polycations such as polylysine (Katchalsky, 1964; Easty and Mutolo, 1960). In these latter adhesions, a bridging mechanism is to be expected in view of the charge relationships involved.

At this point we can see that experimental results are still insufficient to establish the correctness of one or other of the theories of cell adhesions even for the adhesion of one single cell type. On the other hand despite this lack the lyophobic colloid approach appears to provide a far more satisfactory explanation of most adhesive phenomena, particularly in living cell systems, than the rival theories. It is the only theory which predicts the existence of two types of adhesion, the close and the 100–200 Å gap adhesion; indeed in general terms, the type 1 mechanism is but a subtheory of the lyophobic colloid theory though some of the specific bonding mechanisms which have been suggested as acting in close adhesion appear to contradict observed results.

POINTS OF UNCERTAINTY

Finally the main points of experimental controversy or uncertainty will be indicated in order to suggest where work is most needed to establish or disprove the various theories of cell adhesion. They appear to be as follows:—

(1) How correct is it that 100–200 Å gaps are found between the plasmalemmae of tissue cells in adhesion? Is there any possibility that electron microscope artefacts may be responsible for these and other claimed details of morphology? Are *zonulae occludentes* etc. adhesive structures?

(2) What is the mechanism by which enzymes act in tissue disaggregation? Can the disaggregation be explained as being due to effects of surface charge density?

(3) Further work is required to definitely establish or disprove the probable relationship between increasing surface potential and decreasing adhesiveness for a wider range of cell types. The extent to which the effect of specific ions on adhesion can be explained by this relationship requires study.

(4) The measurement of the actual strength of adhesions under a variety of conditions would provide a useful tool for diagnosis of adhesion mechanism.

(5) How do molecules such as peptides and proteins affect cell adhesion?

Thus, it can now be seen that although precise details are still uncertain the treatment of the problems of cell adhesion in terms of the interaction of colloidal particles is one which provides a coherent and integrated approach.

MECHANISMS FOR ALTERATION OF CELLULAR ADHESIVENESS IN VIVO

Later in this book, particularly in chapter 7, morphogenetic processes will be described which appear to depend on the development of differences in the adhesiveness of two or more cell types. At the present we can only conjecture about the nature of the changes in the cell surface required to alter adhesiveness, but it seems apposite to consider possible mechanisms. Alteration of surface charge density should provide an effective general method of changing cellular adhesiveness. This might be brought about by increasing or decreasing the amount of mucopolysaccharides synthesised in or for the plasmalemma.

If cell division were to proceed faster than the synthesis of new plasmalemmal material the surface charge density of the cells should drop at each successive division. Obviously this process cannot continue indefinitely. Alternative methods of altering cell adhesiveness are for a cell to produce substances such as the aggregation-inhibiting protein which has marked effects on cell adhesiveness. Such substances may be of general effect on all the cells in the embryo or may conceivably specifically affect one cell type. Such substances presumably act in the cell surface.

Up to this point it has been supposed that changes in cellular adhesiveness would affect the whole surface of the cell equally. However, it has been suggested (Curtis 1962) that one part of a cell may become more adhesive than another if one part of the surface becomes expanded (low surface charge density) relative to another part of the surface. Mayhew and Nordling (1966) and Nordling and Mayhew (1966) have attempted to test whether the surface charge density of flattened (spread) cells was different from that of the same cells when rounded up. The mobilities were measured after fixation of the cells to preserve their shapes. It was supposed by these authors that the rounded up cells were less adhesive than the spread ones (though this was not established). Since they found no difference in surface charge density they concluded that the general hypothesis that surface charge density affects cell adhesiveness must be incorrect. These conclusions seem unwarranted mainly because (i) they failed to test the adhesiveness of cells and (ii) Curtis (1962) suggested that only a small part of a cell might become adhesive by reduction of its surface charge density locally at the expense of increasing the charge density elsewhere. Consequently, the overall charge density would not change even though the cell became more adhesive in character. Interestingly Mayhew (1966) showed that cells of a human sarcoma line became of higher surface potential towards mitosis, a time when cells may possibly be less adhesive. Patinkin and Doljanski (1965) claimed that the electrophoretic mobility of Landschutz ascites cells did not alter when the cells were swollen by immersing them in hypotonic solutions. The concentration of ionic salts was kept constant in these experiments, the tonicity being altered by varying the concentration of sorbitol. Unfortunately although differences in mobility were found no test was carried out on the significance of such differences and there are moreover numerical mistakes in the published data.

CHAPTER 5

The Behaviour of Single Cells

(1) CELL MOVEMENT

The extent of the involvement of the cell surface in movement is uncertain. Several theories of the mechanism of movement make no mention of the plasmalemma, but other theories attach varying degrees of importance to plasmalemma action in cell movement. Of course adhesion must play a part in all cell movement on solid substrates, because a certain degree of adhesion is necessary in order that the cell can thrust and pull forwards on its adhesive contacts. On the other hand too great an adhesiveness will cause the cell to be so sticky that its locomotor mechanisms will be unable to move it. Almost no attention has been paid to these factors except by Ambrose (1961). The plasmalemma may act as the main site of energy conversion to provide the locomotive force for movement or as the structure controlling the application of locomotive forces to the substrate: in addition plasmalemma properties may control the direction and speed of cell movement. Unfortunately studies of cell locomotion are at present at a very controversial stage and it is impossible to do more than indicate in outline those theories of cell movement which specifically implicate the plasmalemma.

It should be pointed out that the term "amoeboid movement" has been used in a wide variety of meanings, at one extreme covering all forms of cell movement and at the other extreme applying to the particular type of movement indulged in by *Amoeba* species and *Chaos chaos*. Since it is unclear whether all cells move by the same mechanism or by a variety of different mechanisms, I do not intend to use this term.

Several theories of cell movement, namely those of Allen (1961, 1964; Allen *et al.* 1960; and Thompson and Wolpert, 1963) place the locus of energy transformation in the inner cytoplasm, and the isolation of contractile cytoplasm from *A. proteus* by the latter

workers (Thompson and Wolpert, 1963) is impressive evidence for
this suggestion. On the other hand, Goldacre (1961, 1964) suggested
that as a result of contact between membrane and plasmagel an
enzyme reaction takes place in which ATP is produced. The ATP
causes contraction of the plasmagel, but the evidence for location of
this reaction at the inside of the plasmalemma rather than merely
near it is not impressive. These two theories are, of course, not
necessarily in contradiction, because the cytoplasm fraction obtained
by Wolpert and his co-workers requires in life ATP for movement,
and it may obtain it from a reaction at the inner side of the plasmalemma. Considerable controversy has been raised over the question of whether the contraction of the cytoplasm occurs at the front
end of the amoeba, as Allen maintains, or at the rear as Goldacre
proposes: this problem, which is at present unsolved, is apparently
rather removed from the subject of membrane behaviour. An
alternative theory to account for the movement of amoebae has been
developed by Marsland (1956) from an earlier theory by Mast (1926).
This theory suggests that (i) plasmasol flow is due to contraction of
the plasmagel, (ii) the plasmagel at the extremities of an extending
pseudopodium is less contractile than further "back" in the cell,
and (iii) the plasmagel sustains pseudopodal shape. Marsland has
provided evidence for this theory by showing that high hydrostatic
pressures cause pseudopodal retraction and solution of the plasmagel
(particularly in the pseudopods) with a disappearance of plasmasol
streaming. On decompression the animals regain normal motility
and normal pseudopodia. Addition of ATP tends to oppose compression effects. This theory does not implicate the surface in locomotion, but the surface obviously must be used as a locomotory
organ. If pressure diminishes cell adhesiveness it might be expected
that the cells would round up: this point should be considered.

As the cell moves the membrane must either move with the cell
or new membrane must be formed with the simultaneous removal
of old membrane. As Wolpert points out (Wolpert *et al.* 1964)
there are the following possibilities. First, the plasmalemma is
rigidly attached to its immediate inner cytoplasm and does not
move as the cell advances. In this case new membrane must be
formed at the anterior end of the cell as it moves forward and be

resorbed at the hind end. This concept is present in Goldacre's theory
(1961, 1962), as well as in those of Bell (1961) and Shaffer (1962).
Second, the membrane is easily deformable so that changes in shape
may take place without formation or removal of membrane. If this
is possible cell movement might take place, as Ambrose (1961)
suggested, by either an undulating movement in the same way that
snails or earthworms move; or as movement whereby the cell forms
an adhesion and thrusts forwards on this to form an adhesion further
on from which a further thrust forwards takes place, and so on.
Third, the surface flows in a directed manner to bring about movement in the same manner that a caterpillar track drives a tractor. To
do this the plasmalemma would have to flow forward over those
parts of the cell surface not in contact with the substratum and backwards in the region of contact with the substratum. These possible
alternatives are shown in diagram in Figure 15.

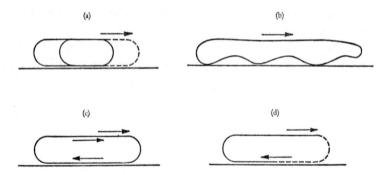

Figure 15.   Possible mechanisms of cell locomotion. (a) Cell grows new
surface (shown by broken line) at its front end and resorbs old surface at
its hind end. (b) No turnover in cell surface; cell moves by undulation of
its lower surface. (c) Cell surface moves actively like the tracks on a caterpillar vehicle. (d) Cell produces new surface which moves back along cell,
            a method which is a combination of (a) and (c).

Two techniques have been used to determine whether the surface
flows or is relatively immovable. First, observation of the behaviour
of particles such as oil droplets or pieces of dirt which appear to be
attached to the surface has been used to determine whether the
surface flows or not (Goldacre, 1952, 1961; Shaffer, 1962). Un-

fortunately there are some difficulties in the interpretation of such observations. Wolpert and O'Neill (1962) pointed out that an observation of the movement or non-movement of a particle refers of course only to the movement of the surface to which it is attached and does not necessarily apply to movement in other parts of the surface. Secondly, since observations are made with the light microscope it cannot be determined whether the particle is firmly adherent to the surface or whether it is movable because it merely lies close but unattached. Lastly, since particle movement is observed on cells which are themselves moving there is the problem of relative movement, which ought to be tackled with quantitative methods but rarely if ever has been.

A vastly more sophisticated method of approaching this problem has been developed by O'Neill (1964), Wolpert and O'Neill (1962), and Wolpert et al. (1964), who prepared a fluorescently labelled antibody against the surface membrane (plasmalemma and mucoprotein coat) of Amoeba proteus. This antibody was directed against the mucoprotein layer, but this almost certainly does not invalidate conclusions about plasmalemma movement which were derived from the experiments, because it is very improbable that the plasmalemma would slip on the mucoprotein coat. When cells were stained with this conjugate, it was found that the half-life of the label on the surface was some 5 hours, the label being taken into the cell, probably by pinocytosis. Small areas of the surface, stained by placing a micropipette containing the conjugate against the surface for less than 1 minute, could be seen to move with ease from one part of the surface to another. It can be concluded from these experiments that the surface is easily deformable and that if new membrane is formed at all during the course of the experiment, the maximum rate of turnover of surface is very slow. The rate of surface renewal is too low for it to play any part in movement.

O'Neill and Wolpert's findings suggest that other functions for the surface in cell movement should be examined. One interesting proposal is due to Ambrose. Ambrose (1961), after examining the movement of fibroblasts using the surface contact microscope, proposed that these cells move (on glass surfaces at least) by the action of a succession of undulations which move away from the leading edge of the cell (see Figure 16). In detail a transverse wave is supposed to travel along the surface of the cell which faces the glass in accompaniment with a compressional wave 90° out of phase with the transverse wave. The undulatory frequency was estimated at c. 4 c/s. and the amplitude up to 1μ. However, using interference reflection microscopy I (Curtis 1964) could obtain no evidence for these undulations using the same cell type as Ambrose, and found that the

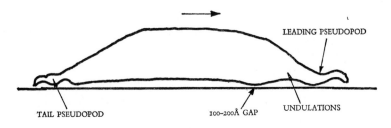

LEADING PSEUDOPOD

TAIL PSEUDOPOD   100–200Å GAP   UNDULATIONS

Figure 16. To illustrate Ambrose's theory of fibroblast movement (Ambrose, 1961). Sectional view of fibroblast.

change in distance between glass and plasmalemma at any point was very slow as the cell moved and was limited to a vertical movement of some few hundred Å. It seems probable that Ambrose may have in fact observed movement of the free surface of the cell (its other side) which may show undulatory movement. The reason for this is that surface contact microscopy depends upon the reflection of light at or near critical angle incidence and conditions for suitable reflection or transmission may occur at the free surface of the cell as well as at its adherent side. Unfortunately interference reflection microscopy has not yet been used to examine the movement of other types of cells.

Further consideration of those theories in which the cell surface is thought to be formed at one part of the cell and resorbed at the opposite end of the cell requires mention of Bell's suggestions and the observations and theories of Shaffer (1962, 1964). Bell (1961) proposed that cell surface may be easily resorbed into the interior by pinocytosis or a similar phenomenon (compare the suggestions of Bennett (1956) about the ease of exchange of surface material with that of the endoplasmic reticulum). If surface can also be formed at one point, movement may take place either by the surface moving backwards along the cell, thus thrusting it forwards, or by the cell interior moving forwards into a freshly made frontal piece of surface. In this latter case the surface need not move relative to the substrate. Shaffer (1962), who worked on the movement of slime mould amoebae, in particular *Polysphondylium violaceum*, observed that attached particles on the side of these cells do not move either forward or backward relative to the ground until they are nearly at the hind end of the cell. This observation suggests that the lateral surfaces of these cells are stationary. These cells tend to develop fine pseudopods called by Shaffer pseudodigits. These project from the side of the cell and may develop to a considerable length; particles adherent to a pseudodigit travel back to its base. From this and related observations Shaffer concludes that the surface is formed at "sources" at the front end of the cell and outer ends of the digits, and is resorbed at "sinks" at the hind end of the cell and bases of the digits. Shaffer concludes that movement occurs by growth of new surface at the front of the cell and resorption at the tail. The various types of cell behaviour discussed by Shaffer will be considered later. One feature of Shaffer's theory is that the surface is stationary on the main body of the cell but moving in the pseudodigits. Of course Shaffer's observations throw no light on the behaviour of the surface on the underside of the cell which is where the locomotive forces act, and conceivably cell surface could move actively in these regions.

In conclusion it can be seen that the cell surface plays an important part in cell movement but it is still uncertain how it acts as a driving surface. Besides the possibilities (i) that it is stationary, the surface growing forwards, (ii) that it moves in a wave motion to drive the cell forwards, and (iii) that it moves backwards along the base of the cell therefore driving the cell forwards, there is a fourth hypothesis. This, due to Mitchell (1956), is that if there is a current flow from one end of the cell to another, the cell will be electrophoresed since it has a charged surface. Bingley and Thompson (1962) found a "front to tail" electrical gradient in *Amoeba proteus* and remarked amongst other suggestions that the cells might conceivably electrophorese themselves.

G

The more complex features of cell movement will be discussed later as examples of cell behaviour, but before finishing this section it will be convenient to discuss the magnitude of the motive force of cells (see also p. 12). Nakai (1960) made a very tentative estimate of the traction force in the filopodia of nerve cells in tissue culture; he obtained a value of 3×10^{-10} dynes. On the other hand L. Weiss (1964b), using Allen's results, gives a locomotive force of $1 \cdot 5 \times 10^{-1}$ dyne for cells of *Chaos chaos*. If Nakai's figure is correct for tissue cells, these very small forces which can break adhesions can only be opposed by weak adhesive forces. The possible relation of the large forces of *Chaos* to contact behaviour has already been discussed. Unfortunately no other measurements appear to have been made.

(2) CELL BEHAVIOUR

To the embryologist and cell biologist, cell behaviour has come to mean the features of the movement of the cells, though of course the term might well apply to a much wider range of phenomena. It is however these details of cell movement which are responsible, as will appear in this and the next two chapters, for the development of definite patterns and arrangements of cells in organs and tissues in the whole body. Cell behaviour is comprised of phenomena of stopping and starting movement, controlling its orientation, frequency of change of direction, mean angle of turn, and speed of individual cells and cell populations. Obviously the basic mechanisms which provide energy for locomotion will play an almost equally vital role. It is clear that the degree of adhesion a cell has with its substrate may control its movement and, as will appear, more complex features of cell behaviour can be well explained in terms of adhesive properties of cells.

Thus it will be appreciated that in moving from the last chapter to this section we are integrating the molecular phenomena of adhesion and movement to give the behaviour of single cells, an integration which will be followed in turn in succeeding chapters: onwards from single cells to cell populations. Unfortunately in following this hierarchy from the basic composition and structure of the surface to morphogenesis in the whole animal, we have now reached that part which is the least satisfactorily understood. For although it is possible to explain many of the observed forms of behaviour of single cells in terms of molecular processes of adhesion and movement there is little concrete experimental evidence to indicate which of the processes at the molecular level actually take place.

Many of these features of cell behaviour are contact phenomena, since they take place only when the cell is in contact with a substrate or meets a new substrate such as another cell. The definition of contact presents some problems. In Chapter 3 we saw that cells are rarely in molecular contact, but it is obvious from Chapter 4 that adhesive reactions may be able to take place with a gap between the cells. Similarly mechanical interactions can take place across the gap and can operate (see Curtis, 1962a) across a gap of several micra. For these reasons we can define contact as occurring when a cell is sufficiently close to another body for physical and chemical inter-actions to take place across the gap. This will take place if the gap is sufficiently narrow to permit the direct transmission of forces or of sufficiently restricted dimensions to provide an environment which can be maintained by cellular activity heterogeneous from the bulk medium. This definition is fairly similar to those given by P. Weiss (1958) and Grobstein (1961) who however ignored the possibility of direct physical interaction of separated plasmalemmae. Grobstein stressed the possible importance of molecular ordering in the gap, but as we have seen there is little evidence for this in the cell-to-cell gap and little further evidence will appear in this chapter. Although some of the contact phenomena may not be adhesive ones it seems that they must all take place through the interaction of cell surfaces, even if the interaction is no more than the transmission of a diffusable stimulus from the interior of one cell to another.

The majority of papers on cell behaviour have dealt with cells in culture for the reason that individual cells in culture are more easily examined in culture rather than *in vivo*. It should be remembered that different cell types are of different morphology under identical culture conditions. And though it is possible to alter the mor-phology and behaviour of a given cell type to some degree, by changes in its culture conditions, towards the morphology and behaviour of another type, those same conditions will cause the second cell type to alter its morphology and behaviour towards a third type. In other words there is a certain but not complete stability of a given cell type in its morphology and behaviour and we must expect to find that not all forms of behaviour are common to even a moderate selection of tissue cell types. Consequently it should be remembered that the various types of behaviour which will be described are limited to certain cell types, even though the same basic mechanisms which control behaviour may operate in all cell types. Willmer (1958) using observational and functional criteria, classed tissue cell types into three varieties. These are: (1) the amoebocyte, frequently a highly motile cell bearing pseudopods all around its periphery (periphery as seen in plan view); (2) the fibrocyte or fibroblast, originally defined because of its ability to lay down intercellular fibrils of collagen, but nowadays more loosely defined as a motile cell type with elongated form having pseudopods at the front (leading edge) and sometimes at the rear too; and (3) the epitheliocyte, a flattened cell of roughly round or square shape without any marked pseudopods but possessing the power of cell movement in the sheets of cells in which it is usually found. The epitheliocyte is usually derived from epi- or endothelial tissues. I feel that with one exception Willmer's classification is still a most valuable subjective one, although it may turn out with the advent of quantitative methods of describing cell morphology that re-classification will be needed. The exception is that a fourth class of cell type should be added, this is the ascites type cell or *spherocyte* which is characterised by being weakly adhesive to normal sub-strates, non-motile except in so far as it floats and moves with the body fluids or culture medium, and lacking pseudopods. This classification now covers all cell types as found in culture, except those which are able to swim actively, such as trypanosomes.

Although cell behaviour is concerned with the morphology of individual cells and cell populations, both as cause and effect; in the main the subject deals with those reactions of cells which control (a) speed of movement (b) stopping and starting, (c) rate of turning, (d) mean angle of a single turn and (e) randomness of a series of turns. These reactions can be described as kineses when only speed of movement or rate of turning are affected, and as taxes when the cells orient themselves and move towards or away from a chemotactic source or other type of tactic source. These forms of cell behaviour can be analysed in terms of orientation, movement and time relation. By providing this quantitative method not only can we describe and compare reactions of cell movement behaviour with a number of ideal models (see for example Patlak, 1953), which helps in the prediction of possible causal mechanisms; it is also possible to make quantitative comparison of surface properties, e.g. potential, and features of behaviour. Unfortunately this quantitative description of cell behaviour is still frequently avoided by authors who take refuge in subjective descriptions. Willmer and Jacoby (1936) and Abercrombie and his co-workers have pioneered the development of quantitative methods of description of cell behaviour. Techniques for these measurements have been described by Abercrombie and Heaysman (1953, 1954), Curtis (1960b, 1961a), Garber (1953) and Curtis and Varde (1964). In discussing behaviour I shall start with an isolated cell in the simplest environment, a uniform one, and then move to more complex situations.

The isolated cell in very simple environments

The most simple situation in which a cell can show some type of contact behaviour is when it has settled on to a plane surface of uniform chemical composition immersed in a homogeneous culture medium. Although it is perhaps not quite certain whether live cells maintained in suspension invariably round up into a spherical form, this appears to be their usual configuration in suspension. Taylor (1961) has examined the process of settling by means of an ingenious microscope which gives a side view of the cell as it approaches a plane surface. He found that the cell flattens and spreads out; this would of course be expected if the cell behaves as a drop of fluid. The periphery of the cell, which is of course most flattened, tends to develop pseudopods in various regions. Rosenberg (1962) examined the number of human conjunctiva cells that became fully spread or partially spread on a variety of surfaces. An index of spreading was measured for each cell population by comparing the proportions of wholly spread and partially spread cells: Rosenberg found that the extent of spreading with time varied with the number of layers of barium stearate–stearic acid monolayers which formed the substrate and with the nature of the underlying support—glass, steel, teflon etc. As in his experiments on adhesion (see Chapter 4) he concluded that these observations indicated the action of long range forces originating deep below the surface. But since statistical tests of the significance of his claimed differences were not given it is impossible to assess the importance of this work at present. It is, of course, possible that the cells are reacting to minute details of the underlying support, but with increasing numbers of layers the structure to which they formerly reacted will become slowly smoothed out. Interestingly enough this effect would be expected if London forces act in adhesion, since the stearate monolayers might screen the London force arising in the glass; the general case has been examined in theory by Vold (1961).

It is unclear from this work whether spreading is solely due to the adhesiveness of a cell to a given substrate. Trinkaus (1963) suggested that the greater the spreading of a cell on a given substrate the more adhesive the cell. However other features of the cell such as the deformability of the whole cell would affect the extent to which the cell can spread.

Unfortunately there are almost no studies of the behaviour of wholly isolated cells, other than protozoans, on which to base a description of what happens next. The process of settling is complete in some hours and the first pseudopods develop, which appears to occur in all cell types except ascites in culture. Formation of pseudopods in culture has been observed for embryonic cells by Holtfreter (1947b), for fibroblasts by Abercrombie (1957), for neurons by Lewis and Lewis (1912) and Nakai (1956), and also for epithelia (see Abercrombie 1957, 1964a,b), for macrophages by De Bruyn (1945), and for leucocytes (De Bruyn, 1946). Several workers (e.g. Tait, 1918; Weiss, 1958; and Abercrombie, 1957, 1964a,b) have suggested that the formation of pseudopods is a fundamental reaction of the cells to settling on a solid surface. It is possible that if the surface is very adherent the cell might spread on it, and that this spreading would be localised in one part of the cell, thus forming the pseudopod. But Gustafson and Wolpert (1961) discovered that a cell adherent on one side to a substrate could form and protrude pseudopods on its other side into the free medium. This finding rejuvenates the problem, for although other parts of the cell are adherent to a solid surface they may do no more than provide a thrust point from which the pseudopod can be protruded. Unfortunately, we have no idea of what molecular processes are involved in the production of pseudopods. Once a cell is provided with one or more pseudopods it is able to move; though it may not, by reason of a variety of processes which will be described. Abercrombie and Lamont (unpublished), using chick and mouse fibroblasts settling on to polystyrene surfaces, found that the cells remain remarkably motionless if they are isolated from each other, although they frequently soon develop two or three pseudopods. However Carter (1965) found that isolated L strain fibroblasts would move in one direction rather actively on a substrate which probably bore a gradient of adhesion. The cells moved up the gradient. Weiss and Scott (1963) reported that cells would not move towards regions of lower pH, which makes them more adhesive, possibly because they stuck too tight to be moved. It is of interest that these pseudopods are nearly always (as subjectively judged) arranged in the motionless cells at just those orientations in which if they had equal locomotory pulls, the forces would balance out. Nevertheless their undulating membranes are active (see earlier) which suggests that the pseudopods are trying to make the cell move. When fibroblasts become motile and move actively, they possess only one large pseudopod at the front end of the cell though a trailing one may be present. This has been observed in fibroblasts (Weiss, 1958; Abercrombie, 1958), neurons in culture (Nakai, 1960), tumour cells (Ambrose, 1961) etc., and probably in macrophages (De Bruyn, 1955), but not in epithelia where movement does not take place until multicellular associations are formed. There is unfortunately no data on how long the cell maintains a single pseudopod or how often it turns. It thus appears that a prerequisite of movement is that a cell bears one pseudopod which is larger (and "stronger") than the others. Weiss (1958) and Weiss and Garber (1952) suggested that the pseudopodia in some way compete with one another, a large pseudopod being able to suppress the formation of others. However their work was carried out in cultures in which wholly isolated cells were exceedingly rare, so that the suppression or stimulation of pseudopods and movement might be due to contact with other cells. Moreover Weiss's hypothesis does not very adequately explain how a cell can become motionless with two or three pseudopods of equal size. Weiss (1947, 1961) suggested that the stimulation of cells to produce pseudopods in a given direction depended on the molecular orientation in the substrate, or as he later put it more vaguely "the molecular ecology". But this does not get us any farther. In culture situations where cells frequently make contact or in which the substrate is curved or oriented on a scale similar to the cells, it is easier, as we shall see, to explain this stimulation of oriented movement. It is perhaps possible that on a plane substrate pseudopods initially form in random positions around the cell (this could be easily confirmed) and that the extent and number is partially a matter of cell type (amoebocytes appear to be able to form them all around the cell) and partly of chance. Weiss's hypothesis is difficult to operate because as we have seen in Chapter 4 orientation on the molecular level over the large regions is not going to affect the adhesiveness of the cell and substrate when they are in differing orientations. Thus there is no reason to believe that the cell surface is in any way patterned to react to such an orientation. However, in conclusion, it is still very obscure what processes are responsible for the establishment, maintenance and replacement of a given pseudopod. As will appear, contact stimulation or repression of cell movement appears in a much more marked way in other situations.

The next stage is to consider isolated cells in a population. In this system any interaction that may occur between cells occurs either by adhesive or mechanical interaction, or by direct passage of substances (and just conceivably electrical stimuli) between the cells, or by cells laying down substances which adhere to the substrate and only affect cells when they come into contact with the substance. Those reactions due to the passage of a substance from one cell to another by diffusion, which result in changes in the position of the second cell, are termed chemokinetic or chemotactic reactions. In chemokinetic reactions there is no orientation of the cells towards or away from the direction of the highest concentration of the substance. Instead a given concentration affects the speed of movement or frequency of turning so that cells tend to accumulate in or to leave the region (see Fraenkel and Gunn (1961) for full description). The cells may respond either to a gradient or to the absolute concentration of the chemical. In chemotaxis there is an actual orientation of a cell's movement towards or away from a source of the substance. Chemotaxis is a more efficient method of producing aggregation or dispersal than chemokinesis. Chemotaxis or chemokinesis may be either positive, leading to aggregation, or negative, leading to over-dispersion. Five systems in which chemotaxis or chemokinesis have been recognised with some certainty will be described; though curiously most workers have not realised the possibility that their systems could be chemokinetic rather than chemotactic.

Chemotaxis and chemokinesis

The first chemotactic system to be considered appears to operate to produce the non-random, overdispersed distribution which monocyte and macrophage cells take up in tissue culture on glass substrates. Jacoby (1944) pictured this phenomenon and noted that the cells are regularly spaced and are almost completely isolated. The cells are elongate or circular in outline and dispersed at some 50 to 150 micra separation. Oldfield (Whitfield, 1963) re-examined this system using adult chick monocytes grown on glass coverslips. She found that cell contacts with another cell were extremely rare, occurring roughly once per 100 cells. Following the technique devised by Twitty (see later), she covered part of the population with a coverslip so arranged that it enclosed part of the population in a narrow open-sided chamber some 80 micra high between coverslips. The cells in the confined space reacted to their changed conditions by first elongating with their long axes normal to nearest edge of the confined space, which suggests that chemotaxis rather than kinesis occurs. Then the majority of the cells migrated to the boundary of the 80 micra deep chamber and moved into the unconfined space outside. A few elongate cells remained widely scattered within the chamber. Before confinement the cells were already in a dispersed distribution, but after treatment they became much more dispersed within the chamber. This reaction of the cells to confinement suggests that a substance which produces chemotaxis accumulates in the enclosed space to higher concentrations than in the open part of the culture, or alternatively that it becomes depleted within the space. Greater concentrations of this substance (or its lack) would direct the movement of cells away from the region of high (or low) concentration. It seemed probable to Twitty and Niu (1948) that such a substance must be one which is easily lost or destroyed in the open part of the culture and this suggested that it must be a gaseous product of cell metabolism. It could however equally well be a gaseous substance which cells require, e.g. oxygen. However no convincing demonstration that it was carbon dioxide, or oxygen, was obtained. Nevertheless Whitefield's results strongly imply that although the actual chemotactic substance may not be volatile, a volatile substance is somehow involved in its production. It should however be mentioned that the lower concentration outside the enclosed chamber might simply be due to the greater depth of medium there.

Presumably the cells react either to the total concentration of chemotactic substance or to the steepness of the concentration gradient. Unfortunately there is little evidence to suggest which alternative is actually used by the cell. The monocytes become regularly dispersed both under the coverslip and outside. The reason for this is that when a steady state is reached (with the substance being either emitted or consumed by the cells) with the concentration gradients being radially arranged around each cell, then there will be no stimulus for any cell to move in any direction because it will experience repulsion or inhibition of movement equally in all directions. As a result, some 12 hours after placing a coverslip over the cells to form a chamber of depth 80 micra and width and breadth 800 micra, Whitefield found that only a very sparse array of cells remained under the coverslip. She suggested that the substance which produces cell repulsion in monocytes acts in a gradient by reducing the number of pseudopods to one, thus speeding up movement, and by diminishing the turning frequency.

Twitty and Niu (1948) made explant cultures of embryonic
Triturus torosus or *T. rivularis* neural crest cells. When the outgrowth
of propigment cells was partially confined by placing a coverslip
over them, the confined population became more dispersed with
cells migrating out into the uncovered part of the outgrowth.
However unlike monocytes no regular dispersion of the cells
occurred outside the coverslip. They continued this work by isolating a few cells in narrow capillary tubes, which led to the cells
moving very far apart. This result appears to confirm their argument
that diffusion would be so restricted in a tube that concentration
gradients would extend over greater distances. It is however unclear
whether the gradients are of equal steepness as in the coverslip
situation for greater distances or whether they are less steep, but
that a certain absolute concentration of substance is found further
from the cell, and it is this which the cells react to. It is not clear
whether this is chemotaxis or kinesis.

The "No-man's-land" phenomenon is an example of cell
behaviour probably resulting from chemokinetic interactions. This
appears in cultures of polymorph cells and was first described by
Carrel and Ebeling (1922). Explants of buffy coat are so arranged
that the outgrowths of polymorphonuclear leucocytes approach
each other. When the outgrowths are separated by a gap of some
100 micra movement of cells in the direction which would lead to
contact ceases. A cell-free space, the no-man's-land, is left between
the outgrowths. Oldfield (unpublished) reinvestigated this phenomenon and found that it only occurs when the liquid layer of the
culture medium is very thin, presumably because only under these
conditions will stable diffusion gradients be set up. She also showed
that when chick polymorphs were confined in a chamber 80 micra
deep they tended to migrate into uncovered parts of the culture.
These observations suggest that polymorph cells produce a diffusible
substance which inhibits movement of cells towards one another.
It is however possible that the substance is emitted by cells etc. in
the original explants.

There is considerable similarity in the behaviour of the three types
considered. All react by migrating away from one another in response to a diffusible substance which is almost certainly produced

by cellular metabolism. With a suitable arrangement of cells these effects can lead to the cells trapping one another in a stationary over-dispersed non-random array.

However chemotaxis and chemokinesis can be positive so that cells tend to move towards one another. Nevertheless positive chemotaxis or chemokinesis have very rarely been proved to act in animal cell systems. It seems to me that the main reason for lack of proof may lie in the nature of the distribution of cells which results from positive reactions. Whereas negative chemotaxis or kinesis tends to lead to ordered patterns of cells being set up, positive chemotaxis, unless very intense, will not result in a distribution of cells which is obviously non-random and highly ordered; and positive chemokinesis results in still less ordering. If chemotaxis is positive and weak the cells will frequently escape its effects so that the non-random distribution of the cells will not be obvious. It should be remembered that the sort of situation in which chemo-taxis is tested for is that in which the area not occupied by cells is greater than that occupied, so that if the cells move randomly they would be expected to be more frequently out of contact than in contact. This effect enhances any negative chemotaxis (if it is weak and random movement can occur) and diminishes the weak effects of positive chemotaxis. The only two examples of positive chemo-taxis which are really well established are ones in which the effect is so marked that it is immediately obvious. For these reasons it might be well worth while searching for examples of weak positive chemotaxis or chemokinesis using statistical methods to examine cell distribution changes with time.

Harris (1961) provides a critique of the various methods used to identify chemotaxis and points out that the demonstration of positive chemotaxis requires the identification of an effect on the orientation of movement of an individual cell, due to a gradient of concentration of a chemical, such that the cell moves or tends to move up the gradient towards the other cell etc. emitting the chemical. Harris devised a very elegant method for tracking the movement of indi-vidual cells using a photographic smear technique. By this means he showed that mammalian polymorphonucleocytes, monocytes and eosinophil leucocytes react chemotactically to a variety of stimuli such as factors derived from bacteria, so that they move towards the source of the chemical. Unfortunately the chemical nature of the agents is still completely unknown.

The second example of positive chemotaxis is that found in the cellular slime moulds, the *Dictyosteliaceae*, in which during one phase of the life cycle the cells (amoebae) are separate and isolated. Subsequently these cells aggregate naturally into multicellular fruiting bodies, which form spores. Initially the isolated amoebae move at random but after a while their movement becomes directed towards certain points termed *aggregate* or *aggregation centres* (Bonner, 1947; Shaffer, 1951a). Bonner also showed that if two coverslips each bearing amoebae were placed side by side under the liquid medium, one bearing cells that were just starting to aggregate and the other a small mass of recently aggregated cells (a centre), then the centre attracted amoebae on the other coverslip towards it, though of course they could not cross the gap between coverslips. This result is strong evidence for a chemotactic mechanism. The chemotactic agent was named acrasin by Bonner (for an amusing discussion of the possible etymology of this name see Shaffer, 1957a). Shaffer (1957b) was able to extract acrasin from the cultures and absorb it in agar blocks; amoebae were then attracted towards such agar blocks containing acrasin. Unfortunately no chemical identification of acrasin has yet been made.

There is considerable reason to think, according to Shaffer (1958), that although acrasin secretion starts at the centre, it soon is being secreted by cells outside the centre. The main reasons for his conclusions are: (i) streams of cells develop which move towards the centre; the shape of these streams and the orientation of cells in them suggest that cells are locally attracted towards the streams themselves, the presumption being that individual parts of the streams secrete acrasin; (ii) at the periphery of the area from which cells are being recruited into the aggregate cell movement is frequently oriented away from the radial course which points towards the aggregate centre. This second observation suggests that acrasin produced far out from the centre might be emitted in gradients which would be unaligned with the gradient from the main centre. However this latter observation and possibly the former could be

explained on other theories such that contact guidance or local
non-chemotactic aggregation was taking place. Shaffer (1961a),
and Gerisch (1961) showed that any cell might start the secretion
of acrasin and so become a centre. It seems probable (Shaffer, 1957c)
that *Polysphondylium* and *Dictyostelium* secrete differing acrasins.
Shaffer (1957a,b) points out that the secretion of acrasin may not be
continuous but pulse-like, since apparent pulsation of aggregation
movements appear to take place. Gerisch (1961) was able to get
aggregation without the action of visible centres by using carefully
defined conditions: this further suggests that cells develop the power
of secreting acrasin.

Samuel (1961) investigated the aggregation of *Dictyostelium
mucoroides* paying particular attention to the behaviour of individual
cells. The rate of movement of these cells falls just before aggregation begins and then rises as streaming towards the centre starts. In
detail the slowing of cell movement just before aggregation begins
occurs within 0·3mm of the future centre; just outside this area the
cells orient centripetally. This tendency for immobilisation and taxis
helps to trap cells in the future aggregation field. Prior to this
preaggregation phase, cycles of alternating drop in cell speed and
centrifugal orientation from the future centre and the converse
take place; random movement occurs at the nodes of this cycling.
Any given phase of the cycle is found at successively later times at
greater distances from the future centre. When the cells were isolated
on a homogeneous standard medium; the locomotory rate was
constant, which demonstrates that the differences are due to intercellular interaction. Non-aggregating cells show negative chemotaxis. Samuel's elegant work ought to be repeated for other types of
cell behaviour in vertebrate and other metazoan tissue cultures, e.g.
where cells such as lymphocytes (De Bruyn, 1945) show alternating
periods of movement and immobility, which cannot at the present
time be ascribed to the intrinsic properties of a single cell or to
interactions between cells. Jeon and Bell (1965) found that *Amoeba sp.*
cells tended to accumulate around *Hydra* or extracts of *Hydra*.
Though they term this chemotaxis there is no evidence from their
published photographs that cells are actually oriented towards the
Hydra. This suggests that positive chemokinesis has taken place.

They suggest that the chemical agent causes the protrusion of a pseudopod in the up-gradient direction.

Thus it can be seen that positive chemotaxis is a powerful means of controlling cell aggregation and dispersal by altering the direction of cell movement. But with the exception of Samuel's work, we are unsure whether positive chemotaxis acts on speed of movement or rate of turning, or whether the prime action is on the determination of where in the cell surface pseudopods form, etc. Samuel's work shows that chemotaxis of *D. mucoroides* cells involves (*a*) orientation and trapping of a field of cells in which the cells are oriented centripetally (presumably up-gradient), (*b*) subsequent stimulation of movement with suppression of turning. It is remarkable that apparently in this system all the possible means by which a cell's movement might be oriented are brought into play in the correct sequence; but this is somewhat confirmed by slime mould aggregation being such a very effective and efficient system. The work of Oldfield suggests that negative chemotaxis involves stimulation of movement and rate of turning when the gradient is established and the converse when the gradient disappears.

Despite the existence of these six examples of chemotaxis or chemokinesis, there are no other known instances in multicellular animals which have been recognised with any certainty, and it should be appreciated that with the exception of the slime mould work and the reactions of amoebae, all the examples are recognised in tissue culture (although there was some reason for suspecting chemotaxis of various white cells to sites of infection etc. from histological sections). It is not that there is any reason to doubt that chemotaxis may not occur *in vivo* but rather, as Harris (1961) points out, that we have not developed the techniques for unequivocally recognising it *in vivo*. Taylor (1962) Weiss and Scott (1963) pointed out that local pH gradients can control the direction of movement of cells; the direction of migration being towards neutrality (pH 7·0) in the gradient. This could be viewed as a type of chemotaxis, both positive and negative. Coman (1965) suggested that leucocytes show negative chemotaxis to each other, but his results could be equally well explained by the operation of contact inhibition of movement.

Roux (1894, 1896) suggested that amphibian embryonic cells obtained by disaggregation are attracted towards each other by a cytotaxis or "cytotropismus" effect. But Voigtlander (1932) and Kuhl (1937) found no observable orientation of such cells towards one another or to a centre but did not use any statistical test to confirm their observations. Lucey and Curtis (1959) obtained a small amount of statistical evidence that there was no effect of one cell on another's direction and frequency of pseudopod protrusion, which argues against a chemotactic effect in aggregation. Unfortunately Kuhl and Roux described the cells they used as blastomeres but it is very improbable that they studied cells from early cleavage stages of development, since these apparently do not aggregate (Curtis, 1957, 1960a). Kuhl found that the cells made long "random" walks. He also was unable to obtain aggregates composed of more than a few cells. It is impossible to completely exclude the action of chemotaxis in this system at present, partly because some workers did not obtain appreciable aggregation but mainly because of the absence of careful statistical tests in earlier work and because of the small samples studied by Lucey and Curtis. Galtsoff (1923) observed no signs of chemotaxis in the locomotion of aggregating sponge cells. As will appear shortly, chemotaxis affords a simple explanation for a variety of features of cell movement in both aggregates and embryos, but with the exception of melanophore dispersal in urodeles no evidence has yet been produced to substantiate such explanations for directed cell movement in embryos. Moreover almost nothing is known about the possible mechanisation of chemotaxis, though Carter (1965) has suggested that positive chemotactic substances make the end of cell up-gradient more adhesive than the other end of the cell, so that the cell moves up the gradient.

The ground-mat or microexudate theory of cell guidance and contact guidance

There is another mechanism by which cells may interact at a distance, but which is a contact phenomenon unlike chemotaxis. Cells may synthesize various macromolecules and release these into the medium, from which they are very rapidly adsorbed on to the surrounding substrate. Since cells react to the substrate in some way,

as we have already seen, it is conceivable that the adsorbed macromolecules on the substrate (the ground-mat or microexudate as P.
Weiss (1940, 1941, 1958) has termed it) might affect cell movement
when cells pass over parts of the substrate bearing the ground-mat.
The evidence in support of this theory is derived from a lengthy
series of observations. Weiss (1929) observed the emigration of
fibroblasts from explants in tissue cultures grown in or on plasma
clots. If two explants are placed close to one another a bridge of cells
frequently develops between them, the "two centre" effect.
Previously this had been explained as being due to chemotactic
direction of cells towards one another, but Weiss showed very
convincingly that the bridge of cells formed because they emigrated
along oriented fibrin fibres between the explants. Weiss and Garber
(1952) and Garber (1953) showed that the rate of migration of the
cells and their morphology was closely related to the texture of the
fibrin clot, highly oriented clot structure favouring rapid emigration
and elongate cell form.

Weiss (1941, 1945) examined the migration of nerve cells, in
culture, along glass fibres. Since most cell types cannot swim it is
very obvious that cells will only move along solid structures as
Harrison (1914) first found, naming the reaction, stereotropism. But
Weiss pointed out that the reaction has some more subtle and
interesting features. These are that the cells tend to show a markedly
rectilinear motion along the fibres, instead of spiralling round or
encircling them. In addition the cells become elongated in the
direction of their motion and probably tend to show faster movement than cells on a plane surface. It is not clear whether isolated
cells show any direction or acceleration of movement on fibres as
compared with plane surfaces. No close study has been made of the
phenomenon but it would seem probably to result from an effect
on turning such that turns which bring the cell into line with the
fibre are preferred. Weiss (1958) later demonstrated that a very
similar if not identical phenomenon occurs when the cells are placed
on a surface bearing grooves. In this case the cells move rectilinearly
but probably more slowly (Curtis and Varde, 1954) than on convex
surfaces. It is not wholly clear from Weiss's work whether the cells
react to the convex edge of a square cut or to the concave groove

itself. Weiss (1941) termed this rectilinear motion *contact guidance*. In the case of movement along grooves the orienting structure may be the groove edge, but it is less immediately obvious what causes rectilinear motion on the smooth cylindrical fibre.

Weiss (1941, 1945) attempted to explain contact guidance as being due to a colloidal exudate, microexudate or ground-mat produced by cellular action which is laid down on the fibres ahead of the cells. As it is laid down it takes up an oriented structure in response to surface tension relationships (according to Weiss, 1945). If this is the case it is surprising that it is oriented in the same direction on both concave and convex surfaces. More recently Weiss (1953, 1958) has not expressed any definite view as to how the orientation occurs. In his 1945 paper Weiss rejected the idea that the cells were simply reacting to the curvature of the surface on the grounds that the curvature was too slight to be of effect on molecules in the cell surface. This idea seems to be inadequate and thus it might be worth reconsidering whether the effect was not due to the topology of the surface and nothing else. Weiss believed that the cells react to the orientation of molecules in the substrate, in other words to their "molecular ecology" (1958). No mechanism by which this might occur to produce cell orientation has ever been suggested, though Taylor and Robbins (1963) observed fine pseudopods which were put out by a variety of cells in culture, which they suggested were used to detect molecular orientation in the substrate. A number of papers describe the identification of microexudates on surfaces on which cells are or have been growing. Weiss himself (1945) noticed a stainable material on the glass fibres on which the cells grow. Although this material may be present in many culture conditions there is no proof that the association of the material and contact guidance is more than incidental at present. Taylor (1961). Rosenberg (1960) and L. Weiss (1961b) produced evidence that cells lay down microexudates on glass surfaces. But the effects that Taylor and Rosenberg describe are directly opposite (see Chapter 4) and technical objections may be raised against these demonstrations (also see Chapter 4).

Rosenberg (1963) found that cells can accumulate preferentially in grooves in behenic acid (C_{22} fatty acid) multilayers as compared with a plane surface of behenic acid. They lie elongated in the grooves. He found that the grooves had to have certain relation-ships between width and depth for this effect to occur. He made no conclusions about this other than the most general one, that cells react to the molecular orientation of the substrate. Spratt (1963) suggested that contact guidance occurs in the migration of chick blastoderm cells over an agar/chick embryo extract surface; the cells responding to a cellular exudate. For criticism see Chapter 7. Johnsson and Hegyeli (1965) isolated from urine a partially purified peptide, called "directin", which appeared to enhance some orien-tation present on glass coverslips so that HeLa cells lined up on the glass in relation to this orientation.

In considering whether contact guidance might not be explained as a direct reaction of cells to the shape of their substrates, Curtis and Varde (1954) pointed out that there is no theoretical reason for expecting orientation of molecules on curved surfaces of appreciable radius of curvature. But it would be expected that curvature would affect packing in the cell surface and this might alter cell behaviour. They also found that features of cell behaviour such as the display of contact inhibition of movement (see later), and the extent of cell spreading were very dependent of fibre diameter in a non-linear fashion. There is no reason to expect that molecular orientation on a surface would vary in this non-linear manner. Curtis and Varde also found that features of cell behaviour were opposite on convex and concave surfaces, which suggests a direct reaction to curvature. But Curtis and Varde were unable to determine the actual mechan-ism of contact guidance.

Consider this example of Rosenberg more closely. Cells are grown on a surface of behenic acid monolayers. When a groove is cut in the behenic acid surface, cells align themselves along the groove. Behenic acid monolayers are built with the molecules oriented with their long axes vertical so that there is no cue in the molecular orientation to align the cells in any preferred orientation. The only orientation paralleling the cell orientation is the groove edge. Although the height of the groove is of molecular proportions, its length is not. But there is evidence that it is not the edge by itself which the cell reacts to, because cells do not accumulate or orient at single edges, such as occur at steps from one monolayer to another. Two groove edges have to be present for this reaction to occur. This fact immediately suggests an explanation which is very different from Rosenberg's ideas. When the base of the groove is glass it will be of different adhesive properties to the behenic acid, and when the base of the groove is behenic acid Rosenberg (1962) presents evidence that it has different adhesive properties from the plane surface outside the groove. If the groove is wide enough for the cell to be able to sag into it without bridging it some 200–300 Å above its base, then the cell will be able to come into adhesion with the bottom of the groove, at say 100 Å separation. This means that if the base of the groove is more adhesive than the surroundings the cells may become trapped there, the shape of the groove leading to cell orientation. Cells however will not trap at a single edge since they can wander away from it on a surface of constant adhesiveness. Until it is shown that this and similar explanations cannot apply, there is no need to invoke the idea of molecular orientation controlling cell behaviour in this instance. Curtis and Varde (1964) suggested that cells react to the topology of the surface on which they adhere. The main difference of this idea from that of molecular orientation of the substrate controlling cell behaviour, is that the former idea supposes that the cell surface reacts on the molecular scale (packing etc.) to features of the substrate whose signal to the cell is by adhesiveness or curvature etc. The concept of control by molecular orientation of the substrate supposes that the signals are from individual molecules to which the cell reacts in an unspecified manner. Neither theory has much experimental evidence in its favour at present. For these reasons it must be admitted that although the idea of control of cell behaviour through the action of microexudates is attractive, there is very little evidence which supports this theory at present.

The contact interaction of cell with cell

Although the reactions of a cell to settling on a plane surface or the phenomena of contact guidance suggest that cells show contact reactions in such comparatively simple situations, it is the interaction of one cell with another which has provided the best evidence for the occurrence of contact reactions. Of course cells which are in contact may interact by chemotactic means, with in some cases the chemotactic gradient being able to operate only over a very few micra. This would not be a contact reaction but it would be easily confused with one. Although very few studies have been made of the fine structure of the contacts between cells, it seems probable that cells do not often come into molecular contact (see Chapter 3). Small regions of molecular contact will occur when *zonulae occludentes* are observed in the cell contacts, but these have been described in only a few instances. Since cells can be as close to one another as 100 Å for appreciable areas of contact, a variety of types of physical interaction can occur. Adhesive reactions have been discussed in Chapter 4 and mechanical ones in Chapter 2. Chemical interactions which result from the closeness of the two cells include the following possibilities: (i) the exchange of diffusible substances between cells which, though not acting in a chemotactic manner, affect the adhesive character or motility of the recipient cell; (ii) the development of a micro-environment of changed composition in the gap between the cells, for example by pH alteration (Weiss and Scott, 1963) or by the accumulation of secreted enzymes (see L. Weiss, 1962a) which affect cell behaviour, or again by the synthesis of intercellular material in the gaps between cells to alter their contact relations (P. Weiss, 1958). It is very improbable (see Chapter 4) that cells can react to the actual orientation or distribution of molecules of various types on one another's surfaces: the reason for this is that there is no means of transmitting any signal specifying orientation or distribution over some 100 Å or more between the cells. It is of course possible, as Grobstein (1961) points out, that intercellular material may be ordered into very specific patterns of structure; but although it is obvious that large scale structures such as collagen fibrils will affect cell behaviour, there is much less evidence that there is any organised structure in the 100–600 Å gap between cells (see Chapters 1 and 3). Electrical interactions between cells in contact leading to activation of movement etc. are of course possible in theory. But although Bingley and Thompson (1962) found that the value of the membrane potential of *Amoeba proteus* affects the locomotory properties of this cell, it is unclear whether two cells can affect each other by electrical discharge across the

THE CELL SURFACE

intercellular gap in such a way that their locomotory properties
change. Conceivably, however, nerve cells, particularly during
development, might be affected in their movements by the electrical
activity of other nerve cells. Consequently the activity or non-
activity or even possibly the pattern of activity of a nerve cell might
control the direction of movement of other cells towards it, thus
helping to build specific connections (see Chapter 7). As suggested
in Chapter 3, if the cell surface is a polarisable electrode then
cellular adhesiveness will be affected by the electrical interaction of
cells, and any alteration of their metabolism resulting from closer
contact might change their adhesiveness by reason of metabolic
control of surface potential. This discussion has unfortunately had
to be almost entirely theoretical since there is at present insufficient
experimental evidence, but its purpose has been to indicate the
various ways in which it is possible that cells might interact to alter
their adhesive and locomotory properties when they make contact.
In the next section the various types of contact interaction of cells
will be examined and their causes considered in the light of these
theoretical considerations.

Contact inhibition of movement

Although features of cell behaviour in populations due to contact
inhibition of movement have been recognised for the last fifty
years, for example the monolayering of cells noticed by Loeb (1921),
the precise description and naming of this phenomenon had to wait
till the work of Abercrombie and Heaysman (1953, 1954). This phenomenon is often and slightly ambiguously referred to as contact
inhibition. Abercrombie and Heaysman studied the behaviour of
embryonic chick heart fibroblasts moving on glass in a medium
composed of serum, embryo extract and saline. Abercrombie and
Heaysman (1953) showed that an increase in the number of neighbours that a cell is in contact with results in a statistically significant
decrease in its speed. These cells were at the edge of the outgrowth.
They also found (unpublished) that the direction of movement of
the cells was strongly controlled by the number of cells they were in
contact with. This can be measured as "contact number" (Abercrombie and Gitlin, 1965). Cells with many contacts show no

preferred direction of movement. Abercrombie and Heaysman(1954) were able to detect by statistical means the phenomenon of cell behaviour which is responsible for these occurrences. Abercrombie and Heaysman considered the marked monolayering of fibroblasts which occurs in these cultures. Explants were placed so that cells in their outgrowths moved towards each other. Some hours after the outgrowths first made contact, cells from each outgrowth should overlap, if their movement were unimpeded in direction or speed, by coming into contact with cells of the other outgrowth. Abercrombie and Heaysman examined the extent of overlap of the cells by counting overlaps of nuclei (because nuclei can be easily seen and their area measured, and because knowledge of their area is needed for calculating the number of overlaps that would be expected if they occur randomly, i.e. if nothing impedes overlap.) On comparing the actual number of nuclear overlaps with those that would be expected on random grounds they found that there were approximately one third as many overlaps as would be expected on random grounds. This result, when considered in conjunction with their earlier results on speed and direction of movement in relationship to contact number, was best explained by the following hypothesis. When two cells make contact, their movement towards each other tends to cease before one cell is able to move an appreciable distance over the other; as a consequence any further movement that the cells may show will be directed away from the axis of contact. This phenomenon they termed contact inhibition of movement. It is clear that the phenomenon will account for the diminution of rate of movement with increasing contact number and will tend to result in cells moving away from the outgrowth, as Abercrombie and Heaysman found. If however a given cell is surrounded on all sides by other cells it will tend to have its movement in any direction inhibited so that it will not show any preferred direction of movement nor much movement. A cell free from contact with others will tend to show more random movement. It is also obvious that this phenomenon accounts for the marked monolayering of fibroblast outgrowths. No slowing of a cell's movement just before it made contact with another was observed: this argues against any diffusable stimulus being responsible for the reaction.

Abercrombie and Ambrose (1958) examined this phenomenon by interference microscopy. They observed that when the leading edge (ruffled membrane) of the main pseudopod of one cell meets that of another, (i) the ruffling tends to cease; (ii) an adhesion appears to form between the two cells; (iii) cell movement ceases; and (iv) after a while a new pseudopod may form in some other part of the cell and tend to move one cell away from the other, if it does not almost immediately contact another cell and become inhibited. The reasons for thinking that an adhesion forms are as follows: (i) Kredel (1927) found that fibroblasts appear to be adherent to one another as judged by the results of attempting to dissect such cells apart, (ii) it can be observed that two cells which have been in contact may break apart as though they were snapping under a considerable strain. This latter observation implies that it is an adhesion which resists the separation of the cells. Weiss (1958) observed the snapping reaction and termed it "contact retraction", but his description which is purely qualitative appears to be of just a part of the contact inhibition phenomenon.

That contact inhibition of movement occurs in other cell types has been established for neo-natal mouse muscle cells in contact with embryonic chick heart cells (Abercrombie *et al.*, 1957), for rat neo-natal heart muscle fibroblasts in contact with themselves (Curtis, unpubl.), for adult chick polymorphs and monocytes in contact with themselves (Oldfield, 1963), and for embryonic heart fibro-blasts from the Japanese quail (Heaysman and Lamont, unpubl.). It is very probable that hamster kidney cell strains show contact inhibition (Macpherson and Stoker, 1962) but these authors have merely subjectively described the cells as failing to overlap without using any quantitative study. It is apparent that epithelial cells show contact inhibition, but the evidence is based only on subjective descriptions (Abercrombie, 1964). Some tumour cell types do not show contact inhibition of movement in culture. Abercrombie and Heaysman (1954) found that mouse sarcoma 37 cells are not contact inhibited by chick 9-day embryonic heart cells. This finding is of considerable interest in relation to the local invasion of tumour cells. Subsequently mouse sarcoma 180 cells (Abercrombie and Heaysman, unpubl.) were shown to display some contact inhibition in culture, and Barski and Belehradek (1965) found that mouse N1 and M6 tumour cells apparently did display contact inhibition. Forrester *et al.* (1964) claimed that hamster BRK malignant cells did not show contact inhibition in culture but this was not upheld on examination (Curtis, unpubl.). Although these findings might appear to question the action of a lack of contact inhibition in tumour malignancy it should be remembered (see later) that chemical conditions in culture media greatly influence contact inhibition.

A number of matters of interest arise from the methods used to make quantitative examination of contact inhibition of movement. It should be realised that the methods used by Abercrombie and his co-workers depend on considering nuclear overlaps and not on cytoplasmic or whole cell overlaps. It is of course possible that there is appreciable cytoplasmic overlap and that the observed pheno-menon is merely due to some mechanism which prevents nuclei overlapping one another. But although this effect might lead to a limited mis-estimate of the extent of contact inhibition it seems unlikely in view of the observations of whole cells undergoing contact inhibition made by Abercrombie and Ambrose to completely negate the existence of the phenomenon. Abercrombie and his co-workers treated nuclei as being circular in plan in their earlier work and as random in plan later, in order to calculate the nuclear area, knowledge of which is required to compute the overlap expected if contact inhibition did not occur. Since nuclei are gener-ally elliptical in shape these two methods of treatment lead to inaccuracy in estimating random expected overlap. A more accurate method is to consider the nuclei as ellipses (see Curtis, 1961 and Curtis and Varde, 1964 for treatment and possible errors). Curtis (unpublished) showed that appreciable errors in calculating expected overlap on the random or circle basis arose if the ratio of long to short axial length exceeded 2.

A more serious unsolved problem is that the greater part of the deficiency of overlap might have quite another explanation. If the rate of cell movement is very much higher for a cell A moving on another cell B than it is when the cell A moves on the substrate or just in contact with the edge of the other cell, and the rate of turning is the same on both surfaces, then cell A would more frequently overlap substrate than cell B even if there were equal areas of cell B and substrate. However, Abercrombie and Heaysman's observations (1953, 1966) on speed of movement in relation to contact indicate that there is a profound effect on speed and direction of motion even for very slight contacts, which would tend to prevent cells ever overlapping.

The degree of contact inhibition of movement shown by cells in contact with their own type is variable from one cell type to another as well as in interactions with other cell types. Abercrombie *et al.* (in press) showed that the degree of contact inhibition between chick heart fibroblasts and neo-natal mouse skeletal muscle fibroblasts was not of the value which would be expected from the values of overlap of the two types of fibroblast each with its own kind. This finding suggests that the reaction cannot be due to a simple linearly quantified cell property, such that interaction between two types would be of a value which is the mean of the two separate values for homologous interaction. The reaction may be of even greater complexity: for example although mouse sarcoma 37 cells are not contact inhibited by chick fibroblasts (Abercrombie and Heaysman, 1954; Abercrombie *et al.* 1957) and so invade these cultures, the fibroblasts are inhibited in their movements by contact with the sarcoma cells so that they do not invade the sarcoma cells. This failure of sarcoma cells to become contact inhibited has also been observed for cells of the mouse MC1M tumour and 311 tumour (Abercrombie, Heaysman and Lamont, unpubl.).

The degree of contact inhibition shown by a cell is very dependent on other factors. Curtis (1961) found that population density has a profound effect on the degree of inhibition displayed by cells of that population. At low densities <5 cells $1000\mu^2$ contact inhibition decreases with rising population density, but this effect disappears and is replaced at densities above nine cells per $1000\mu^2$ by an increase in inhibition with population. The existence of this indicates the need to take population density into account in comparing the incidence of contact inhibition in two populations. Statistical methods for doing this were derived by Curtis and Varde (1964). Medium conditions affect the display of contact inhibition markedly. In crude experiments Curtis (1961) found that increased embryo extract concentration in the medium resulted in increased overlap. Heaysman and Lamont (unpubl.) found that serum concentrations affect it, and Curtis (1963) found that the pure protein which inhibits aggregation can abolish contact inhibition presumably by diminishing cellular adhesiveness, if applied at twenty times its concentration in crude serum. Todaro *et al.* (1966) also found that calf serum increased mitosis and perhaps diminished contact inhibition as well. Nucleotides (c. 0·004M) added to the standard culture medium weaken the display of contact inhibition (Curtis, unpublished). The nature of the substrate may also affect contact inhibition markedly, for example colloidion (mixed cellulose nitrates) substrates result in a diminution of contact inhibition as compared with plasma, glass, collagen or polypropylene substrates (Heaysman and Lamont, unpubl.). The topology of the substrate can control the display of contact inhibition to a considerable degree (Curtis and Varde, 1964), contact inhibition being enhanced on concave surfaces and increasing with diminishing radius of curvature. Contact inhibition is diminished on convex surfaces and lessens with diminishing radius of curvature.

A number of workers have used the term "contact inhibition" in a sense which may be rather different from that which Abercrombie used. One of the predictions which can be made is that at high population densities any one cell on making contact with another and then another will soon find itself confined in a space which is elongated parallel to the cells on either side. Consequently cell alignment might build up. Macpherson and Stoker (1962) as well as Berwald and Sachs (1963) used the subjective appearance of cell alignment as a criterion of the presence of contact inhibition. Despite the prediction from contact inhibition, it has to be added that the past history of a culture, in particular the rate at which cell population builds up, would considerably affect the area over which cell alignment would develop. Until cell alignment and nuclear overlap are compared in the same cultures we cannot be sure whether the cell alignment is wholly due to contact inhibition or whether it is due to some other type of behaviour. Curtis (unpubl.) found no evidence that the alignment reported by Macpherson and Stoker was due to contact inhibition.

A variety of mechanisms have been suggested to account for contact inhibition of movement. The majority of these hypotheses are based on direct contact reactions between cells but, as Abercrombie (1961b) pointed out, short-range chemotactic reactions may account for contact inhibition. However, although simple chemotactic hypotheses and micro-environment theories such as Taylor (1961) has put forward (in terms of the pH of regions between cells inhibiting movement) fail to account for the rhythmic changes in cell behaviour during contact inhibition, this does not preclude the setting up of more complex hypotheses. The majority of theories presuppose that adhesive properties control cell behaviour in contact inhibition. The simplest hypothesis is that when one cell runs into another either the adhesive energy or the inertial energies of both cells are such that both cells cease to move. One simple example of this hypothesis would be that when two cells collide they have insufficient energy from movement to continue movement. However Curtis (1961) found that although contact inhibition decreases with increasing embryo extract concentration in the medium, speed of movement declines in the higher range of extract concentrations. This result is directly contrary to the expectations of the collision hypothesis. A second simple hypothesis is that contact inhibition occurs because cells are more adhesive to the substrate of the culture vessel than to one another, so that they are unable to move over one another (see Carter, 1965). Steinberg (1964a,b) suggested that this phenomenon occurs in a variety of aspects of cell behaviour without specifically implicating contact inhibition. Curtis and Varde (1964) produced evidence against this hypothesis. They found that sheets of cells could be grown suspended free in the medium like sails. These sheets were anchored to fibres at either side. In these sheets the only possible substrate a cell may have is another cell. Contact inhibition is very marked in these sheets. Since there is no competition between a cellular and a noncellular substrate for the adhesion of the cells in this situation, the occurrence of contact inhibition disproves Carter's simple hypothesis. In addition cell behaviour in this culture situation is of interest in connection with the question of the time spent in overlap. Since these sheets are composed of coherent cells which show little overlap and since any movement a cell can make must never take it out of contact with another cell, we cannot account for the deficiency of overlaps by the hypothesis that cells in overlap move so rapidly that they rapidly move out of overlap. Nevertheless this hypothesis might still apply if we restate that only over the nuclear region of the cell does fast movement of another cell occur. Observation of cells in contact (Abercrombie and Ambrose, 1958) does not substantiate this alternative hypothesis to account for the deficiency of overlap.

It has been claimed in several papers that contact inhibition of cell movement is basically an adhesive phenomenon (Abercrombie, 1957, 1961a, 1962, 1964a,b; Abercrombie and Ambrose, 1958) and only secondarily one of cell movement. Yet the inhibition could be well explained if the occurrence of contact between two cells resulted in a paralysis of cell movement due perhaps to some chemical or electrical interaction. The evidence that contact inhibition is basically an adhesive reaction is supported by the following evidence. Kredel (1927) found that fibroblasts in contact were in adhesion as judged by the reactions of the cells to microdissection of their contacts, and the snapping reaction "contact retraction" points to a similar conclusion. We have already seen that it is impossible to explain contact inhibition in terms of a simple competition between substrate and cell for the adhesion of another cell. But it is possible that more complex theories of adhesion may account for the phenomenon, which still need to be tested. In general, these theories would propose that although the periphery of the cell as seen in plan view and the pseudopods have a very adhesive surface, the rest of the surface is of low adhesiveness. As a result cells meeting one another would tend to adhere at their edges but not to overlap. A small amount of circumstantial evidence in favour of this theory was discussed by Curtis (1964).

One observation which suggests that any adhesion theory is insufficient, is based on the observation that the interaction of tumour cells of the mouse sarcoma S37 in culture are not impeded in their movements on making contact with a fibroblast, but that the movement of the fibroblasts is impeded by contact with the sarcoma cells. Obviously this cannot be due to adhesion between the two cells.

There thus appear to be two general alternative theories left to account for contact inhibition of movement. First, that it is a reaction which has nothing to do with cell-to-cell adhesion, being caused by one cell being able to paralyse the mechanism of cell locomotion of another directly, possibly chemically. Nevertheless this type of hypothesis seems a little unlikely in view of the circumstantial evidence that adhesive processes are involved in adhesion. The second type of general theory supposes that when cells meet they interact so that their adhesive properties change. An elaborately detailed example of such a theory was published by Curtis (1960a): though this particular example still remains a speculation it will serve to illustrate the main features of the alternative general theory about contact inhibition. Curtis made two premises: first, that as adhesiveness increases so a cell's potential motility increases up to a certain degree of adhesiveness and then declines with increasing adhesiveness; second, that the adhesiveness of a cell is affected by another cell's movement across it. He suggested that when one cell moves across another both cells shear one another with their actively moving pseudopods, and that their cell surfaces react to the shear by changing their properties, in particular surface potential. A cell may react either by increasing its adhesiveness on shear, or by decreasing it, according to its type. Consequently contact inhibition of movement of two fibroblasts would be thought to be due to the increase in adhesiveness of the cell surface, consequent on shear when two pseudopods touch (see Chapter 2, page 70). It is possible that only the regions sheared show an increase in adhesion, while other parts of the surface actually show a decrease (Curtis, 1960a, 1962a). The resulting adhesion would stop pseudopod movement, so that shear would vanish and the adhesion in turn, with the result that the surfaces would return towards normal and attempts at movement would be renewed to be followed by another period of adhesion and so on. It is very interesting that Abercrombie and Ambrose (1958) found that when two cells make contact, periods of apparent adhesion of the contacting pseudopods are alternated by periods of renewed pseudopodal movement. Similarly such a hypothesis explains the result observed when sacroma cells meet fibroblasts; the fibroblast becomes more adhesive but the sarcoma cell less adhesive, so that the sarcoma cell is little affected in its movement over the fibroblast. The reason for the contact inhibition of fibroblast movement in this case is that as a cell's pseudopod becomes more adhesive on shear, those parts of the pseudopod in contact with the substrate adhere to it firmly enough to prevent movement. Of course, this theory is almost completely unsubstantiated at present but it does provide a testable hypothesis. A very similar hypothesis has been put forward by Shaffer (1964) to account for the phenomenon called "contact following" (see later).

Contact inhibition of cell movement has a number of interesting consequences for the behaviour of a cell population: some of these have already been mentioned (see also Abercrombie, 1957, 1961a,b). Although contact inhibition is not displayed by all cells of a culture at all times. it can often be displayed by the majority of cells at any one time. Abercrombie and Heaysman's method of counting overlaps and comparing expected and actual overlap provides a very exact measure of the phenomenon, by which it is possible to recognise populations which show only say 10 per cent less overlap than expected, to ones in which 90 per cent less occurs. The following phenomena are typical of populations in which at least 60 per cent less overlap than expected is found. It is obvious that appreciable contact inhibition will lead to monolayering since overlap is impeded. The fact that movement is inhibited on contact and then restarted in a new direction will result in cells tending to continually move away from contacts. This explains the emigration of fibroblasts from an explant in a radially oriented outgrowth. Abercrombie and Heaysman (1953, 1966) have shown that the direction of movement of cells at the edge of the explant is strongly dependent on the extent of contact with other cells, which suggests that chemotaxis is unlikely to act in producing the radial outgrowth.

Thus it is possible to explain to a considerable extent the features of outgrowths in tissue culture in terms of the operation of contact inhibition of movement without having to have recourse to phenomena such as chemotaxis of cells. It might be expected that contact inhibition would operate to produce outgrowth (emigration) of cells from an explant. The reason for this is that contact inhibition would prevent cell movement in the edge of the explant in any direction save that which takes a cell out of the explant. However, treatment of explants with the aggregation-inhibiting protein, which diminishes contact inhibition (Curtis, 1965b), accelerates cell outgrowth. This result would not be expected if contact inhibition played a major part in producing the emigration of cells. It seems more likely that the diminution of cell adhesiveness as a result of treatment of the protein allows the cells to escape from the explant. Presumably in any outgrowth from an explant the cells have to be of low adhesiveness or else they would not escape from the explant. The possible operation of contact inhibition in the sorting out of tissue types in reaggregates, embryonic cell movement and tumour invasion will be discussed in the next two chapters. It can now be seen that contact inhibition of movement is a phenomenon which leads (a) in the presence of free space into which the cells can move, to an increased rate of turning; and (b) in the absence of this space, to the trapping of the cell. By this means contact inhibition provides a very precise method of stopping cell movement into an area when a population occupies it as a monolayer. Curtis and Varde (1964) described a contact phenomenon which may be closely associated with contact inhibition. They found that the spread area (plan area) of a fibroblast growing in a culture was very closely related to the proportion of the cell's perimeter in contact with another cell. The relation was a curvilinear one, the greater the contact length the smaller the spread area; but areas below c. 700 μ^2 per cell were not found.

The finding that some forms of tumour cell at least did not show contact inhibition in culture, and the possible relation of this lack of contact inhibition of movement to malignancy, stimulated considerable speculation. Macpherson and Stoker (1962) and Stoker (1964) found that polyoma-transformed hamster cells showed a different form of colony morphology in culture from that shown by their normal equivalent cells. Stoker based a theory of the mechanism of contact inhibition on these results (but see below). Normal cells tended to form colonies in which the cells were markedly aligned over considerable areas whereas the malignant cells tended to be randomly arranged. Since it is clear that parallel packing of cells can be a consequence of the operation of contact inhibition of movement it is

tempting to conclude as Stoker (1964) did that the polyoma-transformed malignant cells lacked contact inhibition. Defendi and Gasic
(1963) found that the transformed cells appeared to have more sialic
acid per unit area on their surfaces than the normal cells, this result
agrees with the finding of Forrester *et al.* (1964) that these malignant
cells have higher surface charge densities than their normal equivalents. Since Ambrose (1961, 1962) had suggested that malignancy
was characterised by an increased surface charge density, which also
probably produced a decreased adhesiveness of the cells, it was
tempting to conclude that malignancy, lack of contact inhibition
of movement, high surface charge density, decreased cellular
adhesiveness and high sialic acid density on the surface were all
correlated and causally connected. However Vassar (1962) and other
workers have shown that high surface charge density is not usually
correlated with malignancy. In my laboratory we have found that
when quantitative methods of measuring contact inhibition are
applied to Stoker's material both malignant and normal cells show
the same appreciable degree of contact inhibition. Finally Kramer
(1966) found no particular correlation between sialic acid density
on the surface and the type of cell behaviour which might be
suspected from colony morphology.

Contact promotion of movement and peripolesis

If we imagine that the contact inhibition of movement reaction
were to be reversed then cells would tend to overlap more frequently
than expected on a random distribution of overlaps. As a consequence cells would move around on "top" of one another. This
reaction does indeed appear to occur in some types of reaggregation
(Curtis, 1960a), and was named (in contrast to contact inhibition
of movement) *contact promotion*. Unfortunately although the
phenomenon has been subjectively recognised there is no quantitative description of it as yet. The reaction would be expected to
comprise the stimulation of movement of cells making contact
and development of an increased rate of turning whenever a cell
started to move off a cellular substrate, on to a non-cellular one.
Viewed at any one moment a population of contact promoting
cells would be indistinguishable from one in which a rather different

H

phenomenon was taking place. This would be the overlap of cells without further cell movement once the overlap was first formed. This might occur if the adhesion of cell to cell was very strong so that two cells on making contact would maximise their areas of contact and then cease movement. This phenomenon is probably equivalent to the clumping of cells observed under conditions in which cell movement is prevented, e.g. low pH or presence of antibodies. Unfortunately workers who have described the clumping of cells in culture (e.g. Saxen and Pentinnen, 1961; Moskowitz and Amborski, 1964) have not observed whether the cells move on top of one another or are stationary.

A phenomenon which may in fact be identical with contact promotion has been named *peripolesis* (Lewis and Webster, 1921; Sharp and Burwell, 1960) in which a moving lymphocyte is restricted in its movement to the surface of a tumour cell or macrophage. Description of the phenomenon seems a trifle incomplete because it is unclear whether movement is restricted to the periphery of the cell as seen in plan view, or whether it can take place over the whole of the free surface of the cell. In the latter case the phenomenon would be identical with contact promotion. No quantitative description of the phenomenon has yet been published. An as yet unnamed type of cell behaviour, which has some of the features of both contact inhibition and promotion as well as peripolesis, was described by Trevan and Roberts (1960) as occurring in cultures of cells of semi-ascites epithelioma. These cells remain monolayered, but when two cells meet, although their active movement continues, they turn so that they do not overlap. Their movement apparently continues in directions such that they tend to remain in contact by their peripheries (as seen in plan view). One cell may exchange its contacts with one cell for contacts with another cell. Unfortunately there is no quantitative description of the phenomenon in terms of frequency and direction of turning or in terms of the frequency with which cells break contacts entirely and emigrate from the monolayered associations of cells in contact. It would seem that very slight contact between these cells is contact promoting, but extensive contact becomes inhibiting. It is possible that these cells individually are insufficiently adhesive to the substrate to move efficiently but small areas of adhesion between two such cells provide enough adhesion for active movement. Large areas of adhesion may bring the contact inhibition mechanisms into action. Curtis (1960a, 1962a) suggested that contact promotion might result from the effects of shear of one cell on another altering surface properties to permit faster movement of one cell over another than on the non-living substrate.

Contact following

Shaffer (1962, 1964), examining the aggregation of slime mould amoebae, noticed that if a previously isolated cell made contact with the middle of a cell in a cell stream, the first cell did not join the

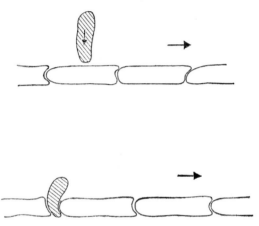

Figure 17. Contact following in slime mould amoebae. Hatched slime mould amoeba does not join train until a junction between two cells passes it.

stream laterally but stayed immobile beside the stream until the end of one cell passed it, when the first cell joined into the cell stream (see Figure 17). This is the phenomenon of *cell following*. The explanation produced by Shaffer for this behaviour was derived from observation of another most interesting type of behaviour in these cells. The cells often form fine pseudopods termed by Shaffer *pseudodigits* which protrude from the side of the cells. Shaffer observed that particles adherent to the pseudodigit were carried down towards its base. Shaffer suggested that this observation means that new cell surface is added to the front end (distal) of the pseudo-digit (presumably derived from the inner cytoplasm) and is removed from its base. The distal end is the "source" of new membrane and the base the "sink". Cell movement is believed to occur by the addition of new surface at the front end of the cell and its resorption at the rear. In order to account for cell following, Shaffer suggested that the source region at the front of the cell and the sink region at the

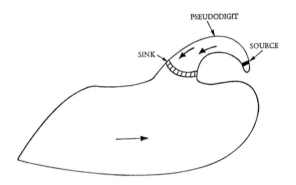

Figure 18. Pseudodigits in slime mould amoebae. Arrows show flow of surface between source at tip of pseudodigit and sink at base.

hind end are specially adhesive regions, or that hind and fore ends of the cell are very adhesive with the source and sink regions slightly lateral to the adhesive regions. The enhanced adhesiveness of these regions ensures that cells adhere only with these regions in contact. Shaffer observed that the front of one cell would adhere only to the hind end of another. One possible explanation put forward by Shaffer for this polarity of adhesions is that when "source" faces "sink" then the movements of membrane on the surface of each cell are opposite (see Figure 17) and shear each other. If source faces source, shear is considerably smaller, as both cell surfaces are moving in the same direction. Shaffer thinks that possibly shear may main-tain adhesion. This theory has considerable similarity to Curtis's theories about the control of cell movement on contact. Indeed the behaviour of fibroblasts suggests that contact following may occur

in these cells (Abercrombie, 1964), for chains of cells form in fibroblast outgrowths arranged head to tail. Some degree of interaction
between the cells occurs in fibroblasts arranged head to tail, as can
be recognised for the following reasons. Fibroblasts in a stream show
a large pseudopod at the front end and a small one at the hind end.
But when the cells become isolated from contact with a cell behind,
the rear pseudopod tends to disappear. This observation suggests
that the leading edge pseudopod can induce a pseudopod in the
hind end of another cell with which it makes contact. As Abercrombie (1957, 1961b) points out, the following of one cell behind
another is predicted on the contact inhibition hypothesis. It is thus
possible that contact following will be recognised as a feature of
contact inhibition.

Parallel packing of cells

The fasciculation of nerve fibres in tissue culture was observed
by Harrison (1910) in which these very elongate cells grow out from
the explants opposed side by side. Nakai (1956, 1960) re-examined
this phenomenon using time lapse filming. These results illustrated
very nicely the observation that the cells lie side by side even when
space is available for them to move into. Weiss (1955) had suggested
that fasciculation takes place by the non-detachment of fibres which
have associated rather than by chemotaxis or other forms of active
association. Nakai (1960) suggested that the filopodia of a fibre, when
they make contact with a fibre, tend to draw their own fibre towards
the other. It is not quite clear whether some form of cell behaviour
similar to peripolesis brings about this association, or whether
another type of contact behaviour, as yet undescribed, is involved.

The other type of parallel packing is that which Macpherson
and Stoker (1962) and Berwald and Sachs (1963) have observed in
cultures of hamster normal BHK cells. These cells in dense cultures
are monolayered and apparently remarkably parallel in their
orientation. Unfortunately it is not quite clear whether the same
behaviour is developed at very much lower cell population densities,
so that it is not certain whether the behaviour is related to peripolesis
or to contact inhibition (see earlier). Stoker (1964) found that polyoma-transformed hamster cells, which grow on polystyrene in an

unoriented array, take up the orientation of normal cells when plated out on top of them.

Pseudopod protrusion

Two observations, although derived from work in whole animals, reveal an interesting facet of cell behaviour and are best discussed here. These observations were made by Gustafson and Wolpert (1961) and by Wigglesworth (1959) respectively. Both papers describe the protrusion of a pseudopod by a cell over a considerable distance, c. 200μ. The pseudopod anchors to a certain site and then the intervening cytoplasm contracts and hauls the cell to a new position (in Gustafson and Wolpert's observations) or hauls in a tracheole towards the cell body (Wigglesworth). Gustafson and Wolpert described this action of a pseudopod in the sea urchin during gastrulation. The remarkable feature is that the pseudopod is protruded free into the fluid of the blastocoel and remains unattached at its end until it reaches the other side of the blastocoel. This work is discussed further in Chapter 7. In Wigglesworth's observations the pseudopod extends between cells and is presumably continuously in adhesion with the surrounding cells as it extends.

The contact behaviour of cells can be classified in the following way.

Interaction with substrate: orienting reactions
   Promoting movement: contact guidance
   Retarding movement e.g. cells transverse to grooves
Interactions between cells
   Promoting movement: contact promotion
   Retarding movement: contact inhibition of movement
   Increasing rate of turning: contact inhibition at low cell population densities
   Decreasing rate of turning: effect of contact inhibition at edge of outgrowth

Contact inhibition of movement has as its opposite type of behaviour contact promotion, and likewise contact guidance has in practice an opposite, though this has not yet been named. The opposite to contact guidance is the failure of cells to move transversely across concave or convex surfaces. Contact inhibition and promotion are cell-to-cell phenomena, contact guidance and its opposite can be cell interactions with non-living substrates. All these phenomena are ones which can bring about the trapping of the cell in one particular environment. Contact following and peripolesis appear to be features of contact inhibition and promotion respectively, possibly combined with other aspects of cell behaviour. It seems possible that the whole range of types of cell behaviour which control cell orientation, movement, turning and trapping have now been described, though aspects of cell behaviour which determine cell morphology are less well understood. These features of trapping, turning, movement and orientation control the distribution of cells in populations and will be discussed in the next two chapters. It is obvious that our attempts to explain these types of cell behaviour in terms of adhesiveness, surface mechanical properties, motility mechanisms etc. are at present remarkably inadequate.

CHAPTER 6

The Behaviour of Cell Populations: Model Systems

The behaviour of single cells integrates to form part of the more complex behaviour of population of cells. In turn, the behaviour of cell populations is built up into those morphogenetic movements which take place in the whole embryo or in regeneration or in various pathological systems. But before the behaviour of cell populations in whole embryos is considered, it is desirable to examine simpler systems in which only one or two cell types are present in the whole population of cells. Such systems are reaggregates, tissue and organ cultures. I term them model systems because it is hoped that they will provide simple and idealised situations in which meaningful experiments can be done. The extent to which this hope can be realised at present is discussed in Chapter 7. The main question to be discussed in this chapter is what mechanisms are available for bringing a cell to a particular position in a body of cells. In Chapter 7 we shall see that cell movements in the embryo lead to the placing of cells in specific positions in the embryo. On the whole, this is done by movements of tissues rather than by the movement of isolated cells, but both types of movement will be discussed in this chapter.

REAGGREGATES

Three main techniques for preparing reaggregates are in use at present. The simplest method was introduced by Wilson (1907), who allowed disaggregated (dispersed) cells to settle on to the bottom of the culture dish. The cells then move, collide and form adhesions, and the adhesions build up larger and larger cell bodies. A variation of this technique has been to place the cells on the chorioallantoic membrane of a chick embryo (Weiss and Taylor, 1960; Garber and Moscona, 1964). Moscona (1961a) used an alternative method of agitating cell suspensions so that in the same manner cells collide and adhere to form roughly spherical aggregate bodies. Although Moscona produced aggregation by using a gyratory shaker to agitate the cell suspensions, other types of agitation work as well, e.g. reciprocating motion (Curtis, 1963). As will appear later in this chapter, there is some slight evidence that the still medium and shaker methods give different results. A third technique for producing reaggregates is to centrifuge a cell suspension to give a cell pellet; this technique was used by Trinkaus and Lentz (1964), and by Hayes (1965).

At present a very wide range of cell types have been aggregated with success using either of the first two techniques, for example adult sponge, coelenterate and echinoderm cells (Wilson, 1907, 1911) embryonic insect cells (Hadorn et al. 1959), embryonic echinoderm cells (Giudice, 1962), embryonic protochordate cells (Scott, 1959), embryonic amphibian and avian cells first by Roux (1894) and Moscona and Moscona (1952) respectively and mammalian cells (embryonic) by Mintz (1964). Dickson and Leslie (1963) claimed to produce the first successful aggregation of adult mammalian cells, but their aggregates were composed of remarkably uncoherent cells. For unexplained reasons aggregations of adult chordate cells have not yet been obtained with certainty, with the exception of adult *Rana pipiens* liver (Ansevin, 1964). Grover (1961) found that the aggregative ability of embryonic chick lung cells declined with increasing embryonic age. Kuroda (1964) reported that the aggregability of cells declines after a period of culture. However cells of human normal and tumour cell lines derived from adults have been aggregated successfully (Halpern et al. 1966; Dodson, 1966).

It is probably easier to comprehend the mechanism of aggregation in agitation (shaker) systems than in still systems. The shaker system was introduced by Gerisch (1960) and represents a considerable technical advance since it made possible the aggregation of cell types which previously had proved recalcitrant. In addition the technique resulted usually in the production of spherical aggregates of uniform size under given conditions. As a result of work by Curtis and Greaves (1965) and Curtis (1967) on the kinetics of aggregation, a physical understanding of the mechanism has been arrived at. It was found that aggregation in most media had kinetics which exactly parallel orthokinetic flocculation kinetics (see Chapter 4). Curtis (1967) was able to measure the probability of a collision between cells resulting in the formation of an adhesion, and showed that the adhesiveness of cells remained constant during the greater part of aggregation for any given cell type, except when whole serum or the aggregation-inhibiting protein was present or when cells had been trypsinised. It was also demonstrated that the cells behave like spherical bodies in aggregation. It is probable that the final size of the aggregates is determined by the balance between shear forces and the cell-to-cell adhesion, and this final size has been used as a measure of adhesiveness (Moscona, 1961a,b). Steinberg and Roth (1964) confirmed the finding made by L. Weiss (1963a) that prothidium bromide increases the size of aggregates; and interpreted this result theoretically in terms of the number of collisions between cells per second. They suggested that prothidium bromide increases the number of collisions per second. They did not suggest that the effectiveness of the collisions might increase, nor did they make any measurements of the kinetics. Steinberg and Roth suggest that there is no change in cell physiology (adhesion) during aggregation in serum but their treatment is too theoretical to detect such changes.

Aggregation in still systems is visibly different from the state revealed in shaker systems by kinetic analysis. Early visual descriptions by Wilson (1907), Galtsoff (1925) and Fauré-Fremiet (1925) supplemented by analysis of time-lapse films (Kuhl, 1937, Lucey and Curtis, 1959 and Shaffer, 1958) show that the collisions result from cell movements or, where the cell bodies do not move, from the protrusion and collision of pseudopods. Although in both types of aggregation adhesion results, as it must from collision, cell movement and pseudopod protrusion appear to play no part in shaker aggregation. In most still aggregate systems (see also later) the cell movements appear to be random, but slime mould cells aggregate under the influence of a chemotactic substance (Bonner, 1947; Shaffer, 1962). Again, in shaker aggregation the cells appear to be spherical until they make their adhesions whereas in still aggregation they are often elongate. It is not yet clear whether contact promotion is the direct cause of aggregation in still systems. If it is, it would be expected that extensive cell movement within the aggregate would take place almost immediately. There is some evidence that this movement occurs in amphibian aggregates after the first contacts have been made (Curtis, unpubl.). Again we do not know whether cell movement starts immediately in newly-formed shaker aggregates or whether there is a lag period before it commences. The time taken for a population of cells to aggregate depends, in the absence of any external inhibitor mainly on the cell type or types (probably reflecting their adhesiveness), their population density, and to a small degree on the temperature (see Chapter 4 for discussion of relevant experiments), though when cell movement is important temperature may play a more obvious part. In shaker systems the rate of agitation is important. At physiological pH the time taken to form an aggregate of 10^5 cells may vary from 2 to 24 hours.

The aggregates formed in still systems are adherent to the substrate and are generally of flattened form. However they are generally several cells thickness in depth. Curtis (1962a) suggested that contact promotion in these systems acts to pile cells on top of one another because they tend to move and adhere to one another rather than to the substrate. The aggregates formed in shaker systems are normally rounded in shape, though Moskowitz and Amborski (1964) found that aggregation in certain sera produced aggregates of more ragged shape: the reason for their results is unknown at present.

Wilson (1907, 1911) made the fascinating discovery that although the cells in a newly-formed reaggregate of a sponge appear to be randomly arranged, a sponge body of normal histology develops in due course from the aggregate. Although the morphogenesis found in other aggregates (see examples shortly) is not as extensive as in sponges a very considerable morphogenesis is frequently found. For example, Townes and Holfreter (1955) found that an aggregate of amphibian neurula cells in which ecto-meso- and endoderm cells were initially randomly inter-mixed, developed so that ectoderm cells lay at the outside, endoderm at the centre of the aggregate with mesoderm between them. Wilson appreciated that there were two mechanisms which might be responsible for the rebuilding of a sponge of normal morphology. Either some or all the cells in the aggregate might de-differentiate and subsequently redifferentiate into the cell type appropriate to their position in the aggregate. Alternatively the cells do not differentiate and lose their type although they may appear to be de-differentiated. The development of normal morphology from the aggregate body would in the latter case take place by the migration of cells to certain discrete places in the aggregate body, each cell type showing a specific site at which it comes to rest (sorting out or segregation). At first biologists inclined to the view that de-differentiation took place (Wilson 1907) but Huxley (1921) showed that there was little evidence for de-differentiation in aggregates of *Sycon* sp. cells, on the grounds that aggregates derived from one or two cell types did not show development to a complete animal but only produced those cell types included in the aggregate. Galtsoff (1923, 1925) made a most thorough histological examination of morphogenesis in aggregates of the sponge *Microciona prolifera* and found that although a certain degree of redifferentiation of archaeocytes takes place, redifferentiation was not a major feature of aggregate development whereas cell movement of cells to specific sites was. Ganguly (1960) maintained that considerable redifferentiation was possible in the reaggregates of the sponge, *Ephydatia*. Trinkhaus and Groves (1955) who studied the fate of chick embryonic wing bud and mesonephros cells mixed together in aggregate bodies, concluded that although sorting out appeared to take place the possibility that redifferentiation (i.e. transformation) had taken place could not be excluded. Weiss and Moscona (1958) found that various mesenchymal types of cell appeared to retain the morphogenetic properties typical of their type. But Moscona (1957b) claimed that chick embryonic retinal cells became lentoid in type in reaggregates. Nevertheless a certain degree of scepticism may be allowed to rest on these results for technical reasons which were unavoidable at the time of the work (see also Grobstein, 1955). In order to demonstrate that either (*a*) no redifferentiation of cells takes place, or (*b*) that tissues which appear to be of one type histologically do not contain cells of another type as well, it is essential to be able to trace the fate of all cells which were of any one type before disaggregation.

The movements of cells can only be traced effectively with some type of labelling of the cells themselves. Although staining properties, e.g. glycogen in heart cells (Steinberg, 1962c) or presence of yolk platelets (Curtis, 1961b), have been used to trace cells, these techniques are a little questionable on the grounds that changes in cellular metabolism may have altered the markers. To a certain extent this may also be held against radioactive labelling but experience with tritiated thymidine as a marker suggests that reliable results can be obtained with this label. Genetic markers can be used (Mintz, 1964) but may occasionally suffer from the disability that the morphogenesis of aggregates of cells of mixed genetic origin may not be the same as that of isogenic aggregates, particularly if antibody reactions are involved in segregation. Okada (1965) used immune fluorescent techniques to identify specific cell types in aggregates of embryonic chick metanephric tissue.

Obviously recognition of cell type in aggregates by radioactive labelling, genetic markers and with immunochemical techniques is superior to recognition by simple histological stain reactions. Nevertheless radioactive labelling has only been used in segregation studies by three workers to date as far I am aware, and genetic and immunochemical markers on one occasion each. Trinkaus (1960), Trinkaus and Groves (1961) and Zwilling (1963) have used radioactive labelling to study the stability of cell type in aggregates and to follow segregation processes if any. An alternative method of studying the stability of cell types in aggregates is to explant the aggregate cells in tissue culture as Umansky (1966) has done. In view of the very small amount of published work in which rigorous techniques have been used to test the stability of cell types we cannot be quite sure at present whether or not redifferentiation may occur in aggregates, or whether there may not be quite a large number of cells of one type hidden as it were in what appears to be a tissue of another type. The general view at present is that a given cell type is stable throughout the processes of disaggregation, aggregation and subsequent morphogenesis, though this concept has been challenged by Umansky. Consequently the appearance of discrete groups of cells of one type in the aggregate some while after aggregation is complete, is attributed to the movement of cells within the aggregate to their segregated positions. As will appear shortly much experimental and observational evidence supports this view, although it cannot yet be regarded as being wholly substantiated. It has been tacitly assumed in the above description that sorting out takes place inside the aggregate body, but before the mechanism by which this takes place is discussed, the question of whether any sorting out takes place during aggregation will be considered.

The segregation of cell types might take place during the initial aggregation of cell types. Galtsoff (1923) showed that the aggregation of sponge cells, after they had settled to the bottom of the culture dish, took place by the random movement of cells of all types, over the glass surface until they formed adhesions. It is not clear from his descriptions whether the adhesions resulted in the cessation of individual cell movement or whether contact promotion took place. Galtsoff found no evidence for the action of chemotaxis leading cells towards one another though rigorous tests on this point were not applied. Roux (1894, 1896) described the chemotactic attraction of disaggregated amphibian "blastomere" (sic) cells to one another, terming this "cytotropism" or "cytotaxis". Voigtlander (1932) and Kuhl (1937) re-examined Roux's experiments using embryonic cells, again described as blastomeres though they were probably postgastrular cells, from Triturus sp., and found that there appeared to be no directed movement of any cell type towards another. These earlier experiments were done in the presence of Mg ions in the culture media. Lucey and Curtis (1959) made time lapse films of the aggregation of Xenopus laevis gastrula cells in Holtfreter medium (free from Mg ions). In the Mg free media the cells show almost no movement of the whole cell body before they aggregate, but appear to aggregate only at population densities which allow the cells to come into contact by their pseudopods. Statistical analysis of the direction of protrusion of the pseudopods showed that the first adhesions were formed in a random manner. Curiously enough the possibility that a sorting out may occur during aggregation in shaker systems does not appear to have been investigated. In view of the results by Curtis and Greaves (1965) and

Curtis (1966a,b) which show that different cell types aggregate at different rates and that this is more marked in the presence of serum (usually used in aggregation studies) it is possible that some sorting out occurs early in aggregation.

Most authors who have studied the segregation of two or more types of cell in an aggregate have hitherto assumed that soon after the aggregate body has formed the various cell types are distributed randomly throughout it. Almost no tests of this assumption have been made although it might perhaps be expected that the mixture of two or more cell types in suspension would give a random arrangement of the various types (but see above). Nevertheless it is possible that even differences in cell density may arrange cell types in a non-random manner as the aggregates form (Ikushima 1959). Much is known about the final distribution of cells in aggregates but again few examinations have been made of the actual redistribution movements during sorting out. For this reason descriptions of the final sorted-out patterns will precede those of the movements.

The final pattern established by the sorting out of cell types is somewhat difficult to define in practice because further normal or abnormal morphogenesis with cell movement may ensue immediately after or even towards the end of the primary sorting out, resulting in the confusion of the pattern established by the initial sorting out. I shall consider first the segregation patterns found in mixtures of cell types from a single species (see Figure 19). Early workers who used sponges, such as Galtsoff, described the first visible sign of sorting out as the accumulation of archaeocytes as a peripheral layer at the surface of the aggregate. But this observation has not been repeated for other species as yet. Although earlier workers had described patterns of sorting out which we now realise to belong to one or other of standard varieties, Townes and Holt-freter (1955) were the first to realise the existence of these patterns and their significance in relation to *in vivo* morphogenesis. They examined the sorting out of ecto-, meso- and endo-derm from amphibian neurulae of various species. They found two main forms of sorting out: (i) cells of one type form a layer enclosing a second type, which may in turn enclose a third type; (ii) one cell type tends to segregate on one side of the other. Other workers have also found a third type: (iii) one cell type segregates into small groups scattered through and on the edge of a continuous mass of the other type. The second form of sorting out has been interpreted in another way, namely that one type partially encloses the other (Steinberg, 1963b, 1964a,b). Steinberg described the cell types which form the inner mass in pattern (i) and which are partially enclosed in pattern (ii) as "internally segregating" or as belonging to the "discontinuous"

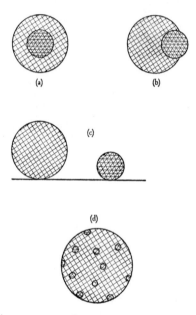

Figure 19.   Sorting out patterns in aggregates. (*a*) One cell type completely encloses another, though the inner cell type is shown as concentric with the outer, this is not necessarily so. (*b*) One cell type partially encloses another. (*c*) Both cell types form completely separate aggregates. (*d*) One cell type scattered in small groups through other cell type (continuous type).

phase. He termed the cell types forming the outer layer as the "continuous" phase. The significance of his terminology of "continuous" and "discontinuous" phases is clearer when pattern (iii) sorting out is examined. In this form of sorting out, one cell type appears as a large number of small groups scattered in and at the edges of a continuum of the other type. It can now be seen that the terminologies "internal/external" and "discontinuous/continuous" do not invariably agree, because pattern (iii) sorting out may show both cell types as external or internal, but one type is discontinuous and the other continuous. Pattern (i) sorting out was described (Curtis, 1962a) as "concentric" sorting out, but although the arrangement is often concentric it is not invariably so; for these reasons his terminology should be abandoned.

A considerable number of papers have described the patterns of sorting out of two or more cell types: patterns (i) and (ii) have been found in sponges by Galtsoff (1925, 1929), in echinoderms by Giudice (1962) in avian tissue by Moscona and Moscona (1952), Moscona (1962), Steinberg (1962c, 1963b) amongst others, and in amphibian tissue by Townes and Holtfreter (1955) and Curtis (1961b). In addition Abraham (1960), Okada (1959), Feldman (1955) and Sobel (1958) have described pattern (i) in amphibian and avian tissue. Pattern (iii) sorting out in which small groups of one type are scattered in a continuous phase with either an internal position of the discontinuous type or both internal and external positions, has been described much more rarely than other patterns of segregation. Moscona (1961b) figures aggregates containing small spheres of cells of one type segregating both internally and externally in a continuum of the other type. It is of course possible that this type of pattern results from one cell type aggregating before a second cell type aggregates. Trinkaus and Lentz (1964) and Trinkaus (1961) described aggregates in which avian retinal pigment cells are scattered as small groups in a continuum of mesonephric cells, and it appears that the pigment cells were also segregated internally. Many aggregates show formation of tubular structures internally when cell types such as mesonephros are included (see Moscona, 1961b, 1962) but it seems probable that this type of morphogenesis occurs after sorting out has been completed. Vesicular structures sometimes appear in aggregates and these may also represent the action of further morphogenesis after segregation has been completed.

Aggregates composed of the tissues of two different species frequently show patterns of sorting out similar to those described above (see Galtsoff, 1925; Townes and Holtfreter, 1955; Moscona, 1957a; Curtis, 1962b), in which either tissue of one species origin will enclose that of another species, or in which the cells sort out according to their histological rather than their species type. For example Moscona (1957a) found that mouse and chick cells of the same tissue type assort together but segregate from other tissue types of both species; this type of sorting out was termed "histotypic". But another type of sorting out is found in mixtures of cells of certain pairs of species. This type of sorting out is characterised by the fact that apparently cells of different species origin either do not enter, or do not remain in the same aggregate, so that in the first case the mixed cell types aggregate into two or more aggregates, each containing cells of one species type only and in the second the aggregates sort out into aggregates of one species type by expelling cells of the other type from the aggregate bodies. This type of specificity was reported first by Wilson for sponge cells; and Galtsoff (1925), Spiegel (1954, 1955), Curtis (1962b) and Humphreys (1963) have all confirmed that it probably occurs in the aggregation of many pairs of sponge species. Unfortunately in all these experiments recognition of cell type (species) has been carried out either by the natural colour of the cells or by staining reactions which are not highly reliable. In consequence even a considerable proportion of cells of one species in an aggregate apparently composed of cells of another species would not necessarily be noticed. De Laubenfels (1928), Sara (1956) and Curtis (1962b) have on the other hand described aggregates of pairs of species of sponge cells in which cells of either species type co-mingle in the same aggregate, although they may show some sorting out within the aggregate.

Although the majority of descriptions of segregation of mixed species of cells have been confined to sponges, we have Moscona's evidence on mouse/chick mixtures (see above) and my own findings (unpublished) that quail (*Coturnix* sp.)/chick mixtures behave in the same manner as the mouse/chick mixtures. It is not clear at present whether vertebrates differ from sponges in that the sorting out of the former is histotypic whereas that of the latter is often by species type. A variety of intermediate types of sorting out might occur, for example one in which all the cells of both species are present in one aggregate with cells of one species segregated into internal and external positions and the cells of the other species again segregated internally and externally to one another but all lying internal to cells of the first species. It has been found (Curtis, 1962b) that it was possible to obtain at least partially fused aggregates when two species were employed, in which although sorting out was species-specific, aggregates of different species remained fused side by side. It should be remembered that in the segregation of aggregates of mixed species some five or six cell types in all may be present when tissue types as well as species types are considered. Most descriptions of segregation and experiments on it have involved at most three types of cell of a single species. It is probable that our lack of knowledge of how tissues segregate in mixed species aggregation (Moscona's work is an exception to this ignorance), is responsible for the uncertainty as to whether mixed segregation in species aggregation can be defined in the same way as Steinberg (see earlier) has done for single species aggregates. But assuming that the same definitions can be applied to both types of sorting out we reach the following scheme:

(1) Cell types separate into distinct non-adherent bodies:
  (a) as aggregation takes place, e.g. sponge species (Galtsoff, 1925);
  (b) after aggregation has taken place, e.g. other sponge species (Spiegel, 1954, 1955; Curtis, 1962b), ectoderm and endoderm in amphibian aggregates (Townes and Holtfreter, 1955; Feldman, 1955).

(2) Cell types separate into distinct adherent bodies, in contact laterally, also interpreted as cell types partially enclosing one another, e.g. sponge species: Curtis (1961b), amphibian medullary plate plus mesoderm on endoderm (Townes and Holtfreter, 1955).

(3) One cell type totally encloses another; the latter may be discontinuous. Many examples already given, including different tissues of different species (Townes and Holtfreter, 1955).

(4) Although the distribution of cell types is non-random, neither type can be described as internal or external in character: e.g. some aggregates (Moscona 1961b), possibly Trinkaus and Lentz (1964).

In addition there are situations where sorting out does not normally occur; see Zwilling (1963) and Curtis (1961b).

Two further points of importance are, first, the question of the reproducibility of these sorting out patterns; and second, the possibility that they may result from inductive interactions between the cells. Most authors have not recorded whether they occasionally obtain aberrant segregation with pairs of cell types, though Steinberg (1962c) found that retinal cells took up rather variable positions. But Abercrombie and Weston (unpublished) studying the sorting out of fused tissue fragments of two cell types (not aggregates) found that a considerable variation in sorting-out pattern was obtained (see also later). Similarly in aggregates of mouse and chick epithelial cells Garber and Moscona (1964) obtained no regular segregation pattern.

The second point is of considerable importance, though we have at present only two papers in which sufficient evidence is put forward to make any decision as to whether inductive interactions between the cells affect sorting out. Townes and Holtfreter (1955) found that if the sorting out of three tissues, epidermis, mesoderm and endoderm, is observed in pairs of these tissues, then in the pair, epidermis and mesoderm, the latter segregates internally; and that in the pair, epidermis and endoderm, it is the endoderm which is found internally. Again in the combination of mesoderm and endoderm, mesoderm segregated internally to endoderm, but when epiderm was added mesoderm then segregated externally to endoderm. Although the action of epidermis may not be a classical inductive one in this situation, it is clear that the sorting-out pattern has been affected. Steinberg (1962c, 1963b) examined the segregation of embryonic cells from chicks. He used 4-day precartilage cells, and 5-day pigmented retina, heart and liver cells. He found that liver cells segregated externally to heart, and in turn heart segregated externally to cartilage. Similarly, as would be expected on these results, liver segregates externally to cartilage. A little discrepancy was found with retinal cells which sometimes segregated externally and sometimes internally to heart, but which apparently always segregated externally to cartilage. Although Steinberg has not reported on the results of all the possible ways in which the four tissue types may be paired, it seems that, with the possible exception of the retinal cells, there is a hierarchy of segregation position, descending as it were from liver to cartilage cell types. In any combination a given tissue always appears to behave in the same absolute way, so that if at the top of the hierarchy it always segregates externally and if at the bottom always internally, if it holds an inter-mediate position it segregates internally or externally according to whether the other tissue type is higher or lower on the hierarchy than it. This constancy of behaviour suggests that we are dealing with a quantitative property inherent in a given cell type which is not affected by interactions between tissues. Steinberg's result is of great importance because if this is the correct interpretation we have only to investigate a single quantitative mechanism to discover the cause of sorting out. But it should be remembered that Townes and Holtfreter obtained results similar to Steinberg's until they tried the combination of three tissues.

Very few descriptions of the sorting-out process have been published by the time of writing. Although observation of dark coloured cells in a rather transparent aggregate is possible, and was used by Galtsoff (1925) and by Trinkaus and Lentz (1964), it is impossible to apply this technique to most aggregates which are opaque. In these cases recourse has to be had to histological sections of aggregates at various times after they have formed, using some means of identifying cell type such as tritiated thymidine labelling. Townes and Holtfreter (1955) described the sorting out of tissue types in aggregates of amphibian neurula tissue. Frequently com-binations of tissues from different species origin were used in order to facilitate recognition of cell type, e.g. *Amblystoma punctatum* and *Triturus torosus*. The aggregates were grown on an agar surface, to which they did not attach. The initial structure of the aggregate was, acccording to Townes and Holtfreter, a random intermingling of different cell types. It is interesting that they and many other authors draw pictures of over-dispersed arrays of cells in order to depict random arrays. No worker has yet described the arrangement quantitatively or made tests for the randomness of the arrangement. It should be remembered that in an aggregate composed randomly of equal numbers of two cell types, any cell will have a 50 per cent probability of three cells of its own type being in contact (assuming a cubic packing), and only 1·5 per cent probability of being entirely surrounded by cells of the other type. Consequently it is probable that long strings of cells of one type can be traced in the random aggregate, and these cells are in a sense already adherent to their own type. Therefore random aggregates must be regarded as showing much contact between cells of like type if the two types of cells are present in similar concentrations.

Townes and Holtfreter describe sorting out as taking place by the migration of externally segregating cells to the outside of the aggregate and the inward movement of the internally segregating types, though evidence for cell movement by individual rather than groups of cells was not provided. It is not wholly clear whether both internally and externally segregating cells move actively towards their goals or whether one type alone moves actively towards the other type, which is being passively pushed around by the actively locomoting cells. These authors also described cell movements in fused pieces of whole tissue; these results will be discussed later.

Galtsoff (1925) observed the segregation of *Microciona prolifera* cells and noticed that the archaeocytes appeared to be the first cells to move from their apparently non-random positions to the periphery of the aggregate. Unfortunately neither in Townes and Holtfreters' nor in Galtsoff's work is it clear whether the cells moved randomly until they reached their correct position where they were then trapped or whether they moved in a directed manner. As will appear later, evidence on this point would be most decisive in testing various theories of segregation. Trinkaus (1961, 1963a) and Trinkaus and Lentz (1964) have described the sorting-out process in chick embryonic tissues. Trinkaus (1961) investigated the stability of retinal pigment cell type in aggregates and concluded that they retain their type and pigmentation. Thus they should provide good markers for segregation studies. Trinkaus and Lentz prepared aggregates of $5\frac{1}{2}$-day chick retinal pigment cells (tapetum) with 4-day heart or mesonephros cells, by centrifuging mixed suspensions into a flat pellet. This provided an aggregate in which the position of pigment cell clusters and perhaps even individual cells could be determined by observation in life. They observed that segregation takes place by the formation of small clusters of pigment cells, probably by individual cell movement. Trinkhaus (1961) labelled chick cells with tritiated thymidine and followed cell movements in the aggregate. He found that tapetal cells become elongate and dendritic, which suggests that they may be undergoing complex contact interactions with other cells. Large clusters show no movement but the smaller clusters may move slightly. Two clusters may join together. But the clusters persist unless they fuse without any disaggregation, so that finally the pigment cells lie scattered in clusters internally in the aggregate. Unfortunately no observations were made which revealed how the internal position was taken up, but it is clear that the discontinuous nature of the pigment cell aggregates resulted from their inability either to break connections once formed or to undergo movement in clusters. The situation appears to be rather different in other tissues, for both Townes and Holtfreter (1955) and Steinberg (1963b, 1964a,b) describe movements by masses of tissue which have not been disaggregated, but there is no direct evidence that once formed, clusters of one cell type may separate into separate cells, which then join other clusters. Steinberg (1962c) described segregation of heart (labelled by their positive staining for glycogen) and retinal cells as occurring by the withdrawal of the heart cells from the aggregate surface, followed by the clustering of heart cells in numerous small foci distributed throughout all but the surface layer of the aggregate. It appears that the clusters move but it is not clear whether clusters also break up into individual cells which then join larger clusters. There is also a curious lack of heart cells at the aggregate centre at the start of aggregation which suggests that they may have been differentially accumulated in the outer layers of the aggregates during aggregation itself. It should also be pointed out that the internal position of heart cells described here does not accord with descriptions in his 1964 paper.

It is clear from Steinberg's work (1964a,b) that when very few internally segregating cells are included in an aggregate, then they do not form one or a few groups at or near the centre of the aggregate but remain dispersed as small clusters or even single cells throughout the aggregate, except that they usually do not appear in the most external cell layer.

An experiment which demonstrates that individual cell movement is necessary for segregation was carried out with *Xenopus laevis* embryonic cells by itself (Curtis, unpublished). Gastrula endoderm and ectoderm were separated before disaggregation and then layers of either disaggregated ectoderm cells or endoderm cells were built up on Millipore filters. The layer of the opposite cell type was laid down on top of the existing layer and they were then cultured. Histological sections showed that soon after the cells were laid down and had adhered together, the uppermost cell type started to intermingle not as streams but as single cells with the lower one, whether the latter was endo- or ectoderm. Subsequently they resorted out with ectoderm uppermost. This experiment shows that individual cell movement takes place early in formed aggregates and that there is some change in cell behaviour after a while in aggregates. Bonner and Adams (1958) and Bonner (1959) found that in aggregates composed of slime mould cells from two species, cells of one species moved more rapidly than those of the other. Cell movement in these aggregates (slugs) is in one direction so that sorting out results.

Unfortunately we know little about cell behaviour in aggregates. Much of our sparse knowledge comes from the thorough histological examinations of morphogenesis in aggregates of the sponge *Suberites suberites* (then known as *Ficulina ficus*) made by Faure-Fremiet (1932a,b). He found that initially the cells were rounded in shape throughout the aggregate but soon afterwards the peripheral cells spread out to form an epithelial-like structure composed of archaeocytes. Collenocytes throughout the aggregate then became fibroblast-like in form but other cell types did not show a change in form. At this stage no other sorting out has occurred, for except at the periphery of the aggregate archaeocytes, collenocytes and other cell types are intermingled. The water canals appear to develop by strands of cells becoming linked up through the aggregate; these cells then de-adhere over part of their contacts to form the canals. The canal pattern undergoes considerable reorganisation after it first appears. It is important to note, that first little if any sorting out into tissue groups takes place. In this respect sponges appear to differ from other groups in which segregation has been studied. However, the segregation of the external layer of archaeo-cytes may resemble segregation in metazoa. Second, the cells take up elongated form. This was also noted by Trinkaus and Lentz in aggregates containing embryonic chick retinal cells. In both exam-ples, cells in the aggregates frequently had elaborate pseudopods. It is impossible to account for the presence of cells of such shape on the differential adhesiveness theory of sorting out, for this theory requires that the shapes of cells be determined by interfacial energies and interfacial conditions lead to simple curved surfaces, unless it is argued that cell movement also plays a part in sorting out, as it is by Steinberg; but one is faced with the problem at the single cell level that one process would tend to upset the other. These cells have surfaces which are not simple curved surfaces.

It is worth examining the possible ways in which the various known forms of cell behaviour might act in aggregates. Contact inhibition might lead to the trapping of cells of one type in one place if one cell type was contact inhibited by its own type, and the other was not, and different types did not contact inhibit one another. Unfortunately we are not yet sure that contact inhibition takes place in three-dimensional tissues. Contact following by cells of like type would lead to movement of trains of cells of one type, which even if they moved randomly would accelerate segregation processes. It should be remembered that, even in the randomly arranged aggregate, long trains of cells of one type are present. These cell trains and cell groups may show co-ordinated mass movement. Contact guidance might play a part in aggregates if intercellular fibres could be laid down before segregation starts.

In conclusion it can be said that these rather incomplete studies indicate that segregation takes place by movement of individual cells and also probably by the movement of cell groups. It should be realised that the randomly packed aggregate will, unless one type is present in very small numbers, contain large extents where a cell of either of the two types is already in contact with other cells of its own type. When one type is present as a small proportion of the total number of cells, then the other type will almost be sorted out already. Although sorting out leads to the external or internal segregation of a given cell type, it is not yet clear how this happens;

although directed movement and trapping are possible causes. Theoretical and experimental evidence on the mechanism(s) will be discussed in the next section. The part played by contact phenomena such as contact inhibition, following, and promotion is completely unknown, and there may be as yet undiscovered forms of contact behaviour which act in segregation.

HYPOTHESES TO EXPLAIN SORTING OUT

There are two features of any theory of sorting out (segregation) of tissue types which are essential for the explanation to be satisfactory. These are, first, that the accumulation of cells of one type in one region should be explained; and second, that the pattern of distribution of one tissue and another should be accounted for. Many theories, as we shall shortly see, appear to explain the sorting out, but predict patterns which are not in fact found, or which are only found in certain rare combinations of cell types. On the other hand any theory which explains the patterning adequately also explains the accumulation of cells of one type in one region.

The earlier theories to explain sorting out all postulated that there was some type of specific adhesion mechanism for each cell type (Weiss, 1958; Townes and Holtfreter 1955; Spiegel 1954; Moscona 1962; Humphreys 1963), so that adhesions could only be formed between like types of cell, or were at least strongest between like types of cell (Steinberg 1958). Both P. Weiss (1953) and Townes and Holtfreter suggested some antigen–antibody-like reaction as the basis for the specificity, but did not advance any experimental evidence in support. Spiegel found that antibodies would prevent aggregation in a specific manner but this provides no argument in favour of specific adhesion by an antigen–antibody reaction. Moscona claimed that extracellular substance or substances appeared to be essential for the aggregation (adhesion) of cells and suggested that specific substances of this type might lead to the sorting out of cells. Curtis and Greaves (1965) have shown that the evidence for the existence of this type of intercellular substance active in adhesion, as suggested by Moscona, is very slight. Humphreys' work (1963) in which he puts forward evidence for cementing substances specific for the cells of one sponge species, has been discussed in Chapter 4.

Although there is at present little evidence that different cell types have specific mechanisms of adhesion (see discussion in Chapter 4), it is possible that further work will provide better evidence for their existence. But it should be pointed out that there are other difficulties with this type of theory. The sorting-out pattern into separate aggregates, which would result from the various cell types each having a specific adhesion mechanism, would depend considerably on the ratio of one cell type to another and the ability of groups of cells to move. But in all conditions, groups of one cell type or the other would develop. If the adhesion mechanisms were completely specific, groups of different type would show no adhesion to one another, so that sorting out would occur as the aggregate bodies built up. Humphreys suggests that this is what occurs during the aggregation of cells of mixed species of sponge. But in all aggregates composed of mixed tissue types of a single species and in many interspecific combinations, sorting out does not occur till after the aggregates have formed. Although groups purely of one cell type would appear in these aggregates because of the operation of specific adhesion mechanisms, there would be no specification of any cell type as being external or internal, some groups of any one cell type lying externally and others of same type internally. But, as has been seen, internal or external segregation is a constant feature of one cell type in a pair. Frequently the internally segregating type is found not as a single mass but as a number of separate bodies. In this case Steinberg (1964a,b) has described the internally segregating type as the "discontinuous" type and the external one which surrounds all internally segregating types as the "continuous" type. It can be seen that if sorting out depended on such mechanisms of specific adhesion the continuous type would be that cell type present in greater number, not as is found in practice that one cell type is "continuous" over all proportion of the two types in the aggregate.

Steinberg (1958) put forward a most interesting theory to explain sorting out, which, although he (1962c; 1963b) appears to have subsequently abandoned, has interesting theoretical possibilities. Steinberg sought to account for the fact that a good deal of adhesion occurs between cells of different type even though they segregate.

He suggested that cells bear patterns of adhesive groupings on their surfaces and that the actual pattern is specific to any one cell type, although all cell types have the same chemical type of adhesive grouping. Incidentally in order to explain the importance of calcium for cell adhesion he supposed that these groupings were carboxyl groups which would act to form calcium bridges. Because the groupings were similar on all cell types, any cell would form at least some adhesive bonds with cells of other types, because there would always be some groups on one surface in register with those on the other cell's surface. Steinberg suggested that the specificity would arise from the fact that two cells of like types would have identical patterns of adhesive groups on their surfaces so that they would form adhesions with more groups in register than unlike types (see Figure 20). Consequently like types of cell would be more adhesive than unlike types. However there is one matter which makes this type of explanation difficult, and this is that one cell may lie against another with their respective patterns of adhesive groups at different orientations so that, although the cells may be of like type, many of their adhesive groups do not link with others on the surface of the other cell. It can thus be seen that there would be a great range in the adhesiveness of like cell types which might overlap the adhesive range of another cell type. However this difficulty would be lessened by certain types of patterning and removed if cells could adjust the orientations of their surface patterns. But this theory does not explain the development of internal or external positioning of cell types.

The two theories which explain sorting out as being due to quantitative differences in cell adhesiveness and not to any qualitative differences are due to Curtis (1960a, 1961b, 1963) and to Steinberg (1962c, 1963b, 1964a,b). It is perhaps easier to understand Steinberg's hypothesis so it will be treated first.

Steinberg suggests that the cells in an aggregate move discretely and that each cell type is of a different quantitative adhesiveness. He then treats the adhesion between cells in terms of their surface energy relationships. The following treatment is not that which Steinberg deduced from surface energy considerations but a more generalised treatment.

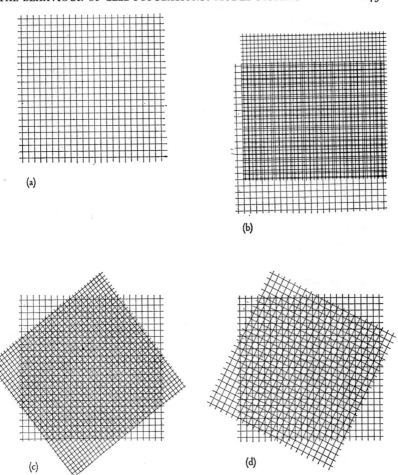

Figure 20. Steinberg's (1958) concept of specific adhesion. The intersections of lines on these grids are taken as the positions of binding groups. (a) Two like cells are in contact and the groups on each surface are in complete register, with 100 per cent of all binding sites active. (b) Two unlike cells are represented in contact in the middle of the picture; since the spacing of groups on one surface is very different from that on the other, fewer bonds are formed between them in unit surface area; in fact about 25 per cent of intersections on the coarse surface lie under intersections of the finer grid. However (b) is little different from (c) in terms of its binding efficiency. (c) Two surfaces of different cells meet at an angle, and binding is c. 20 per cent efficient in terms of the coarser surface. (d) Two like cells are pictured overlapping each other at an angle, and here the binding is only 15 per cent efficient even though they have like surfaces.

Consider an aggregate body containing two cell types in equal proportions arranged randomly. Let one cell type be more adhesive than the other, without any specificity in the adhesions. Take a cell of the more adhesive type which is surrounded by cells of the less adhesive type. As this cell moves about in a random fashion, it will eventually encounter another cell of its own type. Since these two cells are more adhesive to one another than to the surrounding cells of the less adhesive type, they will tend to remain together rather than move away from each other into regions where they would be surrounded by cells of the less adhesive type. Suppose this group of two cells of the more adhesive type encounters a third cell of the same type: it is more probable that this third cell will remain to join the group rather than leave it. As further cells of the more adhesive type join this group it will enlarge. When about six or seven cells are in the group, the addition of a further cell will result in at least one member of the group being wholly surrounded by cells of its own type. When this happens the central cell will be in contact on all sides with cells of the more adhesive type, and it will be held there more firmly than the peripheral members of the group which have adhesions one side to the less adhesive cell type. As a result, these groups of cells of the more adhesive type will tend to enlarge as they capture single cells of the more adhesive type and also perhaps small groups (if these be motile). At the surface of an aggregate there is in a sense a contact of negligible adhesiveness, that of peripheral cells and the medium. If any cell of the more adhesive type is found in the periphery it will not be trapped there, because the very weak adhesion against the medium will always allow it to move on to other sites which it has less probability of leaving, namely inside the aggregate, either in contact all round with cells of the less adhesive type or even more probably in contact with cells of its own type.

From what has just been said one can deduce that the probability of a cell of the more adhesive type leaving a contact with its own type of cell is less than the probability of a cell of the more adhesive type leaving a set of contacts with the less adhesive type of cell. In turn the probability that a cell will leave a contact with cells of the opposite type around it is less than the probability that a given cell of the less adhesive type will leave a particular set of cells of its own type. This can be rewritten as Steinberg does in terms of the energy of adhesion of the various contacts. Suppose that the more adhesive cell type is called A and the less adhesive B.

Then if $W_{AB} < \dfrac{W_{AA} + W_{BB}}{2}$, where W is the energy of

adhesion and the subscripts refer to the cells involved in adhesion, the cells sort into an external type B enclosing an internally segregating type A. This treatment applies also to the formation of separate or partially separate aggregates by two cell types. When $W_{AA} > W_{BB} > W_{AB}$ cell type B will not mix with type A, either partially enclosing it or demixing.

One of the features of this hypothesis, which is of great attraction, is that it predicts that one would find for mixtures of two cell types of different adhesiveness (a) complete intermixing of the two types, (b) one type spreading over the other or (c) one type partially spreading over the other or separating from the other completely, depending on the adhesiveness of the two cell types. Such arrangements are found in sorted-out aggregates. The spreading of one cell type over another would result in the simple enclosure of say type A by type B, and the position of the inner type A need not be concentric with the outside of the aggregate but might be eccentric. Since the surface energy relationships would merely specify that one cell type should be internal and the other external, if there were few cells of type A relative to the number of type B, it would be expected that type A cells would sink into the aggregate body, assuming the correct surface energy relationships, and because of their small numbers, fail to collect into one group, but instead form a number of scattered smaller groups. This type of sorting out probably appears in the pigment cell systems studied by Trinkaus and Lentz (1964), though of course its occurrence does not prove the Steinberg hypothesis. Steinberg's hypothesis was foreshadowed by Tait (1918) to a small extent.

This hypothesis would also predict, as Steinberg points out, that two pieces of tissue of different cell type would behave in a manner similar to segregation in aggregates, namely that one cell type would wholly or partially enclose the other or separate from it. Steinberg (1964a) describes examples of this behaviour for fused pieces of tissue and mentions similar findings by Holtfreter and Townes. Steinberg (1962c, 1964) points out that much if not all of the cell movement involved in sorting out would result from the interfacial relationships of the cells tending to minimise their surface free energy. He also suggests that active cell movement might tend to oppose movement directed by interfacial relationships.

At present the proofs which Steinberg offers for his hypothesis do not as yet include measurements of the adhesiveness of cells. It would be exceedingly difficult to do this. The proofs he puts forward are that (a) the sorting-out patterns and (b) the kinetics of sorting out can only be explained adequately by his theory. To do this he discusses the "timing" theory of sorting out (Curtis, 1961b, 1962a,b, 1963) and the chemotactic theory (see below). Therefore I propose to examine these other theories before describing and discussing Steinberg's proofs.

The so-called "timing hypothesis" is similar to Steinberg's theory in that it is supposed that quantitative differences in cell adhesiveness and related properties are responsible for sorting out. The hypothesis was foreshadowed by Curtis (1960a), stated in a special form (Curtis, 1961b) and in a general form (Curtis, 1962a). The hypothesis states that during the segregation period any given cell type changes its behavioural properties so that it becomes trappable in a fixed adhesion with other cells. It is also presumed that different cell types develop these trappable properties at different times; before they become trapped they are supposed to be freely motile. Consequently cells of like type tend to trap together forming adhesions together. In order to explain the segregation into external and internal phases it is supposed that trapping is initiated at the periphery of the aggregate. Further trapping takes place against the cells which have already been trapped. Consequently cells of the type or types which develop trappable properties later segregate internally.

Experimental evidence in favour of this type of hypothesis was obtained by attempting to confuse this supposed "timing" of the changes in behaviour of the various cell types. This was done by allowing one cell type to aggregate several hours before a second cell type was added. Consequently it might be expected that by delaying the addition of one type it would reach the trappable state some hours later than usual. If the trappable states of both cell types then coincided, one would expect no segregation of the two types. In other situations it might be possible to reverse the order in which two types reached the trappable state, in which case it would be expected that the internal or external position of the two types would be reversed. Both these experimental situations have apparently been realised, with amphibian embryonic cells (Curtis, 1961b), and in interspecific sorting out between sponge species (Curtis, 1962b). But objections may be made on the ground that identification of cell type was not sufficiently reliable, and in the case of the experiments on embryonic material the sorting-out patterns obtained might have been due to differences in developmental age between cell types introduced at the start of aggregation and after some hours' interval.

It should be pointed out that explanation of the various patterns of segregation by the "timing" hypothesis differs somewhat from that put forward by Steinberg. According to the timing hypothesis the situation where one type completely surrounds another is due to trapping starting at the outside of the aggregate. Partial enclosure may perhaps be interpreted as side-by-side aggregation due to trapping starting in the centre of one region. Trapping might start in the centre of an aggregate if it is a process controlled either by the preferential accumulation of a substance there or if it is a process dependent on the accumulation of a certain number of cells of the same type (more likely to occur in the body of an aggregate than at its surface). When a cell is trapped it is presumed to cease movement. It is not clear whether Steinberg assumes that cells stop movement when sorting out is complete. If interfacial energies alone are responsible for sorting out, movement should cease when sorting out is complete, but since Steinberg presumes that active cell movement takes place in sorting out there is no reason why it should not continue afterwards. Evidence from organ culture experiments tends to suggest that cells cease to move after segregation is complete. Centre trapping provides a good explanation of the beaded-

I

chain like aggregates of sponges found by Curtis (1962b) in which a block of one species type was adjacent to blocks of the other species type on either side.

Steinberg suggested that the timing hypothesis could not explain the occurrence of many small discontinuous regions of the internally segregating type in some aggregates. I feel that this can be explained by the "timing hypothesis". These aggregates contain relatively few internally segregating cells. Cell migration in them has to be more extensive than in other aggregates in order to (a) move all the internally segregating cells to a central position and (b) give the rarer internally segregating cells a sufficient probability of collision with their own type to segregate as one mass. Consequently it might be expected that the sorting out of such aggregates would take longer than in aggregates with more equal numbers of each cell type. But if the first cell type became trappable before this segregation is complete, it would immediately or very rapidly form a continuous phase of non-motile cells, which would prevent the second cell type from migrating to a single central site. In consequence sorting out to form a single central mass would only be expected if the numbers of both cell types were fairly equal. As the proportion of the internally segregating type falls, it would be expected that cells of this type would be found in smaller and smaller clumps and that the region containing the clumps would extend more peripherally. The reason for this is that as the proportion of the externally segregating type increases, the smaller numbers of internally segregating tend to be more isolated from each other. Consequently it is less probable that two internally segregating cells will meet in a given time. The space between them will be filled with the extra cells of the externally segregating type. When this cell type changes its properties it will trap the internally segregating type in isolated pockets before it has had time for all its cells to make contact. This is in fact found to be the case. This argument appears to meet the objections put forward by Steinberg (1962c, 1964a,b).

The timing hypothesis would predict that, when the numbers of cells of each type are fairly similar, the internally segregating types would be found to move towards the centre of the aggregate during segregation. As mentioned earlier, this appears to occur. Slight variations in the distribution of the various cell types in the aggregate at the start of aggregation, or a variation in the time at which trapping starts in various parts of the aggregate, would lead to eccentric segregation of the internally segregating type, as has been found (see Steinberg, 1964a,b).

Another type of theory which explains sorting out was put forward by Townes and Holtfreter (1955) and by Stefanelli and Zacchei (1958). They suggested that chemotactic mechanisms lead to the sorting out. It should be pointed out that neither these nor any other workers have yet obtained any clear evidence for sorting-out being due to chemotactic mechanisms. It has been somewhat naïvely supposed that the only chemotactic gradients which could be set up in aggregates which are nearly spherical would be radial ones, so that segregation would always result in concentric spheres. But more complex situations are imaginable. For example chemotactic action might appear only when a certain number of cells of the same type meet together by chance. This might happen at one or more sites in the aggregate leading to one, two, three or many groups of cells of one type. Townes and Holtfreter (1955) appear to have been a little uncertain as to the mechanism controlling sorting out, for they also made suggestions that surface-tension differences might exist between the cells leading to the inward segregation of one cell type (compare with Steinberg, 1964a,b), that specific adhesive bonds formed between cells of like type through a certain non-specific adhesiveness is never lost, and that the surface coat (see Chapters 1 and 7) plays some part in controlling segregation. They suggested that the fact that mesoderm remains adhesive to both endoderm and ectoderm in developed aggregates indicates a specificity of adhesive bonds on ectoderm and endoderm such that mesoderm can interact with either type. As has already been seen, the aberrant position of mesoderm in various combinations can be explained as being due to inductive interactions. The weak- or non-adhesion Townes and Holtfreter found between endoderm and ectoderm is explained on both the differential adhesiveness hypothesis and the timing hypothesis, as being due to considerable differences in adhesiveness.

Steinberg (1962c, 1963b, 1964a,b) sought to establish his theory of segregation by demonstrating that the timing and chemotactic hypotheses predicted patterns and kinetics of sorting out which were not in fact found. He claimed that with few cells of the internally segregating type, the timing and chemotactic hypotheses would predict a central location for these cells, whereas in fact they are located in a number of discontinuous subsurface groups. He pointed out, as we have seen, that his differential adhesion hypothesis would predict their discontinuous subsurface location. But we have also seen that in fact the timing and chemotactic hypotheses can also make the same correct prediction. Steinberg also suggested that the eccentric position of the internally segregating cell group found with a greater proportion of internally segregating cells would not be predicted by the timing and chemotactic hypotheses, but this is not borne out when these hypotheses are examined. Finally he argues that the timing hypothesis predicts an "inward" herding of the internally segregating type which is not found in practice; but as explained here, this is only expected on the timing hypothesis when the proportion of each cell type is fairly similar, and in fact does appear to occur according to the experimental evidence.

There are a number of consequences which arise from the assumed random packing of the unsorted-out aggregate. These are of importance in consideration of the mechanism of sorting out. Consider a randomly packed aggregate containing equal numbers of cells of both types. Assume that the aggregate is packed in a cubic fashion so that each cell has six nearest neighbours (we shall ignore contacts at corners for the sake of simplicity). Then the probability that all these six cells are of the opposite type to the central cell is 1/64, whereas the probability that one of its neighbours is of the same type is 63/64. This neighbour of the same type may be part of a group of cells of the same type or of just one of two cells of the same type surrounded by cells of the opposite type. If we consider the longest extent of linear and non-linear chains which run randomly through three dimensions there is a 66 per cent probability of the cells of one type being in chains of length 13 cells long, and a 32 per cent probability of chains being 35 cells long, the average length being 19 cells. (Formation of loops has been ignored in this calculation.) Parts of these chains will form nodes several cells thick. These chains will of course be branched, and if we consider the main branches (i.e. first order branching) there is a 16·6 per cent chance of a branch of one or more cells starting from any one of the cells of the primary chains. A chain of 75 cells in length would on average bear $12\frac{1}{2}$ main branches. In turn these branches bear secondary, tertiary and further sub-branches, there being on average a point of bifurcation for every six cells in a chain; the average length of each branch would be 19 cells. If we now consider the average basic chain and its branching out to tertiary branches, there are 247 cells of the same type jointed together, and an average of 760 cells if we include quaternary branches. Obviously an important question is whether the majority of all the cells of one type in the aggregate are members of the network which is built up by the branches. In other words do the branches of one region join with those of others?

Consider a branching system up to quaternary branches: this will occupy on average a volume of 2143 cells of which 1076 are cells of one type. Of these 760 are in long branched chains; 17 of these cells are single, and the remaining 299 are in small groups or as fifth and sixth order branches. If we represent this branching system by a sphere, it will have on a average surface of 804 cells of which 283 cells will be cells of one type attached to the branching system. If we imagine this surface to be contacted by a randomly packed array of cells of the same two types in equal proportions, 12 per cent of the chains of one cell type in one region (tertiary branching) will contact those in another region. Consequently the majority of cells of one type in a randomly packed aggregate composed of equal numbers of cells of two types are in contact with cells of the same type through a network structure.

This result may seem intuitively obvious to some readers but it is clear that many workers have failed to appreciate the consequences of random packing. Consequently the process of sorting out is not due to the trapping of single cells into groups, because there are hardly any such single cells. We can hardly suppose that cells frequently separate themselves from groups of their own kind and move to new positions, because if this happened at all frequently it would mean that the probability of a more adhesive cell leaving a group of its own kind was so high that the trapping process would be very inefficient. For example if there was a 45 per cent chance that such a cell would leave a group and a 55 per cent chance that it would stay, segregation would be very slow and inefficient. So we must suppose that groups of sorted out cells move together to reach their final segregated positions. This in itself is a little surprising since it has on the whole been supposed that this is not the mechanism by which segregation takes place. It is reasonable to suppose that the cells in a chain might tend to move towards a node because in a node they would be packed together with maximal contact area amongst themselves, and so would be most strongly held together. But in doing so it is improbable that a chain of the more adhesive cell type would frequently be broken, because to do so would mean exchanging two adhesions of the more adhesive type with themselves to two of the weaker adhesions.

These considerations have a considerable bearing on the mechanism of segregation. Obviously individual cell movement must be rare if the Steinberg hypothesis is correct. Consequently in order to preserve his hypothesis one must either presume that movement of clusters of cells of one type occurs, or alternatively that some other mechanism is used to move cells to a certain location. One mechanism can be imagined which might do this and which would be affected by differential cell adhesiveness. This is the protrusion of pseudopods by cells which are then used to haul the cell body to a new site. Examples of this action in embryos have been described by Gustafson and Wolpert (1963). If the pseudopods protruded by the inner cell type make contact with a more closely packed region of their own type, they will tend to move the cell to one of these nodes; provided of course that they are sufficiently adhesive. Chains of the more adhesive cell type will have initially a large number of "free" ends (ends against cells of the outer type) near the periphery of the aggregate. Cells of the more adhesive type at the ends of chains will have much less restriction on their migration to more central sites than a cell in the middle of a chain. As a result the internally segregating cells will soon leave the peripheral regions of the aggregate.

It is more difficult to explain the complete elimination of the externally segregating type from the centre of the aggregate, if indeed this ever happens. A single cell of the less adhesive type near the centre of the aggregate, originally surrounded by cells of the more adhesive type, would have no reason to leave this site on the differential adhesiveness hypothesis. Again, sorting out would be an inefficient process on the differential adhesiveness hypothesis, since cell movement would be towards local concentrations of cells at first, rather than towards the centre of the aggregate. But it should be remembered that the results with aggregates containing relatively few internally segregating cells suggest it is a rather inefficient process. These problems do not exist on the "timing" or chemotactic theories but since they are not insuperable they do not disprove the differential adhesiveness theory.

An argument by Steinberg (1964a,b) against the timing hypothesis is based on the fact that the timing hypothesis predicts that cells change their properties during aggregation. Steinberg pointed out that tissue fragments which had never been disaggregated apparently behave in the same manner as aggregating cells in sorting out. He considers it unlikely that cells in the tissue fragments could be altered by excision so that they subsequently changed their properties back to normal in any way similar to that predicted by the timing hypothesis. Curtis and Greaves (1965) found that the adhesiveness of cells increases during aggregation in medium containing serum (normally used in aggregation studies and for culturing tissue fragments) which suggests that changes of the type required by the timing hypothesis take place in both systems.

In effect, I feel it is difficult to summarise and evaluate the various theories of sorting out at present. Steinberg has made a major contribution in putting forward his ideas on differential adhesiveness. Unfortunately although his theory fits nearly all known observations it has not yet been experimentally tested by altering the adhesiveness of the cells in a segregating system. Nor have measurements of cell adhesiveness in aggregates been made yet. It should be remembered that both these matters are difficult to undertake. Steinberg's theory explains the tissue dispositions found when fragments are fused (see later in this chapter), whereas the timing and chemotactic theories do not; but this does not disprove these other theories, since the similarities between the two systems could be fortuitous. At present the timing hypothesis has little good experimental evidence in its favour, but it also explains the observations which have been made on aggregates as does the chemotactic theory. The timing hypothesis should be easy to test rigorously but this does not seem to have been done (the author is partly to blame for this). Obviously evidence from experimental alteration and interference with segregation processes is required. When we have it we may find that some other type of theory as yet unimagined is required.

ORGAN CULTURES

When pieces of two or more tissues are placed side by side in organ culture, they frequently adhere and undergo re-positioning. This rearrangement often results in one type wholly or partially enclosing the other, so that it resembles the segregation patterns found in aggregates. Holtfreter (1939) was the first to realise that some type of specificity was shown, for when fragments of tissue of several types were placed together they tended to rearrange themselves so that like types of tissue fused—he termed this tissue affinity. The mechanisms by which these artificial organs are formed may bear a closer resemblance to normal morphogenesis than do segregation mechanisms in aggregates for the following reasons. It is conceivable that disaggregation leads to abnormal cell behaviour or that intercellular materials, such as collagen which influence cell behaviour may be lacking in aggregates, whereas cell-to-cell and cell-to-collagen relationships are not disturbed in tissue pieces. If on the other hand it can be demonstrated that the same mechanisms operate in aggregates and organ cultures, then because organ cultures so closely resemble the situation in whole embryos, it is more reasonable to assume that aggregates provide a satisfactory model for normal morphogenesis.

Although most authors appear to have assumed that the tissue fragments lie immobile in the culture medium and only adhere where they make contact, some workers claim that the tissue fragments make directed movements towards each other (Alescio and Cassini, 1962; Wolff and Marin, 1957). However it is very important to discount the possibilities that surface tension or vibration may lead to centripetal motion of fragments. Unfortunately there have, at the time of writing, been very few studies of morphogenetic movements in organ cultures, and the majority of these have only described the final pattern of tissues set up. Wolff (1954) described the associations of embryonic chick, mouse and duck tissues in organ culture; the tissues were combined in pairs, either from within a species, or in pairs from different species. In the latter case both similar and different organs from two species were combined. When an organ such as a testis was cut into fragments and placed in culture, the fragments readhered to form a body which rounded off its external contours. If the organ contains epithelial or endothelial cell types, they form an external membrane around the mass. If these cell types are not present other types may be used to form an external layer. When fragments of tissues such as testis or mesonephros which normally contain tubules and cords of cells are cultured, these structures are maintained. Wolff found that one tissue type often wholly or partially surrounds the other, but that there is no constancy in which tissue of any pair of types takes the external position. Connective tissue cells appear to show considerable powers of movement as individuals. When two organs which both contain connective tissue cells are placed side by side, considerable interpenetration (invasion) of each organ by connective tissue cells from the other organ takes place. However this invasiveness is much less marked with other cell types, although mixed epithelia of differing origins may form. For example Wolff and Haffen (1952) found that when duck testis and ovary were combined in organ culture, almost no interpenetration of cells from either organ took place.

Bresch (1955) combined fragments of embryonic chick (6 to 10 days' incubation) liver, lung, thyroid, mesonephros and spleen in organ culture. He found that one organ tended to spread over the other, and that the most external cells of the fused fragments were flattened to resemble an epithelium. Lung tissues might lie either external or internal to liver tissues, and though liver usually lay external to mesonephros and mesonephros to thyroid, the last might lie external to liver. He noted that the relative size of the two tissue fragments appeared to affect their relative positions. Like

Wolff he found that connective tissue cells tended to invade the other tissue fragment. Bresch remarks that the connective tissue cells show no tissue-specific behaviour. Wolff and Weniger (1954) noted that in associations of duck testis and mouse ovary, connective tissue cells from both tissues intermingled throughout the explants.

These findings that there was no constancy in the relative external or internal position taken up by any two tissues in combination were confirmed by Bermann (1960), who found that in combinations of embryonic chick mesonephros and mouse testis, either tissue might be external.

Townes and Holtfreter (1955), besides examining the sorting out patterns of disaggregated amphibian neurula cells, also investigated the morphogenetic patterns set up when whole tissue fragments of ectoderm (epiderm), medullary plate, mesoderm (somite or lateral plate) and endoderm were combined together in pairs or triplets. In general it can be said that they found that the same morphogenetic patterns were set up whether the cells were disaggregated or present in whole fragments. However when medullary plate was combined with endoderm, with both tissues in the form of whole fragments, the medullary plate finally appeared side by side with the endoderm; whereas when both types were disaggregated the final position of the plate was wholly surrounded by endoderm. Townes and Holtfreter noted that the sorting-out patterns appeared to be constant for any pair of tissues combined together in fragments. Vakaet (1956) appears to have obtained a similar constancy of pattern in his combinations of embryonic chick lung and thyroid and liver and mouse lung. But his results are contradicted by those of Bresch, who specifically tried the effects of combining different relative amounts of each tissue (using the same ones as Vakaet had) and found that this affected the pattern. McLoughlin (1963) reported the effects of combining 5 day chick embryo epidermis with various forms of mesenchyme in organ culture. She found that in many cases the epidermis spread over the mesenchyme to form a cap two cell layers thick. Whether spreading occurred or not appeared to be a constant of the substrate mesenchyme. Dameron (1961) describes an apparently constant spreading of lung and certain other mesenchymes over pulmonary epithelium.

Grobstein (1953) separated epithelium and mesenchyme from mouse submandibular gland rudiments and cultured them *in vitro* in a plasma medium. He found that the mesenchyme spreads over the epithelium and thus stops the epithelium spreading on the glass substrate. This of course could be well explained in terms of quantitative differences in the cell adhesiveness of the two types. Golosow and Grobstein (1962) observed in a similar experiment on 11-day mouse embryonic pancreatic epithelium and mesenchyme, that mesenchyme spread over the epithelia, and that lung, kidney, stomach and other mesenchymes behaved in a similar way.

Steinberg (1962c, 1964a,b) fused pieces of embryonic chick liver and heart, heart and limb pre-cartilage, retina and limb pre-cartilage and retina and heart in hanging drop culture followed by culture in a shaker. He found that one of these tissues invariably surrounded the other and that the tissue type which was externally segregating in any pair was also the externally segregating type in combinations of the same pairs of types using disaggregated cells.

Steinberg (1962c, 1964a,b) pointed out the similarity in sorting out patterns obtained by himself and Townes and Holtfreter with both disaggregated cells and tissue fragments. He suggested that this similarity would be expected if the surface free energy of cells acted in the adhesion of cells to control the relative positions of two types according to his differential adhesiveness hypothesis. But it appears that there is little clear evidence that this similarity in fact exists. Further work is required to elucidate whether the relative amounts of tissue present and the method of culture affect sorting-out patterns of whole pieces of tissue. The lack of similarity in some systems between sorting out in aggregates and in fused tissue fragments suggests that at least additional factors may act in morphogenesis in whole tissues by comparison with aggregates.

The mechanisms by which fused tissue fragments arrange themselves into a definite if somewhat variable pattern involve either movements of whole groups of cells as single masses or individual cell movement. Two alternative routes might be taken by the cells or cell groups in order to reach their final positions. Either tissues might move around one another, or cells or groups of cells might invade the other tissue and move through it to their final positions.

Unfortunately there is comparatively little information on these points. Wolff (1954), Wolff and Weniger (1954) and Bresch (1955) all found that connective tissue cells in two fragments would intermingle with one another in a way that could only result from the individual movement of cells. However intermingling of other tissue types did not occur, with the possible exception of epithelia. Abercrombie and Weston (unpubl.) examined fused pieces of liver and heart. In one set of combinations they combined pieces of liver in pairs, one of the pieces being labelled with tritiated thymidine. They found that no intermingling of like tissue occurred, and that one piece frequently spread around the other even though of like cell type.

In all the work so far described on the rearrangement of tissue fragments, no parallel with the association of cells of like type found in aggregates has been mentioned. If each tissue fusion had been prepared from more than one fragment of each type it could be argued that cells of like type associate together, but unfortunately authors do not appear to have reported whether this was done or not. However Rerolle (1959) fused fragments of embryonic and early adult rat skin and young adult mouse skin in homoconfrontations and hetero-confrontations. She found that the *strata basilare* of each explant match up into contact; similarly the *strata spinosa* join in contact. Since the *strata basilare* are comparatively thin, the matching-up suggests a considerable specificity. Chiakulas (1952) found that if grafts of epithelia of various organs of urodeles were placed on the flanks of the same species they would spread out and meet the skin epithelium of the host. Fusion with the host epithelium took place if the graft came from ectodermal epithelia, fusion did not take place if the graft was of endodermal origin. However oral epithelium was anomalous in its behaviour if it is regarded as endodermal. Chiakulas suggested that these results show a specificity of cell segregation. Somewhat similar results were obtained with paired explants of epithelia of the same or different type. It is of interest that Garber and Moscona (1964) were unable to obtain any regular segregation pattern with 15-day mouse embryo skin cells and 8-day chick cells in reaggregates (grown on the chorio-allantoic membrane).

Easty and Easty (1963) combined pieces of mouse sarcoma 180 and Harding Passey melanoma tumour and liver, kidney and lung tissue in organ culture. They found that considerable penetration of the normal tissue took place. This compares with Leighton's *et al.* (1956) work with tumour and normal tissues grown in culture in a sponge matrix. Both workers found that tumour cells tended not to invade tissues in which connective tissue was oriented at right angles to their path. This may indicate that contact guidance plays a part in cell invasion. Wolff and Schneider (1957) and Schneider (1958) combined mouse sarcoma 180 with various embryonic chick organs in an extensive and painstaking study. Schneider (1958) found that in all combinations (liver, lung, mesonephros, metanephros, gonads) of organs, from ages between 6 and 18 days' incubation, and in hatched animals, four types of reaction could be observed. There were (i) encirclement of the explant by tumour tissue, (ii) migration of tissue cells into tubules in the tissue (e.g. bronchioles), (iii) invasion of the connective tissues, and (iv) invasion of all tissues. Schneider observed that with increasing frequency of mitosis the type of association of the tissues tended to change from (i) towards (iv). Complete invasion was commoner in younger tissues. Wolff and Schneider also observed that the tumour cells increased in number as the number of normal tissue cells fell. Thus the tumour may invade the host tissue partly by eroding it away. When Abercrombie and Weston combined heart and liver, no interpenetration appeared to occur. In summary, it appears that interpenetration of one tissue by another as a result of individual cell movement can occur in these systems, although this does not appear to lead to development of re-segregated tissue patterns. In other cases where the tissues remain separate and merely rearrange their mutual positions, it seems probable that the cells move as masses which are incapable of interpenetration. McLoughlin (1961a,b) has described the movements of individual basal cells which take place as a piece of embryonic chick epidermis spreads on mesenchyme. Initially those basal cells in contact with the mesenchyme orient with the cells in lateral register to form a basal layer. This orientation then spreads into the next layer of cells, and then the basal cells more than one layer away from the mesenchyme insert processes between the basal layer to the mesenchyme and pull themselves down into the basal layer. In this way a layer of basal cells one cell thick is formed, and as the basal layer approaches this thickness it necessarily spreads, probably pushed outwards by the pressure arising from cells being pulled into the basal layer. A layer of periderm of one cell's thickness covers the basal layer. McLoughlin suggests that the cells attach themselves to a basement membrane on top of the mesenchyme and that their strong adhesion to this membrane drags them down towards it. It should be noted that the cells develop an elongate shape during their movements.

A very interesting study of morphogenetic movements in an organ culture system was reported by Wartiovaara (1966). Mouse embryonic metanephrogenic mesenchyme rudiments were grown in culture. Mouse dorsal spinal cord fragments were used to induce tubulogenesis. The formation of tubules starts by the ingrowth of a branching epithelial ureteric bud. The cells are separated by 100–150 Å gaps. Cell protrusions down to 1000 Å diameter are found but these do not appear to "zipper" cells together or make molecular contact between cells (cf. Pethica, 1961). Initially the cells move actively about and large intercellular spaces between cells are formed frequently. When aggregation of mesenchymal cells around the epithelial branches occurs, the contact relationships remain the same save that the large intercellular spaces disappear from between the aggregated cells. The minimisation of surface area of the aggregates as they develop would be expected on Steinberg's hypothesis. Zonulae occludentes develop in the cell contacts on the outside of the aggregate: this would be expected since the adult kidney possesses them (Farquhar and Palade, 1963). Interestingly a number of cells fail to aggregate. Obviously there are many interesting and unexplained features in this system: for example, what ensures that the cells aggregate around the epithelium: is this due to chemotaxis or to some other process? It seems improbable that the whole structure of the kidney can be explained as arising from some process such as the operation of the differential adhesiveness mechanism.

The mechanisms which lead to the rearrangements of fragments of two or more types of tissue after they have been fused together, may as Steinberg (1963c, 1964a,b) suggests, be the same in their basic nature as those which operate in the segregation process in aggregates. Steinberg found that identical patterns of tissue arrangement appeared in aggregates and in fused fragments of the same two types of tissue. This similarity suggests that the basic mechanisms in each process are the same, though it should be remembered that the movements of tissues in fused fragments over or under one another may result from the exchange of homonomic adhesions (between like cells) for heteronomic adhesions (between unlike cells), whereas in the segregation process in aggregates the reverse process occurs to a considerable extent. In each case the resulting similar structure is that which is least likely to allow disrupting cell movement. If tissue fragments are considered to move in some way similar to that of cells and to move in single units, then the explanation given for segregation in aggregates can be applied to this case. Unfortunately we are unsure about the mechanism whereby tissue fragments move and whether there may not be migration of cells as individuals to and from the fragments during the process. Moreover there are reports of regular failures to obtain constant patterns for the segregation of fragments of tissue of two types after their fusion: obviously further work is needed to discover the cause of these apparent anomalies.

It seems a little improbable that the morphogenetic movements found in tissue fragments can be explained on the "timing" hypothesis, mainly because it would be necessary to suppose that the cells emigrate individually to a considerable extent from the explants and mix before resegregating: this may of course happen, but it seems unlikely to us at present. The timing hypothesis specifically explains segregation resulting from individual cell movement, whereas it is clear that most if not all of the morphogenetic movements in fusions which result in re-patterning of the tissues are due to movement of cell masses. Since they are already sorted out, and are only rearranging their mutual positions, there is no need of a trapping and segregating mechanism.

In conclusion it may be said that the rearrangements of tissue position in fusions in organ culture do not present a very coherent picture. Apparently cell interpenetration or individual cell movement does not occur in those tissue combinations in which the various tissue types finally appear separate. The theory that the relative sizes of the individual pieces of tissue that are fused together affects segregation appears the best way to explain the facts when individual cell movement does not occur, but very few workers have even incidentally produced any evidence on this point. It is unfortunate that no clear instances of the action of individual cell behaviour in these systems has come to light, though various speculations will be made in the next chapter. Although it was mentioned at the start of this section that intercellular materials might play an important part in these systems, by comparison with aggregates, this possibility has not been considered by those interested in their morphogenesis. Similarly the possibility that induction might interfere with tissue rearrangement in organ cultures and segregation in reaggregates has rarely been considered. These points will be considered immediately.

THE EFFECTS OF OTHER TYPES OF TISSUE AND SUBSTRATE INTERACTION ON MORPHOGENETIC MOVEMENTS IN VITRO

Up to this point in this chapter, we have considered cellular interactions in organ and tissue culture and aggregates only in terms of direct cell surface interactions between cell and cell. However, there may also be adhesive and other interactions with intercullular connective tissue elements, as well as alterations in cell surface properties due to diffusible factors which may spread from one cell type to another in aggregates. Evidence on these points will be considered here.

Recombinations of epithelia and mesenchyme in organ culture, similar in some respects to those described above, were made by McLoughlin (1961a,b; 1963) using 5-day chick epidermis and various mesenchymes. On many of the mesenchymes the epidermis spread over the mesenchyme, often entirely surrounding it. When epidermis is cultured by itself she found that the presence or absence of cell-free dermal intercellular material controlled the orientation of the epidermal cells in a radial pattern. This would suggest that the intercellular material is able to control cell form and those local movements required for orientation to take place. However in detail her papers show that the apparent effect of the material is to increase the extent of orientation around any one centre. She points out that the intercellular material may act only as a carrier of the agent responsible for the orientation, and that the presence of intercellular material is only a concomitant and not a cause of these morphogenetic changes.

However Wessels (1962) showed that PAS staining mucopolysaccharide is associated with the epidermis and can be present on isolated epidermal explants which do not show the orientation behaviour described above. In 1962 Wessels found that intercellular materials may be formed after tissue rearrangement in between layers of embryonic chick dermis and epidermis. Wessels also demonstrated that dermis produced a factor which would pass through Millipore filters to orient cells in the epidermis. Chick epithelial tissues were used and both mouse and chick mesenchymes were active, whether dermal or other mesenchymes were used (Wessels, 1964a). Wessels (1964b) was able to induce similar effects with embryo extract (chick) and succeeded in partially characterising the factor as of proteinaceous nature.

One protein which may affect cell behaviour is the Nerve Growth Factor (NGF) (Levi-Montalcini, 1965; Cohen, 1959). Although Levi-Montalcini (1965) does not claim that the NGF protein affects cell behaviour, examination of her results on the outgrowth of nerve fibres from explants of sympathetic ganglia in tissue culture, which outgrowth is stimulated by the presence of the protein, suggests that emigratory behaviour is influenced by the protein. This stimulation might be due to the NGF resulting in sufficient increase in cell number for cells to be available for emigration, but Lefford (1964) has shown that cell number in explants is not related to the extent of emigration. Instead it seems probable that emigration is due to certain specific cell behaviour (Abercrombie, 1957) or to reduced cell adhesiveness (see Chapter 5).

The aggregation-inhibiting protein (Curtis and Greaves 1965), which is of molecular weight 150,000, affects cell behaviour profoundly. At low doses between 125 per cent and 500 per cent of the amount of this protein present in a normal serum medium, contact inhibition is affected, being reduced in incidence so that invasion takes place in tissue culture. At higher doses cell detachment and rounding up of undetached cells takes place. This protein also accelerates and enhances the rate of emigration of fibroblasts from an explant, possibly due to lowering of cell adhesiveness, and accelerates their speed of movement. A further point of interest is that the aggregation-inhibiting protein affects different cell types to various extents. Unfortunately at present we do not know whether this protein or similar ones occur in embryonic fluids. However it is clear that if proteins with such specific effects on cell behaviour were produced in one or more parts of an embryo, a very precise control over morphogenetic movements could be effected in this way.

Other examples of the control of cell movements by extra-cellular factors in culture are the possible role of intercellular material in controlling and impeding emigration from explants in tissue culture (Lefford, 1964), the effects of pH on cell movement (see Chapter 5) and the examples of factors affecting adhesion described in Chapter 4.

Summary

In summary, we can see that there is as yet no clear understanding of the mechanism or mechanisms whereby a given cell type can take up a specific position in these comparatively simple systems. It is clear that cell movement either of individuals or in groups is of considerable importance, but the mechanisms whereby a given cell type is trapped at a given place are as yet unclear. There is some suggestive evidence that cell adhesiveness plays an important part, and that at least in certain instances cell interactions of an inductive type control the morphogenesis (e.g. some of Townes and Holt-freter's results and those of Wessels). At the present time these model systems do not serve by themselves to explain morphogenesis in the whole animal; but taking their results together with those on whole embryos, a moderately coherent picture of the possible mechanisms of morphogenesis will appear in Chapter 7.

CHAPTER 7

Morphogenetic Movements:
The Cell Surface in Embryo

Movements of cells, as individuals and as groups, take place during embryogenesis, regeneration, in pathological conditions and in some instances normally in adult life. The cells may move by their own motile mechanisms or be swept along in a body fluid such as blood. It is intended to discuss the probable mechanisms by which these movements are brought about and the ways in which they are controlled. This will involve discussing the methods by which cell behaviour may affect specific movements, and in turn the means by which properties responsible for cell behaviour are specified. It will appear that there are considerable changes in the surface properties of a given cell line during development. There has been much speculation and a little evidence put forward on the manner by which cell behaviour controls morphogenetic movements, but surprisingly little interest in the development and changes of surface properties during embryogenesis has been aroused yet.

The first type of morphogenetic movement to be discussed will be the migration of cells in the vascular system; thereafter movements in which cells move on a solid substrate will be examined.

THE SPECIFICITY OF CELL DISSEMINATION BY THE VASCULAR ROUTE

Although comparatively little work has been carried out on this topic it is of considerable importance because it has been claimed that the mechanisms whereby cells settle from the bloodstream and attach themselves to certain organs are those which act in segregation in aggregates. Weiss and Andres (1952) stated that embryonic chick pigment cells injected into the yolk-sac circulation of chick embryos, became specifically sited in the feather germs, which are, of course, the normal site for these pigment cells. Similarly, it has been observed that many tumours form secondaries in specific sites (see Cowdry, 1955) and that apparently during avian development germ cells migrate from the extraembryonic mesoderm to the gonads (Willier, 1937; Simon, 1957a,b, 1960; Meyer, 1964). In all three cases the migration is routed through the vascular system, and there is apparently a remarkable specificity in the localisation of the cells after their migration. If this specificity occurs it could perhaps be explained as being due to the specific adhesion of like cell types, so that cells finally stop moving only when they contact their own type (or in the case of tumour cells, types with some similarity to the tumour cells).

However, does this apparent specificity really exist? First, it is possible that the cells are disseminated randomly through the body but that only those cells which settle in the correct site survive or alternatively show their correct differentiated type. Second, it is conceivable that mechanisms other than specific adhesion would account for the specific siting of cells after their journey through the blood vessels. For example, if cells are released from one particular site into the vascular system and then settle out rapidly, but at a constant rate, then they will be localised close to their point of origin. Again, if the cells are of a certain size they may become trapped specifically in regions where the blood vessels become narrower than the cell size. Finally, certain regions of the body may secrete substances like the adhesion-promoting-protein (Chapter 4) to cause cells to settle down locally. It should be remembered that there are two stages to the localisation of cells in a certain tissue: the cells must first pass through the vascular wall at some point and then move into a given tissue to finally adhere there. It seems probable that if there is any specific mechanism in the settling out of vascularly born cells then this must act at least in part on the passage of the cells through the vascular wall.

Weiss and Andres (1952) considered these points in some detail. They estimated that they were injecting 500 pigment cells per animal. The hosts were White Leghorns and a variety of pigmented breed were used as donors. When hosts, which had received pigment cell injections while embryos, developed their adult plumage, patches of feathers pigmented like the donor type were found. One, two or three such patches might be found on a bird. These patches were of considerable size and represented the activity of thousands of pigment cells. Obviously as Weiss and Andres point out mitosis must occur in the injected cell strain. However, do these pigmented patches arise from a small proportion of the cells injected or from the greater proportion? Weiss and Andres considered that all or nearly all the cells injected would be required to produce the pigmented patches. If this is so then, as they claim, the cells must become specifically located in the feather germs. But if sufficient mitosis can occur then a very few of the injected pigment cells would need to be localised in the skin for the patches of pigmented feathers to appear. In this case there would be no evidence for the specific localisation of cells in this system. Unfortunately we do not know about the extent of mitosis of pigment cells. In conclusion it can be said that it is not yet clear whether specific localisation of pigment cells occurs.

Three main examples of cell migration through the vascular system are known to take place naturally; (i) tumour cells during metastasis, (ii) migration of primordial germ cells in birds, and (iii) the homing of lymphocytes in mammals.

Metastasis is believed to take place largely by the vascular dissemination of tumour cells (see review by Cowdry, 1955). Certain types of tumour appear to form secondary tumours selectively in only a few of all the possible sites. It is not intended to review this data in any detail but rather to describe experimental material which appears to explain this phenomenon. Cowdry (1955) pointed out that the vertebral venous plexus and the liver are regions where the rate of blood flow is reduced and that metastases tend to be found in these areas. Coman and DeLong (1951) found that the dissemination of the rabbit V2 carcinoma (viral) was such that secondaries tended to form in the venous plexus but that the siting of tumours could be changed to the lungs by applying abdominal compression, which presumably results in faster blood flow through the plexus. A variety of workers have injected tumour cell suspensions into animals. Coman *et al.* (1951) injected Brown-Pearce rabbit tumour into the left ventricle (?) of rabbits and found that the cells accumulated, preferentially in the iris, pituitary, adrenals and kidney. They suggested that this distribution was in accordance with the mechanics of blood flow. Green and Lovincz (1957) found that if the mouse Krebs 2 tumour was injected as a cell suspension into the circulation of 10–11-day chick embryos then the cells were selectively accumulated in the liver, brain and heart. Dagg *et al.* (1956) placed a variety of human tumours as cell suspensions on the chorioallantois: some of these tumours formed foci in the chick embryo, particularly in the brain, liver, heart and kidneys. They found, as did Green and Lovincz, that the tumour cells tended to disappear after hatching, possibly due to some type of immunological reaction. Humphreys (1961) injected chick embryos with Ehrlich ascites tumour cells and then discovered the location of the cells which were recognised by their appearance. Cells were found in the chorioallantois, lung, brain, yolk sac and in other tissues, though some tissues (unnamed) were free from cells. Four days after injection the number of ascites cells decreased in all tissues except the brain and yolk-sac.

Zeidman (1961) has observed the settling of tumour cells in the vascular system in fine capillaries. It should also be pointed out that the lymph system can provide a pathway for the movement of metastasing cells.

These examples show that liver, kidney and brain are preferred sites for metastasis of a wide variety of tumour cell types, and this result suggests that it is not a reaction such as specific adhesion between tumour cell and site which determines the location of metastases, but rather some non-specific factor such as mechanical properties of blood flow.

Wood (1964) examined the effect of injecting ascitic cells of the rabbit V2 carcinoma into rabbits. The cells were injected into the central artery of the ear and their movement was examined by filming the vascular system in an ear chamber. He states that the initial sticking of the cells to the capillary endothelium was of short duration, and was independent of capillary diameter and rate of blood flow. He concluded that these capillaries do not act as simple mechanical filters. Fibrinolytic agents broke up thrombi of tumour cells which had formed in the capillaries. If Wood's conclusion about mechanical trapping of tumour cells can be extended to other situations, then the theory that the distribution of the tumour cells is controlled by simple mechanical factors is untenable. But Wood refers only to initial settling which is then lost, and it seems probable that the final settling down is under control of mechanical conditions. Unfortunately, even those workers who have searched the tissues for the presence of disseminated tumour cells have never done this within a few hours of the injection but have only examined the tissues after longer periods. For this reason we cannot yet be sure whether the initial distribution of cells is not completely random, with tumour cells in some tissues dying off very soon after their arrival there. If this were the case the mechanical hypothesis would have to be abandoned.

Passage of tissue cells through the blood stream has been noted in a few pathological conditions and in one instance in normal development, namely the location of germ cells in the gonads.

As early as 1914 (Swift) (see also Firket, 1920) it was suspected that the primordial germ cells in birds first appeared in the extra-embryonic tissues, and that they were then carried or moved themselves to the gonads. Although it seemed very probable that a vascular route was involved, decisive proof of this point was not obtained until 1957 (Simon, 1957a,b, 1960). She produced sterile embryos by excision of the extraembryonic region, and then parabiosed them with normal embryos. She found that primordial germ cells then appeared in the previously sterilised embryo as well as in the other embryo, but only if the vascular systems fused. This result established that the germ cells are transported by a vascular route, but there still remains the problem of whether the cells are dispersed randomly throughout the embryo or whether they settle in one specific site.

Swift (1914), Firket (1920) and Willier (1950) suggested that initially the primordial germ cells are distributed randomly throughout the embryo. But Goldsmith (1928) and Simon (1960) have claimed that these cells are concentrated in the gonad. Meyer (1964) made a most thorough histological study of the distribution of primordial germ cells in the embryonic chick between stages 11 and 18 (Hamburger-Hamilton stages). He found that by stage 13 cells were found in the head mesenchyme, heart and various other regions, but there were none in the gonads. By stage 15 the germ cells had begun to appear in the gonad as well as in the head mesenchyme and other tissues; by stage 16 larger numbers were found in the gonads, and to some extent in the head mesenchyme, than previously, but the numbers in other tissues were little different from the previous stage. At stage 17 only the gonads showed an increase in number of cells over the previous stage. Meyer points out that a number of cells will be found extragonadly at stage 18 particularly in blood vessels. He concludes as Swift (1914), Witschi (1948, 1950), Dantschakoff (1931) and others have done, that there is either chemotactic attraction of the cells to the gonad, or some kind of mechanical trapping. Willier (1950) suggested that some kind of specific adhesion might trap the primordial germ cells in the gonad. However his data are also compatible with the explanation that the germ cells divide only when in the gonad, so that the basic distribution is always wholly random. Van Limbordh (1957) suggested that mitoses of primordial germ cells occur only in the gonad. Meyer favours the hypothesis that the cells are trapped by mechanical features of the circulation in the gonad, and points out that these germ cells possess pseudopods when situated extravascularly which may suggest the mechanism by which they invade tissues. In birds the left gonads develop much more extensively than the right ones. Dantschakoff (1935) claimed that this asymmetry was due to more germ cells settling in the left gonads than in the right. Meyer (1964) and many other investigators (see Meyer for bibliography) have however found that both gonads contain equal numbers over the first few stages in which germ cells are found in the gonads.

Before summarising these results the so-called homing of lymphocytes will be discussed. The possibility that lymphocytes accumulate specifically in certain tissues arose from two sources. First, many lymphocytes (small) are found in the lymph which flows from a thoracic duct fistula. The rate of flow suggests that in the intact animal a lymphocyte must pass into the thoracic duct several times daily. Gowans and Knight (1964) discussed these earlier experiments and re-examined the system very meticulously; their work will be discussed shortly. Second, injection of bone marrow cells into rats and mice which have received bone marrow, thymus, lymph node or whole body irradiation has shown that when the radiation dose is sufficient to destroy all lymphocytes, the bone marrow etc. is recolonised by cells of the injected type. (See Till and McCulloch, 1961; Harris and Ford, 1963a,b amongst others.) These findings suggest that lymphocytes, blast cells and haemopoietic cells can become specifically located in tissues in which they are normally found. Gowans and Knight (1964) labelled small and large lymphocytes with tritiated thymidine (large lymphocytes only) and tritiated adenosine (both cell types) and injected them into the femoral vein of rats. After one or more days large lymphocytes were concentrated in intestinal and mesenteric lymph nodes and in the stroma of the villi. Labelled small lymphocytes were recovered from the lymph duct about 2 hours after their injection. Roughly 98 per cent of the transfused cells were recovered over 4 days. In addition, after 24 hours many, if not all, lymph nodes contained many small lymphocytes. They were also found in the post-capillary venules within 15 minutes and appeared to be penetrating the endothelium of the venules. These results appear to demonstrate that the small lymphocytes accumulate rapidly in the lymph nodes where they then enter the lymph circulation. It should be remembered that the rate at which the labelled cells emerge from the thoracic node is so slow that unless other lymph nodes are behaving in the same manner and competing for the supply of lymphocytes, there is no need to suppose that there is any homing of these cells to the nodes, but rather a random trapping of cells in the thoracic lymph node as they pass through it in the vascular system. It is unfortunate that Gowans and Knight did not make quantitative studies on lymphocyte distribution in other sites.

Marchesi and Gowans (1964) made electron micrographs of lymph nodes of rats and detected leucocytes in the endothelia of the post-capillary venules. These cells were described as passing through the endothelial cells and not through their intercellular junctions. Unfortunately the electron micrographs published were of insufficient resolution to judge whether the leucocytes really are "phagocytosed" by the endothelial cells or whether they push through a deformed intercellular contact between two endothelial cells. Gesner and Ginsburg (1964) found that if P^{32} labelled rat lymphocytes were treated with *Clostridium perfringens* glycosidases they appeared to accumulate in the spleen to a lesser degree than untreated lymphocytes. L-fucose and N-acetyl-galactosamine partly reversed this effect of the enzyme. Enzyme-treated cells appeared to accumulate preferentially in the liver. However since most of the experiments appear to have been carried out only once, it is uncertain how much importance is to be given to this result. Weisberger (1951) and Bainbridge *et al.* (1965) found that lymphocytes tend to be trapped in the lungs and liver respectively to a considerable extent, though they also accumulate in the lymphoid tissue.

In conclusion it must be admitted that there has been no unequivocal demonstration that cells are either attracted chemotactically to, or are trapped by some specific mechanism in, one or two of all the possible sites in the body which they might reach by a vascular route. Although the advocates of mechanical trapping have been able to put forward a considerable body of circumstantial evidence, they, like those of other views, have not carried out sufficient quantitative work to establish that even the phenomena they are trying to explain actually occur. Again those who hold the view that the cells are disseminated entirely randomly are unable to support their view with quantitative work. In this thoroughly unsatisfactory state of affairs all that can be said is that if all cell types behave in the same way, which is unlikely, it seems probable that distribution by the vascular route leads to a non-random but not highly specific distribution of cells in certain sites. The fact that the lungs and liver appear to be preferred as sites for the settling of a variety of cell types suggests that one of the main factors in this process may be the existence of suitable conditions of blood flow.

CELL MOVEMENTS WITH CONTINUOUS ADHESION

Although cell migration by the vascular route is of great importance in malignancy and in a few special processes of normal development, the majority of cell movements in embryogenesis take place with the cells continuously involved in adhesion either to one another or to a substrate. The purpose of this section is to examine how far these morphogenetic movements can be explained in terms of the behaviour of cell populations and of individual cells. Ultimately this will lead, one hopes, to an explanation of these movements in terms of molecular changes in the surface structure and composition, but at present it is impossible to do this. Morphogenetic movements are not here described in great detail, except in those few cases for which no description will be found in standard textbooks of embryology. Rather, after a brief summary of the features of each main type of movement the book will concentrate on the experimental analysis of their causes. The majority of experimental work has been carried out on vertebrate embryos and each of the main morphogenetic movements which vertebrate embryos undergo will be discussed in the order in which they occur in development. However there are three theories which have been used in an all-embracing way to explain morphogenetic movements which are, I feel, exceedingly improbable, and they will be discussed before examining movements in detail.

THE ROLES OF CELL DIVISION, CELL SWELLING, HYDRO-STATIC PRESSURE DIFFERENCES AND "SURFACE COATS" IN MORPHOGENESIS

The first of these three theories is that movements are due to pressure differences between various embryonic cavities. The pressure differences might arise because of osmosis, active transport of water, or hydrostatic pumping by cells. This theory has been put forward by Zotin (1962), Lovtrup (1962), and Tuft (1961a,b; 1962). Very little direct evidence has been found to support the existence of such pressure differences; indeed Tuft (1957) found that there was no difference in osmotic pressure between the blastocoel and archenteron of amphibian embryos. Again Brown *et al.* (1941) found no evidence for water exchange between various parts of the amphibian embryo. Two observations show that this theory is improbable. First, morphogenetic movements continue almost unimpeded when the archenteron, blastocoel or other cavities are open to the surrounding medium (see Stableford, 1949; Moore and Burt, 1939), also even small fragments of embryonic tissues will continue to show morphogenetic movements (Schechtman, 1942 for example). Second, the 100–200 Å gaps between cells make the maintenance of any appreciable pressure difference between different parts of the embryo very difficult, since flow along these gaps will be little impeded. It can be argued that very few electronmicrographs have been made of embryonic cell contacts, but the low adhesiveness of such cells (Chapters 3 and 4) suggests that 100–200 Å gaps are present between embryonic cells. Gustafson and Wolpert (1962) have pointed out that hydrostatic pressure differences could lead to the stretching of cell sheets.

The second theory is that the vitelline membrane, hyaline layer, surface coat etc. of embryos plays an important role in providing either a mechanical support against which cells can thrust to obtain their movement, or more vaguely and grandiloquently, a mysterious "supracellular" structure controlling morphogenetic movements. (Holtfreter, 1943a; Lewis, 1947). Unfortunately there is considerable confusion in this field due to problems of terminology. All eggs normally possess vitelline membranes outside their plasmalemmae after fertilisation and most embryos also possess these membranes during the earlier part of their development. It seems that the vitelline membrane has little morphogenetic importance since morphogenetic movements are normal in all embryos after removal of the membrane. Echinoderm eggs possess a hyaline layer between the plasmalemma and the vitelline membrane; but this structure appears to be unique to this group—its significance in the morphogenesis of these embryos will be discussed shortly. Holtfreter (1943a) postulated that a surface coat occurs in amphibian embryos and that it consists of a structure which is continuous over the whole surface of the embryo. As we have seen in Chapter 1 the evidence for such a structure is now negligible. Spratt (1963) proposed the action of such a structure in avian embryogenesis, but his evidence was purely circumstantial. Similarly Lewis (1947) suggested that a similar structure—the gel layer—was responsible for features of the development of teleost fishes. Dollander (1962) was unable to find any surface coat in electron micrographs of amphibian embryos, and Bellairs (1965) could not discover such a structure in electron micrographs of the avian embryo. There is of course evidence that the outer parts of the amphibian embryo are important in morphogenesis (see p. 295), and Holtfreter used this evidence in support of his hypothesis, but there is no necessity to suppose that this proves the action of a surface coat; it is as reasonable that it is due to different properties of the surface layer of cells as compared with deeper layers of cells.

The third hypothesis to account for morphogenetic movements supposes that they arise partly or wholly from differential volume changes in various groups of cells of the embryo, as a consequence of differential mitotic rates combined with enlargement of the daughter cells through synthesis or by osmosis (see also p. 308). Of course division by itself does not produce any increase in the volume of a tissue, other than a very slight increase due to the formation of gaps between the daughter cells. This latter effect would for example produce an 8% increase in volume over the division of one cell of radius $10\,\mu$ into 64 cells assuming that 200 Å gaps were formed between the daughter cells. Pasteels (1942) and Gillette (1944) investigated whether differential changes in the volumes of tissue, consequent on differential mitosis, occurred in amphibian embryos. They found no evidence for such an effect. The failure of Brown et al. (1941) to find evidence of relative changes in cell density would seem to argue against the possibility that differential swelling of cells might occur, but it should be remembered that this could be brought about either by differential rates of breakdown of yolk or by differential synthesis of macromolecules using carbonate from the medium (Flickinger 1954) to offset density changes that would otherwise arise from relative swelling of cells. No studies have been made on other groups, except that by Kessel (1960) on the development of the fish, *Fundulus*. In no case has a sufficiently precise series of measurements of cell swelling or the converse been made on individual cells or small groups of cells to provide a rigorous test of this hypothesis. It should be remembered that morphogenetic movements may have their origin in the behaviour of small numbers of cells, so that volume measurements on quarter embryos etc. are almost useless. However, at present the evidence, such as it is, argues against differential cell swelling as being a cause of morphogenetic movement.

Gustafson and Wolpert (1962) pointed out on theoretical grounds that cell swelling could be an important morphogenetic agent. If a cell swells without growth of new cell surface, the surface will come under greater tension. The increased tension would tend to decrease the area of cell contact and thus weaken cell adhesions.

The magnitude of morphogenetic forces

Accurate measurement of the magnitudes of the various morphogenetic forces ought to provide a useful means of identifying the mechanisms involved in morphogenetic movements. A further check would be provided by considering whether the embryo's metabolic processes could provide the energy required. Unfortunately it is difficult to measure or even estimate the likely values of the forces. In the main, two techniques have been used: (i) to measure the force just necessary to prevent the movement, e.g. Moore (1951), Selman (1958) and Waddington (1939); and (ii) to attempt to determine the mechanical properties of the system and thence to calculate the force required to bring about the movement. The first method is fraught with the difficulty that preventing the movement with an opposing force may alter the magnitude of the morphogenetic force. The second technique requires a knowledge of the theory of complex non-isotropic structures which we have not got, so that our calculations are at best inspired quasi-guesses. For these reasons there are only a few features of morphogenetic movement which have been investigated in this way, and on the whole they are not very informative. Some of the most promising work has been done by Gustafson and Wolpert (1963) on the process of gastrulation in echinoderms (see page 292) and by Selman (1958) on the mechanism of amphibian neurulation (see page 31).

THE CELL SURFACE IN FERTILISATION REACTIONS

Before considering cell movement in embryos, the behaviour of the cell surface in embryonic development during those early stages when no movement occurs, will be discussed. The origin of the cell surface in an egg shortly before and just after fertilisation is somewhat uncertain. Has the cell surface of the primary germ cells of the parent been incorporated into the developing oocyte, and does this surface of the oocyte persist substantially unchanged to form the surface of the fully formed egg? If the surface persisted throughout oogenesis, it might form a system by which surface properties could be inherited from one generation to another extranuclearly (Curtis, 1965b). Bellairs (1965) has suggested that the plasmalemma of the fowl oocyte breaks up into vesicles during oogenesis and fails to surround the egg cell continuously, on the basis of her electron micrographs. Possibly the plasmalemma of the new embryo is derived from these vesicles. During fertilisation the sperm may make a contribution of plasmalemma to the egg according to Colwin and Colwin (1963).

In many groups of animals fertilisation is followed by more or less obvious surface changes before cleavage and mitosis begin. These changes are referred to as being part of the process of activation of the egg. As a result of this reaction a vitelline membrane lifts clear from the cell surface, and the latter undergoes a series of rapid changes in its properties. Fertilisation reactions (cortical reactions) are perhaps most obvious in echinoderm eggs. Visually they appear as the spread of a colour or turbidity change over the surface of the egg (Runnstrom, 1928; Just, 1919) starting at the point of sperm entry. In *Psammechinus* sp. this change takes about 25 seconds to spread over the whole egg (Rothschild and Swann, 1952). This change marks the formation of the fertilisation membrane, which is composed of the vitelline membrane together with material derived from the cortical granules. Before formation of the fertilisation membrane the cortical granules lay just exterior to the pigment granules and contiguous with the plasmalemma of the egg, See Figure 21. As the membrane is formed it is "elevated" off the egg so that an appreciable space appears between it and the plasmalemma of the egg. Electron microscopy has shown that before this reaction the plasmalemma is probably infolded to surround the cortical granules (Wolpert, 1960; Wolpert and Mercer, 1961). After the cortical granules are released and broken down, the plasmalemma remains as a highly convoluted structure. High resolution electron micrographs do not yet appear to be available to show whether or not the cortical granules are completely surrounded by plasmalemma, which then breaks at the "top" to release the granule, or whether the granules are sunk in open pits in the plasmalemma. A very considerable amount of data has accumulated as to the mechanism whereby the fertilisation reaction takes place (reviews by Runn-strom, 1952, 1957; Runnstrom and Krizat, 1960; Perlmann, 1959), but it is not clear whether this activation involves anything but passive changes in the plasmalemma. Other groups of metazoa do not show such a complex set of fertilisation reactions.

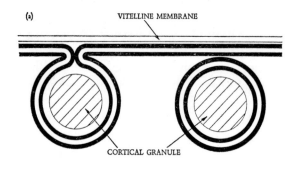

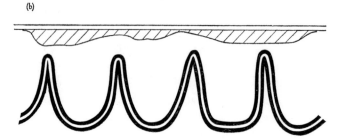

Figure 21. The effect of fertilisation on the sea urchin egg surface (*a*) before fertilisation, (*b*) after fertilisation and egg activation. Plasmalemmae shown as double black lines. It is not clear whether the membrane surrounding the cortical granules is continuous with the plasmalemma (as shown on left) or not (see right). After activation the membrane either alters shape to release the cortical granules or a fusion of cortical granule and plasmalemmal membranes takes place. The vitelline membrane lifts off the plasmalemma as a result of activation.

However several other groups of metazoa have eggs which show phenomena somewhat similar to the fertilisation reaction of echino-derms. In mammalia, fertilisation is followed by a change in the refractivity of the outer parts of the egg in at least some species, e.g. hamster (Austin, 1956), which is due to the disappearance of

cortical granules. It has been suggested (Austin, 1961) that the disappearance of these granules is involved in the reaction of the zona
pellucida around the mammalian egg whereby the zona becomes
refractory to penetration by further spermatozoa. However, this
reaction is not the only block to polyspermy. The vitelline membrane of mammals is formed before fertilisation (see Chapter 1).
In both sea urchins and mammals this vitelline membrane is involved
in the other and main block to polyspermy which spreads rapidly
across the membrane in sea urchins (Rothschild, 1956) and perhaps
more slowly in mammals (Austin, 1961). As in sea urchins activation
of the egg in mammals is accompanied by the elevation of the
vitelline membrane from the plasmalemma, probably due to a
contraction in size of the egg cell.

THE CELL SURFACE DURING CLEAVAGE AND
BLASTULA STAGES

After fertilisation and its accompanying reactions have finished,
the normal embryo settles down to cleavage. As cleavage proceeds,
more and more surface is formed. If a cell divides into equal spheres
without volume change, the two daughter cells have a combined
surface area 26 per cent greater than the parent cell, or 50 per cent
greater if they form two hemispheres. Obviously at some point
during cleavage new cell surface must be synthesised, for otherwise
the surface derived from the egg would become too thin to function.
Selman and Waddington (1955) suggested on the basis of visual
observation and measurements of the deformability of the cell, that
new surface is formed in the cleavage furrows of the newt egg,
starting at the first division. At the moment, we do not know for
certain whether synthesis of new surface is exactly matched to
division, or whether it lags behind or alternatively "overtakes"
division so that the surface pressure of the plasmalemma falls or
rises over the cleavage period.

One approach to this problem is to discover whether new surface
is formed, by measuring the mechanical properties of the surface.
Expansion of the surface would lead to altered mechanical properties.
Unfortunately contraction and expansion could take place in the
plasmalemma without the addition of new surface. Mitchison and

K

Swann (1952, 1953), Mitchison (1952) and Wolpert (1966) examined the surface mechanical properties of sea urchin eggs during cleavage stages. Some of the problems associated with the interpretation of such measurements have been discussed in Chapter 2. Mitchison and Swann (1954a,b) used an "elastimeter", a device by which a micropipette could be placed against an egg to apply suction to it. The extent of deformation of the cell under this suction provides some measure of a complex of mechanical properties, some of which probably are surface properties. Mitchison and Swann noted that the egg became more resistant to deformation as the first cleavage approached. However the interpretation of these results has proved too ambiguous (Wolpert, 1966) for any definite conclusions to be made about changes in the mechanical properties of the cell surface. The main reason for this is that assumptions have to be made about surface thickness, internal cell pressure, plasmalemma surface pressure and the non-anisotropy of such mechanical properties. Similar difficulties attach to the interpretation of Selman and Waddingtons' results.

Unfortunately these experiments failed to solve whether the cell surface undergoes expansion or contraction prior to cleavage and whether new cell surface was formed before or after cleavage. Another technique which has been applied to the solution of these problems is to observe the movement of particles attached to the cell surface. Dan (1947a) and Dan and Dan (1940) examined the movement of kaolin particles attached to the surface of cleaving eggs of the sea urchin, *Mespilia*, and concluded that the surface starts expanding at the poles and then the expansion spreads equatorially. The surface in the cleavage furrow itself starts by contracting and then expands greatly. Similar results were obtained with eggs of the sea urchin, *Astriclypeus* (Dan et al. 1938) with those of the mollusc *Ilyanassa* (Dan and Dan, 1942) and with those of the echinoid *Paracentrotus* (Mitchison, 1952) (see also Dan, 1954). Scott (1960) however observed movements of echinochrome granules in *Arbacia punctulata* and concluded that the plasmalemma actively contracts in the cleavage furrow. Ishizaka (1958) showed that rings of the cell surface on either side of the furrow remain stationary and do not expand or contract during division in eggs of the sea-urchins, *Mespilia globosus* and *Hemicentrotus pulcherrimus*. This observation is difficult to correlate with the theory of cleavage proposed by Mitchison (1952), according to which cleavage is due to surface expansion in which no part of the surface would remain stationary. However, the use of particles as markers is somewhat unsatisfactory, as they may slip or fail to move with the surface. Buck and Krishan (1965) claimed that cell surface synthesis was located in the cleavage furrow of epithelial cells of the tadpole tail tip on the grounds that desmosomes surround the interphase cell on all sides but are absent from the furrow. Such a conclusion supposes that desmosomes are formed irreversibly (see Petry et al. 1961) and cannot move on the surfaces of cells; moreover this observation might be accounted for as well if the surface in the furrow expands.

It appears from these results that we have at present very little idea as to what happens to the cell surface during the earliest stages of development. It would be of great interest to know how and when synthesis of new surfaces starts and whether it plays any part in mitosis. Labelling experiments might be of great value in this field. Similarly the part played by the surface in the activation reactions remains unclear.

Blastulation

The first phase of development during which morphogenetic movement begins is during the blastula stage. In a holoblastic egg, such as an amphibian one, a space forms inside the embryo, when a number of cleavages have taken place. This space is referred to as the blastocoel and the embryonic stage as a blastula. In embryos with blastoderms such as the avian or mammalian embryo where the embryo is flattened the blastocoel forms as a flattened cavity underneath the embryonic presumptive ecto- and mesoderm. The cells underlying the cavity are known as hypoblast cells and represent presumptive endoderm. In mammalian embryos somewhat earlier, a layer of cells, the enveloping layer or trophoblast, separates from the main body of the embryo except for its attachment by its circular periphery to form a cavity over the embryo. A variety of processes might give rise to the appearance of these two cavities. For example the blastocoel might be produced by a loss of adhesiveness by those parts of the cells facing the centre of the holoblastic embryo, or those facing the middle of the blastoderm in avian and mammalian embryos. The gap between the trophoblast and the main part of the embryo in mammals might arise by a similar process. However such a process also requires an additional force or forces to separate the now non-adherent surfaces. This force might arise in one or other of the following ways: difference in cell adhesiveness in various parts of an embryo would lead to tensional forces under suitable conditions which would tend to pull the blastocoel open (as suggested by Gustafson and Wolpert, 1961, 1962)—the weight of the cells might also contribute to this effect. Again, an increase in the internal pressure of the cells concerned would lead to their tending to assume a spherical shape—this might break down adhesions and form gaps between cells. Similarly high surface tension (low surface pressure) might tend to round cells up and thus break adhesions. It is more difficult to explain the formation of a parallel-sided cavity (like that between hypoblast and the ectomesodermal roof in avian embryos). If cells in a flat sheet become more adhesive to one another, they may increase their areas of mutal contact so that the sheet becomes thicker and of less lateral extent, or alternatively the increase in adhesiveness on each cell may be graded in a direction normal to the extent of the sheet so that the sheet then becomes curved. The second effect might explain the formation of a cavity in a blastoderm.

Perhaps the most complete experimental and observational work on the processes of blastulation has been carried out on echinoderm embryos. In these embryos the blastula is a hollow sphere with a wall one cell thick; as the blastula develops the blastocoel enlarges in volume. The egg is surrounded by the hyaline layer which consists of a clear viscous colloid, immediately outside the plasmalemma. This layer forms from the cortical granules (*vide supra*) and consists of sulphated polysaccharides (Immers, 1956). During the blastula stage it tends to split into two layers and apparently incorporates protein. Gustafson and Wolpert (1963) observed, from their time-lapse film studies of the development of the echinoderms *Psammechinus miliaris and Echinocardium cordatum*, that cleavage planes are almost always radial and that blastocoel volume increases at each cleavage. At the end of each cleavage the volume decreases slightly. During blastulation the cells are very adherent to the hyaline layer.

Dan (1952, 1960) and Devillers (1955) considered that preferential adhesion of the cells to this hyaline layer caused the cells to form a hollow ball with a wall one cell thick. Dan suggested that the swelling of the blastocoel is due to the presence of colloid in the blastocoel which produces osmotic swelling. Monne and Harde (1951) demonstrated the presence of such substances. However it is very hard to account for the transient reduction in blastocoel volume on Dan's theory. Gustafson and Wolpert point out that the volume changes can be accounted for on simple mechanical grounds. When cleavage occurs the cells elongate tangentially thus tending to increase the blastocoel volume.

Balinsky (1959) and Wolpert and Mercer (1963) have examined the nature of cell contacts during blastulation. Balinsky used *Tripneustes gratilla*, *Toxopneustes pileolus* and *Salmacis bicolor*, and the latter group continued to use *Psammechinus miliaris*. Septate desmosomes were found at the distal part of cell contacts, and Wolpert and Mercer noticed that they formed a band about 1μ wide surrounding each cell at its distal end. These desmosomes appeared during blastulation. The outer surface of the cells is highly convoluted and this may aid their adhesion to the hyaline layer just exterior to the plasmalemma. In the species Balinsky used there was considerable interdigitation of cells but *Psammechinus* apparently frequently shows deep clefts between cells on the inner side of the cell layer. The adhesion of cleaving cells by their distal portions and the position of the clefts suggests that the septate desmosomes are strong adhesive structures. It seems probable that the hyaline layer is of considerable mechanical importance in the formation of the blastula—not only, as already mentioned, do the blastomeres remain adherent to the hyaline layer during cleavage, but also destruction of it in a medium free from divalent cations results in dispersion of the embryonic cells (Herbst, 1900). However this evidence is not conclusive, because this treatment would be expected in any case to destroy cell-to-cell adhesion.

Gustafson and Wolpert (1963) point out that during the blastula stage in echinoderms the amount of cell surface increases geometrically about a hundredfold, but they suggest that new surface only appears in the cleavage furrows because the external convuluted surface of the embryo remains in firm contact with the hyaline layer. They consider that the radial plane of the cleavages helps to produce a blastula whose wall is one cell thick. Blastocoel formation is accounted for by them by the action of the strong adhesion of the cells to the hyaline layer and strong lateral adhesion of these cells to each other by separate desmosomes at their distal (peripheral) ends. Consequently the cells after cleavage tend to separate except at the distal contacts and the tangential elongation of the cells expands the blastocoel. This theory of course has not yet been tested experimentally but it does provide an attractive explanation of the observed phenomena.

In those animals such as birds or fish where the embryo forms as a blastoderm on top of a mass of yolk the early stages of embryogenesis are very different from those we have just examined. In the avian embryo the blastoderm expands (a form of epiboly) to surround the yolk in the first few days of incubation. The blastoderm is composed of two cell layers by the time the embryo has reached the equivalent of the blastula stage, and a space forms inside the embryo by the delamination of these two layers. This space is homologous with the blastocoel of other types of embryo. Almost nothing appears to be known about this delamination, other than its observable features, which suggest that a loss of adhesion on the "central" ends of the cells takes place so that a gap can then form. However these embryos undergo elaborate morphogenetic movements at the same stage. New (1959) has studied the expansion of the blastoderm in chick embryos, using experimental means. By detaching the periphery of the expanding blastoderm from its contact with the vitelline membrane, New was able to show that adhesion of the edge of the blastoderm to this membrane is essential for blastoderm expansion. If the vitelline membrane was inverted in *in vitro* culture and the blastoderm placed on its outer surface, the embryo failed to expand and instead folded back on itself, though it formed a strong adhesion with the outer side of the membrane. He also found that the cells forming the edge of the blastoderm are adhesive to the membrane, whereas those placed more centrally on the blastoderm surface are non-adherent to the vitelline membrane, and that blastoderm expansion is greatest along lines of tension in the membrane; so much so that blastoderms one cell thick may form. Two blastoderms can form strong adhesions to one another by their edges.

Spratt (1963) confirmed New's findings, and extended them by showing that if the blastoderm was inverted so that the cells of its lower layer were in contact with the vitelline membrane, blastoderm expansion was accelerated. Expansion of a normally arranged blastoderm would only happen if the upper cell layer or at least its lower half were present. He suggested that the expansion is due to "growth" (mitosis, presumably together with synthesis) at a "growth centre" in the blastoderm, probably situated in the area opaca. He had introduced this idea earlier from other evidence which will be discussed shortly, but in many ways this (1963) paper provides the most complete introduction to this idea. Spratt argues that since any procedure which interrupts the continuity between this centre and the rest of the blastoderm, stops expansion in the isolated portion, then the growth centre is the causal source of expansion. However some of his results could only be explained on this theory by suggesting that additional growth centres can form in blastoderm fragments under some conditions so that expansion of isolated pieces is not impaired. This addition to his theory runs close to admitting that there is no evidence for a growth centre. All these observations which will shortly be discussed, do not satisfactorily demonstrate the existence of a growth centre, because it may be that the blastoderm expands, as New demonstrated, under the control of mechanical tensions by the activity of its margin and that cutting the blastoderm up destroys the tensions necessary for its proper expansion.

Earlier evidence for the existence of a growth centre in the chick blastoderm was obtained in the following way. Spratt and Haas (1961a) examined the movements of carbon particles placed on isolated parts of the upper or lower surface of the very early and later stages onward of the blastoderm. They observed that the direction of movement of these particles on the lower surface of the blastoderm was centred on the posterior marginal region of it. Since it is fairly reasonable to assume that the movement of these particles represents cell movement (see p. 218 for criticism), Spratt and Haas concluded that this result demonstrated that there was a centre, the growth centre, which causes and controls these expansive movements. This conclusion is however not wholly convincing, since exactly the same result and observations would be expected on the hypothesis that the edge of the blastoderm directs these cell movements. Spratt and Haas (1961b) observed that the movement on the lower surface of the blastoderm was directed radially but with a marked preference for greatest movement along the anterior radii, while movement on the upper surface of the blastoderm was directed in a more equally radial manner. Spratt and Haas (1961a,b) fused portions of blastoderms and obtained, in some cases, single embryos when two portions containing the "growth centre" were put together, whereas when two portions which did not both contain the "growth centre" were fused twin embryos resulted. They also fused up to four whole blastoderms and found that integration to form single embryos would happen with up to three embryos. Fragments lacking the growth centre could still produce normal embryos after fusion. This last result argues strongly against Spratt's hypothesis because these embryos should have been unable to undergo the normal expansion which is apparently essential for development. Spratt and Haas postulated that normal development depends on the growth centre whose cell division produces extra cells which can then migrate radially. In those embryos in which regulation, the formation of a single embryo from two or more blastoderms, took place, they concluded that one growth centre had come to dominate the others, whereas those experimental embryos which showed little or no regulation they interpreted as being due to a failure of this interaction. Regulation was interpreted as being due to the integration of cell movements in the fused blastoderms. This hypothesis provides a superficially simple explanation of the results. Although it is obvious that development is impaired if cell movements are interfered with, this is no proof that the movements are the main cause of development and of regulation in particular. None of Spratt and Haas' experiments clearly demonstrate that the "growth centre" is anything more than the geometrical centre about which cell

movements are centered. This particular centring of the movements of expansion could be due to the action of the blastoderm
edge. In order to substantiate the hypothesis that the growth centre is
(a) a region of rapid mitosis and enlargement of cells (probably by
synthesis), and (b) a system which in some way integrates and
controls the expansion, very many more experiments will have to
be carried out. It is clear however from the work already published
that the chick blastoderm presents fascinating problems for the
student of cell movements.

As far as I can tell no experimental work has been carried out on
morphogenetic movements or on changes in cell adhesion during
pregastrular stages in other types of embryo. In consequence it is
impossible to come to any clear conclusion about the changes in
cell behaviour or cell adhesiveness responsible for blastoceol formation or blastoderm expansion. Although such changes in cell
adhesiveness probably play an important role in blastoceol formation,
as Gustafson and Wolpert have suggested, there is as yet no experimental work to test this idea. It seems likely that the adhesiveness
of embryo cells to the hyaline layer may be important in blastocoel
formation in the echinoderms but in other groups there is no hyaline
layer, nor does the vitelline membrane play any part in blastocoel
formation since it can be removed without effect on morphogenesis.

GASTRULATION

The definition of gastrulation is surprisingly difficult. To the
amphibian embryologist, gastrulation seems to be the process of
invagination of endoderm and mesoderm; to the older avian embryologist gastrulation is the process of ingression of mesoderm; yet
to invertebrate embryologists such as Hyman (1940), gastrulation is
the process of endoderm formation, irrespective of whether the
endoderm finally ends up inside or on the outside of the embryo.
Either we can adopt Hyman's definition, allowing that we might
alter it slightly to include the facts that in most groups of animals
endoderm is positioned internally after gastrulation and that in
higher metazoa mesoderm is also involved in the process. If this
definition is used we have to allow that avian gastrulation starts
with the delamination of the endoderm from the mesectoderm.

Alternatively we can adopt the conventional definitions of gastrulation appropriate to each group. Here the latter will be chosen, not because it is necessarily the more defensible attitude but because it avoids fruitless argument.

Four basic types of gastrulation have been described. The best known type is invagination or emboly, which occurs in embryos with a substantial blastocoel. The wall of part of the blastula folds

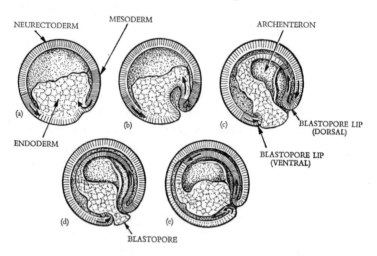

Figure 22.   Gastrulation in an amphibian, urodele. The embryo is sectioned sagitally. (a)–(e) Successive stages of the process, from earliest gastrula stage (a) to late gastrula (e). In (a) invagination of mesoderm starts at dorsal lip of blastopore, invagination of mesoderm at ventral lip begins in (c). Arrows show direction of cell movements. Redrawn from C. H. Waddington's *Principles of embryology* by kind permission of the author and publishers.

inside the other part of the wall; in animals other than in certain sponges the vegetal part of the blastula folds into the blastocoel. The space inside the invagination is known as the archenteron (see Figure 22). The second type of gastrulation is known as epiboly. The ectoderm cells move over the outside of endo- and mesoderm cells which are thus placed inside the embryo. Delamination is the third type of gastrulation in which the presumptive endoderm and mesoderm detach from the inner side of the blastula wall to form an inner mass inside the embryo. The fourth type of gastrulation is known as ingression, in which cells migrate individually into the interior of the embryo. The migration may involve cells over all or part of the blastula wall, and the migration may be routed in one particular direction or spread widely through the embryo.

Ingression and delamination produce gastrulae which can be entirely without any cavity, whereas invagination and epiboly produce a cavity, the archenteron, within the embryo. Again, invagination results in the production of a pore through which the invagination took place, and epiboly can leave a pore if the outer coat of cells failed to completely cover the endoderm and mesoderm. This pore is termed the blastopore. Metazoans can be classified as Deuterostomes or Protostomes. In Protostomes the blastopore gives rise to both the mouth and the anus, whereas in the Deuterostomes, including vertebrates, the blastopore gives rise to the anus, the mouth arising elsewhere. It should be realised that embryos do not invariably gastrulate by only one of the four methods but may do so by a combination of methods.

It is not intended here to describe gastrulation processes in detail for representative members of each class of multicellular animal, for this material is well covered in many textbooks. Instead, experimental work on the mechanisms of gastrulation will be discussed. Once again experimental work and detailed analysis of observational results is almost entirely confined to studies of gastrulation in echinoderms and vertebrates.

INVAGINATION

Invagination has been extensively studied in echinoderm and amphibian embryos. Nevertheless, little more has been done than to provide a detailed description of the process.

In echinoderms such as *Psammechinus miliaris* (see description by Gustafson and Wolpert, 1963) there appear to be three stages to gastrulation, first the ingression of the primary mesenchyme cells, second the primary invagination in which the vegetal plate part of the blastocoel wall folds in to bring itself about a third of the way across the embryo and third, action of the secondary mesenchyme cells which have attached themselves from future tip of the archenteron to the blastocoel wall and which subsequently move their attachments to the blastocoel wall towards the animal pole. By doing this and by their contraction they pull the archenteron tip till full invagination has taken place. It should be noted that Gustafson and Wolpert do not consider that the ingression of the primary mesenchyme cells is part of gastrulation. However I intend to consider this ingression as part of gastrulation since similar cell movements in other embryos are considered part of gastrulation. In normal echinoderm embryos gastrulation starts with the primary mesenchyme cells, derived from the vegetal micromeres, migrating from the wall of the blastula (vegetal plate) into the blastocoel. These cells appear to leave the vegetal plate because they lose adhesiveness and because their pseudopodal activity seems to push each other apart. After these cells have entered the blastocoel individually or as a group they pile up "on top" of the vegetal plate. The primary mesenchyme cells then protrude pseudopods freely into the blastocoel. If these pseudopods contact the wall they contract, moving the cells to new positions. At first the cells migrate away from the vegetal plate. Later, they form a ring-like accumulation of primary mesenchyme cells in a sub-equatorial position. The ring-like accumulation of cells develops two ventrolateral columns which extend towards the animal pole. The pseudopods of several cells may apparently fuse in these columns to form cable-like structures which climb up the inner wall of the blastocoel. Later on the cells in these structures give rise to the skeleton. The pseudopods give the impression of searching with their tips (up to 10) for the most adhesive site on the blastocoel wall at which they can form firm adhesions and it is for this reason that a certain pattern of these cells is produced. Owing to the difficulties of judging the exact position in space of an object from time-lapse films it is a little uncertain as to how precise is the searching and final siting of the pseudopods. By the time their pseudopods are about two-thirds of the way up the inside of the blastocoel the invagination movements which infold the blastocoel wall have started. The primary mesenchyme cells appear to play no part in gastrulation movements other than their own. When the archenteron tip is between one-third and two-thirds of the way across the blastocoel, presumptive secondary mesenchyme and future coelomic wall cells at the tip of the archenteron protrude pseudopods which anchor themselves on to the animal pole region of the blastocoel wall. By their contraction they pull the archenteron tip towards the blastocoel wall. It appears that the oral ectoderm region of the wall is most adhesive so that the archenteron wall finally touches this region. Invagination is completed by the contraction of the secondary mesenchyme cells which thus draw the invaginating wall of the blastocoel close to the animal end of the embryo.

Micromanipulation experiments (Moore and Burt, 1939) showed that the mechanisms responsible for at least part of the invagination are confined to the vegetal plate of the embryo. Gustafson and Wolpert observed that at the start of invagination the cells in the vegetal plate round up and reduce the area of their contact with one another while still remaining adherent to each other and to the hyaline layer. This rounding up of these cells causes the vegetal plate to increase in lateral area, and as Gustafson and Wolpert point out, this expansion would cause the plate to buckle inwards or invaginate, since the hyaline layer prevents it bulging outwards. Moore and Burt found that the isolated vegetal plate would show invagination; which at first sight disagrees somewhat with Gustafson and Wolpert's theory, since on this theory the isolated vegetal plate would merely be expected to expand without folding. If one supposes that a ring of cells round the vegetal plate remain firmly adherent then the expansion would produce folding, but this might again be expected to produce equally invagination and evagination. But Gustafson and Wolpert point out that the cells of the vegetal plate are most adhesive at their outer sides so folding takes place inwards and not outwards. Moore and Burt's experiment also rules out the action of any mechanism such as hydrostatic pressure differences between the blastocoel and the outside. Gustafson and Wolpert suggest that the second movement, completion of invagination (third stage) is due to the contraction of the pseudopodal cells which bridge from the invaginated vegetal plate to the animal side of the blastocoel. They suggest that most or all of the contraction occurs in the pseudopodal part of the cell. Dan and Okazaki (1956) found that in echinoderm embryos in which pseudopodal activity had been suppressed by removing calcium from the medium the second stage of invagination failed. However, removal of calcium would have other effects including weakening of cell adhesion so that it is impossible to interpret these results as necessarily giving support to Gustafson and Wolpert's theory. Gustafson and Kinnander (1960) observed that the pseudopodia explore the inner side of the blastocoel wall and suggested that the orientation of invagination was due to the pseudopodia forming strongest adhesions at the ventral side of the blastocoel.

This analysis of echinoderm gastrulation has led Gustafson and Wolpert to suggest that changes in cellular adhesiveness are responsible for the process. However, it should be realised that no experiment or measurement of cell adhesion has been carried out on these embryos which gives any information as to whether changes in adhesion are responsible for gastrulation. The magnitude of the morphogenetic forces involved in the completion of invagination appears to be $0 \cdot 1$ dyne, according to Gustafson and Wolpert (1963) who arrived at this value on the basis of the force required to elongate a cylinder of appropriate dimensions. This is a maximal value, and they consider a force of c. 1×10^{-2} dyne sufficient. Such a force could be produced by pseudopodal contraction. The total energy expended on gastrulation might be 6×10^{-5} erg over 4 hours; an energy which could easily be provided by respiration. The adhesive force required to change the curvature of a sheet could easily be provided by the adhesive forces due to the London force.

Again work on the mechanism of gastrulation in amphibian embryos has been confined almost entirely to analysis of non-experimental data. Gastrulation in amphibia involves the invagination of a tube of cells. This is very different from the situation in echinoderms in which an entire plate (the vegetal plate) is invaginated. In amphibia the invaginating tube of cells surrounds the endoderm which protrudes at the blastopore as the yolk plug. It is usual to consider that gastrulation begins when the first external signs of invagination appear, namely the development of the dorsal lip of the blastopore. Nevertheless, a number of cell movements take place before the blastopore appears and Nieuwkoop and Florschutz (1950) consider that gastrulation movements begin in *Xenopus laevis* before the blastopore appears. Anurans and urodeles gastrulate in rather dissimilar ways.

Gastrulation will not be described here in great detail, the reader is referred to the detailed descriptions of anuran gastrulation by Nieuwkoop and Florschutz (1950) and of urodele gastrulation by Vogt (1929), (see also Holtfreter, 1943b). In both groups the vegetal area, usually noticeable because of its lack of pigment, diminishes in area before gastrulation (Vogt, 1929) due to the migration of superficial cells towards the vegetal pole. This movement appears to be a type of epiboly. The first cells to invaginate begin to change shape shortly before they invaginate, and this change spreads to other cells just before they invaginate. The cells elongate slightly in the direction which runs toward the blastopore so that they appear to be oriented towards it. A little while later these cells elongate again in a direction perpendicular to the surface of the embryo with their blunt ends facing inwards in the direction of invagination in the embryo. Holtfreter (1943a,b) termed these cells "bottle cells". Both endo- and mesoderm contribute to this cell type. Holtfreter also suggested that the "surface coat" (see page 44) is almost wholly responsible for cell adhesion in the late blastula and early gastrula, so that the cells have a very slight adhesiveness. He deduced that the "surface coat" holds the bottle cells together at their peripheral ends by their rear pseudopods which are joined into a "cable". This "cable" compares with the rather similar structure in echinoderms. The bottle cells advance into the embryo as compact groups. Holtfreter suggested that (i) the shape of individual cells reflects their direction of movement, (ii) invagination is connected with a change of cell shape and (iii) the surface coat holds the cells together, may aid invagination owing to tangential stresses in it, and integrates the movements of individual cells. These suggestions do not really reveal a mechanism for the invagination of the cells even if the surface coat is connected with control of the movements of individual cells.

In this almost entirely theoretical study of gastrulation mechanisms, Holtfreter suggested that the interfacial tensions of different cell surfaces and the surface coat against each other and against the external and blastocoel fluids were different and such that the ectoderm tends to surround other tissues due to its lower surface energy than that of other tissue surfaces. He postulated that the fluid in the blastocoel had a lower surface tension than that externally, and suggested that the high pH of the blastocoel fluid would bring about this effect. These suggestions are similar to those made by Steinberg (1964a,b) to explain segregation in aggregates. Holtfreter (1944) showed that invaginated endoderm tends to spread over the inner surface of the blastocoel in explants of endoderm and blastocoel roof of amphibian embryos. This result argues against any osmotic explanation or an explanation based on the mechanical stresses in the whole intact embryo. Again Holtfreter observed that balls of cells from the blastoporal lips would invaginate into explants of endoderm, forming a depression reminiscent of the blastoporal groove. Similarly endoderm explants will engulf the mass of meso-derm they are placed on. These results would be expected if Holt-freter's hypothesis were correct although it would be necessary to postulate that the cells formed adhesions with no gap between them. However, other hypotheses predict similar experimental results, and these will be considered shortly. Holtfreter also suggested that the surface coat is under considerable tangential stress, since his experiments showed that isolated fragments of embryo which were believed to be covered with surface coat, tended to round up. Holtfreter's evidence cannot be regarded as conclusively proving or disproving his hypothesis. Holtfreter (1944) showed that isolated cells from amphibian gastrulae in tissue culture showed marked motility and a tendency to spread out on the glass surface. These two features are reminiscent of cell behaviour in the intact gastrula and may indicate that gastrulation is wholly or mainly due to individual cell behaviour and not to the action of such structures as the surface coat or to such processes as the actual distribution of stresses in the sphere which represents the blastula. Holtfreter's hypothesis is very similar to that of Rhumbler (1902).

Schechtman (1942) is responsible for a very elegant set of experi-ments on the mechanism of gastrulation in the embryo of the anuran *Hyla regilla*. These experiments bear on the question of whether or not gastrulation is due to the orderly but independent behaviour of individual cells, acting in the right way at the right time or whether it is due to an integrated action of the embryo, an action whose component parts cannot be recognised in the behaviour of individual cells. Schechtman isolated various regions of the early gastrula and found that (i) the dorsal lip when explanted by itself produces a club-like protrusion of cells but does not show invagination movements and (ii) the lateral marginal region of the blastopore lip in isolation produces blastoporal grooves and shows signs of invagination movements. He also reported that if the dorsal lip were isolated from the rest of the blastopore margin in the intact embryo by the insertion of strips of cellophane or of presumptive ectoderm on either side of the dorsal lip, then invagination of the endoderm failed. These results indicate that isolated parts of the blastoporal lip continue to show at least part of the behaviour which they normally display during gastrulation in the intact embryo. The third result shows that interaction between the various parts of the embryo is essential for the complete appearance of gastrulation but the first two results appear to show that very small numbers of cells in isolation display the same type of behaviour as they do in the intact embryo. For example the dorsal lip *in vitro* appears to be inherently capable of extension along the midline as it does immediately after invagination. However, in order that the dorsal lip can invaginate it has to be in contact with the marginal lip; the fact that ectoderm strips inserted between the two will stop invagination of mesoderm suggests that some process more complex than the transmission of mechanical stresses between the two is essential for gastrulation. In many ways these results can be made to fit well into Holtfreter's scheme but according to it there would be no difference in surface tension between nearby mesoderm cells so that small fragments of lip would not be expected to undergo invagination, and even more cogently the rod-like structure formed by the dorsal lip would be impossible. Stableford (1949) showed that the blastocoel fluid had the same surface tension as the external medium and that gastrulation took place perfectly well when the blastocoel was open to an external medium so that the same pH and surface tension occurred inside and outside the embryo. This result strongly argues against Holtfreter's theory.

Curiously Schechtman appears to be almost the only person

other than Holtfreter to have approached the problem of amphibian gastrulation mechanism in an experimental manner. A number of observations about gastrulation in this group have supported one or other theories about the mechanism. Waddington (1942) made rough measurements of the surface properties of amphibian gastrula cells by finding the strength (and surface tension) of saponin solutions required to lyse the cells. The strength of such solutions required for lysis increased during gastrulation which might possibly mean that the surface tension of the cells decreases during the process. Waddington (1939) made a measurement of the force produced by the embryo during the invagination by opposing it with an iron ball in the embryo attracted by an electromagnet and measuring the magnetic field required to just prevent invagination. He found values of c. $3 \cdot 4 \times 10^{-5}$ dyne. Shapiro (1958) measured the pH of the blastocoel fluid of amphibian gastrulae and found that it was very alkaline.

The dearth of experiment on amphibian gastrulation has been accompanied, perhaps happily, by an absence of speculation. However, very recently Lovtrup (1965) has re-analysed the existing data to produce a very detailed theory of the mechanism of gastrulation based in terms of cell behaviour, cell adhesion and difference in hydrostatic pressure between the blastocoel and the archenteron. Lovtrup's complex terminology for various cell types and forms of cell behaviour will not be used. Lovtrup suggested that the ingression or invagination at the blastopore is due to a transformation by which the invaginating cells become more adhesive. Once inside the embryo these cells move between the less adhesive endoderm and ectoderm. The mechanism for this migration is not given by Lovtrup. Cell spreading (enlargement of area of a cell in contact with others) occurs in the endoderm so that it tends to be internal. To enlarge on this, peripheral cells of the endoderm can obtain greatest area of contact with other cells by spreading over the inside of the mesoderm. Lovtrup suggested that the archenteron is formed by water being pumped into it by the amoeboid action of cells. This hypothesis suggests that movements are due to changes in cell adhesiveness and hydrostatic pressure difference. Since no one has measured the adhesiveness of all or any of the different types of cells in a gastrula such a suggestion is more than daring. Shapiro (1958) suggested that difference in pH between the outside and inside of the gastrula were responsible for the movements, since pH would affect cell adhesiveness. Schechtman's experiments and the fact that gastrulae will form even if the blastocoel is open to the medium disprove this hypothesis.

It can now be seen that our knowledge of the mechanism of gastrulation in amphibia is very unsatisfactory. Holtfreter's hypothesis is incorrect since the surface coat does not exist (see Chapter 1). However, strong adhesions in the outer part of the most peripheral cells of the embryo might produce the same effects. Schechtman's experiments disprove this idea. Those parts of other theories which postulate that differences in hydrostatic pressure between the blastocoel, archenteron and exterior are effective are also incorrect. More serious objections to both Holtfreter's and Lovtrup's arguments can be directed at their explanation of the forward migration of the invaginated mesoderm cells as being due to an increase in their adhesiveness. If the invaginated mesoderm cells become more adhesive it would be expected that they would round up into a ball. One could escape this argument by claiming that the inner side of the ectoderm and neurectoderm cells are still more adhesive than the mesoderm cells and that there is a gradient of adhesiveness from front to rear. However, in view of Waddington's measurement of the force involved in gastrulation this gradient would have to run over an improbably large range of adhesiveness. Ascribing any role to mechanical tensions in the surface of the embryo is rather futile because Schechtman's experiments disprove it and in any case they would tend to prevent invagination of the mesoderm rather than aid it. Finally, any explanation of gastrulation in terms of surface free energy differences between the various cell types involved is unsatisfactory. The reason is that the surface free energy relationships between two bodies would be such that those cells with a lower surface free energy would tend to form spheroidal or subspheroidal masses inside the cell type with a higher surface free energy. It would not be expected that the inner cell type would form a thin lamina as the mesoderm does. It could be argued that the endoderm and mesoderm behave like one unit in this process, thus forming a spheroidal inner mass, but Schechtman's experiments and the fact that the mesoderm moves forwards over the invaginating endoderm disprove this possibility.

Schechtman's experiments suggest that every hypothesis put forward so far to explain amphibian gastrulation is incorrect. Schechtman's work shows that movements can take place in isolated pieces of the embryo, which would be impossible on grounds of differences in adhesiveness or surface free energy between different parts of the embryo; for example, the extension of the dorsal lip after invagination. If small groups of cells all of the same type (or supposedly so) are able, as Schechtman showed, to undergo in isolation the same movements, foldings and extensions as they perform when in the intact embryo, then all theories which suppose that invagination is mainly or wholly due to differences in surface free energy between tissues, or to differences in hydrostatic pressure between one part and another of the embryo, or to mechanical forces in the *shell* of the embryo, or to differences in adhesiveness between one part of the embryo and another, are wrong. It is not proposed to offer any new theory of gastrulation mechanism here, but merely to indicate those features of gastrulation which have not been investigated experimentally and which ought to be worth studying.

We have no good information on the following points:

(a) Do the ectoderm and mesoderm cells change shape in such a way that local lateral contractions and expansions occur? If this happened, mechanical forces, like those which may occur in echinoderm gastrulae, might play an important part in the movements of small regions in the intact gastrula.

(b) Do changes in contact area between cells take place?

(c) What are the values of cell adhesiveness during gastrulation for each cell type?

(d) What mechanical stresses are there between groups of cells?

(e) Do contractile processes play any part in gastrulation?

(f) Do such cellular processes as contact promotion, or inhibition take place in the gastrula? Can such processes be used to explain gastrulation?

EPIBOLY

The third type of gastrulation process to be considered is that of epibolic expansion of the outer layers of the embryo over mesoderm and endoderm. Teleost embryos show this process. Gastrulation has been most thoroughly examined, in the following genera *Salmo*, *Brachydanio* and *Fundulus*. The teleost embryo forms as a blastoderm on the surface of the yolk, which is often spherical in shape. The blastoderm expands and a large subgerminal cavity forms between it and the yolk sphere except at the margins of the blastoderm which adheres to the yolk. Gastrulation continues by the margins of the blastoderm folding and moving inwards. Endoderm is invaginated before other tissues. Cells invaginate all round the margin of the blastoderm but do so more rapidly at a dorsal point lying on the future axis of the embryo. While this invagination is taking place the blastoderm expands to cover the whole of the yolk. The margin of the blastoderm is often thickened into the germ ring. Much of the more ventral part of the blastoderm becomes extraembryonic. As the germ ring closes round the yolk it forms a structure which is perhaps equivalent to the blastopore although the yolk within the blastopore is unsegmented.

Pasteels (1940), Devillers (1951) and Trinkaus (1951) showed that the blastoderm undergoes this considerable expansion and that the mesoderm cells in the blastoderm stream backwards towards the region of invagination, converging on the midline as they do so. Mesodermal cells at the margin of the blastoderm also move rapidly posteriorly to the main region of invagination. Morgan (1895) described cell thinning and expansion in the cells of the extraembryonic part of the blastoderm of *Ctenolabrus*.

The teleost egg has been described (see Trinkaus, 1951) as possessing a surface coat or surface gel layer, equivalent to Holtfreter's surface coat. Trinkaus claimed to be able to dissect the gel layer off the embryo as a continuous supracellular sheet, though as with Holtfreter's observations this may mean no more than that it is composed of very adhesive parts of superficial cells which can be torn off the remainder of these cells. Devillers (1959) and Devillers and Rajchman (1960) denied that the surface gel layer existed in any way equivalent to the surface coat of amphibia. It should be mentioned that there does not appear to be any evidence from electron micrographs as to whether this structure exists or not. The yolk is surrounded by the perivitelline pellicle (surface gel layer of the yolk sphere) which, according to Lewis (1949), plays an important role in gastrulation. Wounding experiments (Trinkaus, 1951; Devillers and Rajchman, 1960) suggest that the outermost part of the embryo has marked contractile powers, and it occurred to Lewis that the contractility might be located in this gel layer and that it might play this part in gastrulation.

Trinkaus (1951) made a meticulous study of the quantitative features of blastoderm expansion in *Fundulus*. He found that the expansion of the blastoderm is not equal in all areas over any short time interval, and that in later stages the marginal region of the blastoderm expands faster than any other region. This suggested to him that differential forces arise within the blastoderm to bring about this type of expansion. He was also able to show that the perivitelline pellicle made no active contribution of contraction to drag the blastoderm over the yolk. Trinkaus was able to get the blastoderm to expand over the periblast without any yolk gel layer being present. These experiments suggest that the blastoderm controls its own expansion. A similar conclusion can be drawn from the demonstration by Oppenheimer (1936) and Devillers (1949) that blastoderm in *in vitro* culture can undergo expansion. Devillers (1959) suggested that Trinkaus was incorrect in coming to this conclusion but no experimental evidence was put forward to justify Devillers' statement. The yolk syncytium, if indeed it be a syncytium, which lies on top of the yolk and underneath the blastoderm also spreads over the yolk during gastrulation and Trinkaus (1951) has suggested that it assists epiboly by helping to pull down the blastoderm over the yolk. Devillers (1959) demonstrated that both the yolk syncytium and the blastoderm co-ordinate to produce normal epiboly, by combining yolk syncytia and blastoderms of different ages.

The invaginating cells at the edge of the blastoderm have an appearance very similar to those of the dorsal lip of amphibian gastrulae. Trinkaus (1963b) has made a detailed study of the behaviour in culture of isolated blastula and gastrula cells from

Fundulus. He found that the cells tended to flatten in culture, and suggested that this was due to increased adhesiveness of the cells, a conclusion which we have seen in Chapter 4 may be erroneous. On this criterion gastrula cells appeared to be more adhesive than blastula cells. The flattening and thinning correlates very well with the changes observed by Morgan at the onset of epiboly in the whole embryo. Trinkaus also observed this change in whole *Fundulus* embryos. In these embryos the cells in the surface layer thin and spread out whereas the deeper cells become more tightly packed, and the deepest cells, those of the hypoblast, also spread out and thin. Trinkaus suggested that epiboly starts with an increase in cell adhesiveness so that the outermost cells flatten and spread laterally. Cell division plays almost no part in this expansion (Kessel, 1960). The periblast is the substratum for cell movement, according to Trinkaus. He also suggested that contractile processes in the marginal ring of the blastoderm and periblast help to pull these layers down the embryo once the expansion has reached the equator. Although Trinkaus' invocation of a mechanism, based partly on an increase in cell adhesiveness, will be very attractive to many biologists it should be realised that (i) his assessment of adhesiveness is entirely subjective and (ii) the spreading of cells on a substrate is very dependent on the population density of cells involved (Curtis and Varde, 1964 for chick fibroblasts). However, Trinkaus makes no correction for this in assessing adhesiveness by the extent of spreading.

INGRESSION

The last group of animals in which a certain amount of experimental work has been carried out on gastrulation mechanism, is the birds. Gastrulation is normally taken to be those movements which result in the formation of a primitive streak (Figure 23). The endoderm is either entirely (Peter, 1939) or partially (Hunt, 1937) placed inside the embryo before gastrulation sets in by the process of delamination (but see Rosenquist (1966) who showed by autoradiographic methods that a pre-endoderm region exists in the epiblast whose cells descend through the streak at gastrulation). The morphogenetic movements associated with gastrulation have been followed by placing markers of vital stain (Pasteels, 1935) or carbon

or carmine particles (Spratt, 1946) on the blastoderm. The movements of the markers have been taken to indicate the direction of
morphogenetic movements. However, various workers disagree
about the details of the movements and this is probably due to the
inadequacy of the marking methods. Vital stain marks tend to diffuse
away and particles may be "kicked" around by the cells during their

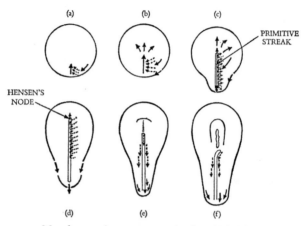

Figure 23. Morphogenetic movements in the chick blastoderm during
gastrulation, at successive stages (a)–(f). Heavy arrows indicate directions of
movement in all layers of embryo, dotted arrows indicate direction of
movement in surface of embryo only. Redrawn from Waddington (1952)
by kind permission of the author and Cambridge University Press.

movements so that they may end up on cells quite far removed from
the ones they started on. Wetzel (1929) and Graper (1929) probably
provided the first detailed description of gastrulation movements in
the chick, and their description has been largely accepted by subsequent workers, such as Pasteels (1935) and Waddington (1932,
1952). At the time this book went to press Rosenquist (1966) published a most masterly investigation of morphogenetic movements
in the early chick embryo. On the whole his results confirm the
reports of Wetzel, Pasteels and Waddington. Unfortunately there
has not been time to include a detailed discussion of his work.

Initially the cells in the midline of the surface of the blastoderm
start moving forward along the midline while those in lateral
areas move posteriorly. The primitive streak then forms in the

midline, first as a thickening at the posterior end of the embryo, which then elongates and develops a slight groove which runs along the midline of the embryo. Wetzel and Graper observed that as the streak appears the movements of cells in the surface of the blastoderm become confined to the caudal end of the embryo and these cells enter the streak. Wetzel and Graper showed that the cells which enter the streak are mesodermal cells. The movement of cells into the streak is not like an invagination in amphibian embryos because only a proportion of cells in the blastoderm surface actually invaginate. In consequence, invaginating cells migrate individually between ectodermal cells to reach the streak. No one appears to have mapped the detailed distribution of the mesodermal cells before they migrate or the exact tracks along which they migrate. Pasteels (1935) demonstrated that to some extent Wetzel's and Graper's descriptions were incorrect since the movements they described of the surface of the blastoderm actually took place throughout its thickness. He also described a tendency for the epiblast to move backward along the streak. Many of the cells also invaginate through the anterior end of the streak, namely at the rear of Hensen's node. Initially Hensen's node moves slightly anteriorly but when the node moves backwards a new streak is added at the hind end so that the streak appears to move posteriorly. Pasteels claimed that the streak "moves" forward as a result of the anteriorwards streaming of cells while Spratt (1955) opposed this idea by suggesting that no actual cell movement takes place in the length of the streak during early stages of its existence, but that changes in the position of the front end of the streak are due to inclusion of extra tissue into the streak as it stretches anteriorly and to invagination of the front end as the streak regresses anteriorly. Spratt, however, did admit that the final stages of the initial anterior movement might be due to cell movement. Spratt and Pasteels agree on the existence of a posteriorly directed movement of cells along the midline of the streak as it regresses. Spratt (1947) suggested that many cells are lost through lateral movement out of the streaks. Invagination through the node and over the sides of the streak continues during regression.

Vakaet (1960) reinvestigated the changes undergone by the streak using a carbon particle marking technique. He found that the regression is probably due to (i) a loss of cells from the node, (ii) an epibolic movement of the whole blastoderm and (iii) a final forward movement of cells at the posterior end of the streak. This description is contrary to Spratt's description (1947).

Balinsky and Walter (1961) have made a careful description of avian gastrulation at the cellular level using electron micrography. They find that cells in the top layer of the blastoderm (the epiblast) lateral to the primitive streak tend to be cuboidal in shape with a 100 Å gap between them. No evidence was found for the existence of any surface coat. The cells in the mesoblast and hypoblast were also cuboidal in shape, but cells in the streak itself were elongated vertically and tended to have a broad base at their inner (and lower) end. Balinsky and Walter suggest that these cells are in the process of invagination and that they are comparable with bottle cells of amphibian cells. They suggest that one of the important processes in invagination is the flow of cytoplasm in these cells from their exterior end to their interior end. It appeared that the cells detached themselves from the surface and migrated inwards, perhaps due to a change in their adhesiveness. No evidence for changes in the morphology of cell adhesion could be detected from the electron micrographs.

All the work described above in detail was carried out on explanted embryos using what we now know to have been rather inadequate culture methods. New (1955) introduced a considerably superior technique for culturing avian blastoderms, which makes it less likely that aberrant morphogenetic movements are taking place. No one has yet taken advantage of the technique to reinvestigate these problems except Rosenquist.

Very little experimental work has been done on the mechanism of avian gastrulation. Shoger (1960) has carried out a number of experiments comparable to Schechtman's on amphibian gastrulation. He isolated Hensen's node and found that it showed the same type of extension as the amphibian dorsal lip. Abercrombie (1950) and Shoger (1960) showed that the orientation of gastrulation movements is controlled by the blastoderm which surrounds the streak and not by the latter itself. Vakaet (1964) showed that the middle part of the streak could "induce" new primitive streaks in other parts of the blastoderm.

The foregoing description and interpretation of avian gastrulation movements have been accepted by a large number of embryologists. But Spratt and Haas (1965) have advanced a radically different description of avian gastrulation. Earlier in this chapter it was mentioned that Spratt and Haas (1961a,b) postulated the existence of a "growth centre" in the avian blastoderm. They suggested that it was the site of particularly intense mitosis, although this was never properly demonstrated. They put forward (1965) the idea that the streak is the centre of a very intense proliferation of cells as was the growth centre which existed during earlier stages of development. Pasteels (1935) has already shown that almost certainly there was no excess of mitosis in this region as compared with the rest of the blastoderm. Spratt and Haas suggested that the primitive streak was in no way comparable to the amphibian blastopore. Particles of carbon placed on the surface of the blastoderm were not normally found to be invaginated through the streak. It was on this basis, namely, "it is difficult to understand why almost all carmine or carbon marked upper surface cells of the streak remain in the upper level of the streak", that they concluded that invagination through the streak did not occur. As an alternative they suggested that the mesoderm and endoderm are already in place as the middle or lower cell layers of the blastoderm. Their conclusion seems somewhat naïve—for if the invagination has certain requirements in the adhesiveness of the cells involved in invagination it might well be that carbon or carmine particles have the wrong adhesiveness to become invaginated. Again, since the cells which are invaginated migrate individually into the streak, it may be that they are less adhesive than the cells which do not migrate, so that the carbon particles tend to stick to the non-motile ones. In the absence of satisfactory quantitative evidence of mitoses and numbers of cells entering the streak together with autoradiographic labelling of the cells to trace their migration, I incline to the view that Spratt and Haas are still far from proving their hypothesis.

It is clear that we have very little evidence on the mechanism of avian gastrulation. It would seem to be a promising field for research. If one feels that gastrulation is due to differences in cell adhesiveness it ought to be possible to collect cells which are about to invaginate, because a gentle disaggregation might release or retain them preferentially. Very telling experiments would then become possible with these cells in tissue culture. Filming the process at high power as De Haan has done with the migration of the chick pre-cardiac mesoderm (see page 320) should also prove useful. Isolation experiments like Schechtman's and experimental interference with the process should be useful.

In summary, our knowledge of the mechanism of gastrulation even in the best known embryos is very slight. Many of the latest theories about the mechanism postulate that differences in adhesive-ness explain invagination but the best that can be said in favour of this is that it looks to our possibly prejudiced eyes like this. Investi-gations of cell movements in gastrulation are almost non-existent and the possibility that the cells can be chemotactically attracted towards the site of invagination has not been studied. All that probably can be said with relative certainty about gastrulation is that it is not due to differences in hydrostatic or osmotic pressure between parts of the egg, that a surface coat plays no part in all or nearly all species, and that individual cells appear to play component but individual parts in the process.

NEURULATION

In vertebrates, gastrulation movements of cells are succeeded by those involved in neurulation, by which a central nervous system is produced in its most simple form from the dorsal and animal side of the embryonic ectoderm. Although the formation of the central nervous system also takes place at a similar time in other higher metazoan phyla the processes involved are little known and will not be discussed here. A brief description of the process of neurulation in vertebrates may be of value before discussing the causal mechan-isms. In amphibia the cells of the dorsal side of the embryo tend to elongate in the radial plane (radial plane of embryo) and in some species extra layers of cells are added to the epithelium. Pigment at the surface of the embryo appears to collect at the margins of the area in which the cells have elongated—this area is known as the neural plate and normally lies symmetrically on either side of the mid-line, dorsally between the blastopore and a point roughly above the front end of the prechordal mesoderm. Then the cells at the edge of the neural plate elongate vertically again so that the plate thickens at this position, and the cells just lateral to the plate push up into folds (neural folds). The folds on opposite sides of the plate then move towards each other, while the plate between them becomes bent with its upper surface concave. Finally the folds meet in the midline and the plate becomes folded into a tube with its former exterior surface now forming the inner side of the neural tube. The folds meet to form the neural crest which later gives rise to a population of migrating cells (see page 311), and a coat of epithelium which then covers the neural tube. Neurulation appears to be very similar in reptiles, birds and mammals, but in fish the neural folds are almost absent and the neural plate sinks as a block into the underlying tissues without rolling up into a tube (although later it forms a lumen within it so that it finally has the same gross appearance as the neural tube in other vertebrate groups).

As in so many other cases the majority of experimental work has been carried out on amphibia, and as with gastrulation the same wide variety of mechanisms have been proposed to explain neurulation. All the following results come from work on amphibian embryos. His (1874) suggested that it results from "growth pressures" due to excessive mitosis in the neural plate, but Gillette (1944) showed that although there is a 23 per cent increase in the number of cells present in the neurectoderm during neurulation in *Amblystoma maculatum* there appeared to be no differential proliferation in various regions of the neurectoderm and surrounding ectoderm. This conclusion of Gillette of an absence of differential mitosis was not based on mitotic counts but instead on the parallel decline in cell size in various regions of the neurectoderm. The various regions which were compared were very large so that localised increase or decrease in cell size in say the folds or plate might not be noted (cell size was calculated from the total volume of each quarter of the neurectoderm divided by the number of cells in a given quarter). Although Gillette was justified in using cell size as a test of His' hypothesis, because the "differential pressures" consequent on mitosis His proposed, could only arise from change in cell size, Gillette's methods were too imprecise to make a definitive test of the hypothesis. Bragg (1938) showed that no differential mitotic activity could be associated with neurulation in a toad, *Bufo* sp. but failed to measure cell size.

The obvious way in which mitosis would be associated with volume changes in parts of the embryo would be if the daughter cells formed by a mitosis undergo enlargement due to synthesis or osmotic effects.

Brown *et al.* (1941) showed that there were almost no density changes of cells during neurulation of amphibian embryos which further confirms Gillette's result, but once again the values obtained were average values for large regions of the embryo. These results do go some way, in their authors' views, to suggest that Glaser's hypothesis (Glaser, 1914, 1916) is incorrect. This hypothesis states that neurulation movements are due to the swelling of cells, in particular the swelling of those parts of cells which are on the inner (ventral) side of the neural plate (see Boerema, 1929). Of course it is certainly correct that the ventral ends of these cells do become larger than the dorsal and superficial ends, but it is not clear that this swelling is the cause rather than the consequence of neurulation movements. Although Giersburg (1929) was able to inhibit neurulation movements with hypertonic solutions it is not clear whether the result was really due to interference with cellular swelling. Moore (1940, 1941) considered it improbable that osmotic effects could account for neurulation movements, as a result of examination of model and experimental systems. The idea that a "surface coat" (Gillette, 1944) or a "gel layer" (Lewis, 1947) contracts locally to produce neurulation appears to have little support as we have seen in the discussion of gastrulation.

Operative techniques have been used but little. However, C. O. Jacobson (1962) has shown that if the entire plate is removed neurulation continues by closure of the folds over the mesoderm and notochord. But if the notochord primordium is removed or if the archenteron roof is cut out, the movement ceases. The latter result is hardly surprising but it is of considerable interest that removal of the notochord primordium stops closure of the folds over archenteron roof considerably lateral to it. These results strongly suggest that the neural plate plays a passive role in closure of the folds, and that the notochord is of considerable importance. Fascinating as these results are they exist too much in isolation for any major conclusions to be drawn from them.

Recently Jelinek and Friebova (1966) revived the idea that mitosis is responsible for the closure of the neural folds, despite Jacobson's experiments (of which they appear to be unaware). They measured the number of mitoses in the neural plate of the chick at various times during neurulation, the surface/volume ratio of the cells and the gross dimensions of the neural plate. They found that the mitotic rate (presumably averaged over the whole plate) is fairly constant and of high value until late in neurulation, and the surface area rises relative to the volume of the cells. They conclude that the total volume of all the cells rises as neurulation proceeds but since they provide no data on total cell number at any stage this is not at all certain. Moreover, the rise in surface area relative to volume suggests that the cells are getting smaller, whereas Jelinek and Friebova believe that each daughter cell of a mitosis is as large as its parent cell. They suggest that mitoses produce the expansion which results in the forces of neurulation. Even if their data supported their hypothesis more conclusively it might be that what is observed is the result of neural fold action rather than the cause of neural fold movements.

Selman (1958) suggested that measurement of the magnitude of the forces involved in neurulation should allow us to discard those theories which make incorrect predictions of the magnitude of these forces. This approach is of considerable interest and originality. Selman placed two iron dumb-bells, one against each neural fold, and arranged that they should repel one another when an electro-magnetic field was switched on. In the experiment the separation of the dumb-bells was held constant so that the neural folds failed to move towards each other, and the magnetic force which had to be applied to do this was measured. Consequently the measurement of this magnetic force gives a measure of the force which can be exerted by the embryo, allowing that no slip occurs between the dumb-bells and the embryo. Selman was also able to take some

account of the elastic compression of the tissues in measurement of
the actual force of neural closure. The maximum force which an
axolotl neurula could exert was about 9×10^{-2} dyne, whereas the
neurula of *Triturus alpestris* could exert a maximum force of $4 \cdot 5 \times 10^{-2}$ dyne. The magnitude of energy requirements was about 10^{-2}
erg for the closure of the neural folds. This value is of considerable
interest in that it is at least 100 times greater than the energy which
could be derived from an increase in either the area of contact
between all the cells of the plate and folds or in their adhesiveness
(assuming a 50 per cent rise in either value and an adhesive energy
of around 10^{-4} erg/cm²). An alternative approach is to consider the
problem in terms of the hypothesis that the energy is derived from
changes in surface free energy resulting from changes in cell shape.
On this basis, assuming as above that the cells involved in neurulation have a total surface area of 1 to 5 cm², an energy of between $0 \cdot 5$
and $2 \cdot 5$ erg would result from a 50 per cent change in surface area.
Tentatively these calculations suggest that this hypothesis does not
apply to neurulation: similarly if neurulation movements are due
alone to an increase in the area of cell contact or their adhesiveness
when cells adhere in the secondary minimum, the calculated energy
is far smaller than that experimentally found. Possibly the solution
to this problem is to search for contractile movements in the cells
involved in neurulation.

Very little biochemical work has been done on the mechanism
of neurulation. Ambellan (1955, 1958) showed that treatment of
Rana pipiens embryos with adenosine triphosphate, diphosphate and
monophosphate at pH $5 \cdot 7$ accelerated neurulation movements.
Since some of these substances probably affect cell adhesion (see
Chapter 4) it is possible that their action is to be attributed to this
effect though other possibilities come to mind easily.

As neurulation proceeds the embryo often elongates in the
anterio-posterior direction. C. O. Jacobson (1962) suggests that this
is due to the action of the notochord underneath the neural plate,
although Jelinek and Friebova (1966) believe it to be due to the
expansion of the neural plate itself. The actual movements of
individual cells in the neural plate (Jacobson, 1962) are more complex
but highly ordered. Cells move towards the axis as neurulation

proceeds, and have greatest posterior or anterior movement when they started in posterior or anterior parts of the plate. The neural plate develops into the central nervous system and the further development of the plate will be discussed under that title; the neural folds give rise to the neural crest cells, whose behaviour will be considered in the next section.

The neural crest

In vertebrates the neural folds (or their equivalents in fish) meet in the midline at the close of neurulation to form the neural crest. Shortly afterwards the cells of the crest disperse by migrating to many different sites throughout the embryo. It is not intended to discuss the vast literature concerned with the identification of those cells which are of neural crest origin because that literature is not concerned with the mechanisms whereby migration and siting of the cells takes place. However it should be mentioned that by 1950, when Horstadius's monograph on the subject was published, it was recognised that the pigment cells (melanophores, xanthophores and guanophores), cells of the fin fold in anamniotes, the Schwann cells, spinal ganglia, the primary and secondary sympathetic nerve chains, the paraganglia, the chromaffin cells, parts of the cranial nervous system, many of the cartilage cells in the branchial skeleton, part of the meninges and part of the subcutaneous connective tissue were probably wholly or in part of neural crest origin. Three recent papers of importance which have dealt with the identification of cells of neural crest origin are those of Triplett (1958), Nawar (1956) and Weston (1963). Little else appears to have been published on the subject since 1950. The various types of neural crest cell appear to migrate to specific sites. Although pigment cells have a rather wide distribution throughout the body of anamniotes they frequently take up spatial patterns in the epidermis and dermis of both anamniotes and amniotes. In anamniotes and some amniotes this distribution is directly responsible for the visible pigmentation of the animals but it seems likely that their effect is slightly more indirect in some of those animals with feathers or hair. In these cases the pigment cells may only synthesise pigment at certain stages of the hair's or feathers' growth leading to a banded or tipped effect. Consequently

L

it can be appreciated that the siting and patterning of all the types of cells of neural crest origin present a number of fascinating problems for study. Unfortunately, nearly all the work on the control of migration and pattern formation by these cells has been confined to pigment, and to nerve cells to a lesser extent. As far as I am aware no studies have been made on the migration and siting of chromaffin cells, the paraganglial and the meningeal cells, the cells of the branchial skeleton and those of the connective tissue.

THE INITIAL MIGRATION FROM THE NEURAL CREST

Surprisingly little is known about the mechanisms whereby the neural crest cells start migration. Although Twitty's finding (1945) that propigment cells tend to repel each other in culture (described in Chapter 5) could be used to suggest that migration starts by the appearance of such a mechanism, further reflection shows that this is unlikely. If the cells move only when there is a concentration gradient of the substance across them then the greater the population density the smaller such concentration gradients will be across one cell, and the less the movement. When the cells are still packed in the crest concentration gradients will be slight even in a direction at right angles to the embryonic axis. However such a gradient might be sufficient to start migration. But Twitty (1945) showed that propigment cells will move from grafts on the flank dorsalwards directly contrary to the expectations of the repulsion theory. Moreover, Twitty (1945) showed that propigment cells will migrate towards a neural crest graft in the flank. Whatever the cause of migration a number of different routes are followed by the migrating cells. It was suggested by Ris (1941) that migration occurs predominantly on the surface of massive structures such as the lateral walls of the neural tube; but Weston (1963), who was the first in this field to take advantage of radioactive labelling, found in chick embryos that migration took place mainly through the somitic mesenchyme with only few cells travelling on the surface of structures. Since few cells travel through the gap between somites a sort of secondary segmentation of the neural crest cells takes place. Weston suggested that this effect is responsible for the segmentation of spinal ganglia. Weston also found that neural crest cells reached the site of the sympathetic chains before spinal ganglia were formed —this result disposes of the earlier claim (Brauer, 1932) that the spinal ganglia determined the location of the sympathetic ganglia. Ris suggested that the propigment cells (melanoblasts etc.) migrate between the ectoderm and mesoderm but Weston observed that the propigment cells probably migrate within the ectoderm (see also Fox, 1949). Weston suggests that the migration involves at least three factors which can tentatively be recognised at the cellular level: (i) a component which orients the cells as they leave the crest; possibly this may be a contact guidance phenomenon, (ii) penetration through an apparently tightly packed cellular environment; this may have relations to the mechanisms of local penetration of invasive tumour cells, and (iii) a directional component which imparts part of the directionality to the migration. Weston put forward the idea that if the crest cells possessed a contact inhibiting property to each other and not to other cells then directional migration would take place.

THE BEHAVIOUR OF PIGMENT CELLS

The greater part of published work on pigment cell behaviour has been directed to the study of these cells in the epidermis of amphibia, particularly urodeles. It should be remembered that nearly all work to date on melanoblast, xanthoblast and guanoblast cell behaviour has depended upon the recognition of these cells when they have pigmented but nearly all experiments have to be carried out before pigment is formed, because pigmentation does not take place until movement of these cells has ceased or nearly ceased. Should cells fail to pigment in certain situations they probably would not be recognised as reaching that position. Lehman and Youngs (1959) were able to demonstrate that melanoblasts in *Tarichia torosa* actually reach the ventral part of the trunk of newts but that they fail to develop melanin in this site. Consequently there are two aspects to the problem: (i) the control of the positioning of pigment cells and (ii) local control of the development of pigment within the cells. Although this latter problem lies outside the strict limits of this book it will be discussed because it is intimately involved with evidence on the former problem.

Twitty (1945) and Twitty and Niu (1948, 1954) were the first workers who approached the subject in terms of the behaviour of individual cells. Their work on cell behaviour has already been described in Chapter 5, but it will now be applied to explain the patterns of pigment cells in the animal body. Twitty (1945) showed that the propigment cells probably mutually repel each other by some negative chemotactic mechanism. Twitty and Niu (1954) showed that the dispersal of propigment cells (melanoblasts only) in culture was in relation to relative rather than absolute concentrations of the repulsion factor, and that the substance appears to be a direct activator of movement. This mechanism would have the effect, as Twitty claims, of increasing the spacing between propigment cells as the population grows in size, or it would also have, slightly contrarily, the effect of tending to prevent the cells from starting migration from the neural crest. The repulsion effect would tend to cause the pigment cell population to migrate and space out into areas previously unoccupied by pigment cells. This suggestion then meets DeLanney's (1941) discovery that the propigment cells from the anterior neural crest migrate out from the crest before other propigment cells and tend to migrate rearwards into other areas unoccupied by pro-melanophores. Other workers such as Rosin (1943) and Holtfreter (1947a) had suggested that propigment cell migration might be due to positive chemotaxis acting in a dorso-ventral direction. If this were so, grafts of crest to other regions of the body should give rise to a ventral movement of pigment cells, but Twitty's finding that pigment cells can move dorsad partly argues against the action of such a mechanism. Negative chemotaxis provides a good, though not wholly satisfactory, explanation of the initial distribution of melanoblast propigment cells evenly over the whole of the flank of a urodele, a distribution which persists in some species with of course the additional factor that melanisation does not take place ventrally. Unfortunately the behaviour of guano- and xanthoblasts has not been worked out.

However, it has been shown that other factors control the appearance and distribution of melanoblasts, xantho- and guanoblasts. Dalton (1950a) investigated the similarities between the melanoblasts of black and white axolotls. He found that melanophores from either mutant would synthetise melanin *in vitro* and claimed that these cells had equal migratory activity in both types of axolotl. However his data suggest that propigment cells from white mutants were less migratory than those from black animals, though he did not measure the speed of cell movement. At this time he suggested that pigmentation in axolotls is partially controlled by the tissue environment. But in his next paper (Dalton, 1950b) he concluded, in view of grafting of neural crest between the two mutants, that his earlier deductions were incorrect and that there is a factor in the epidermis of white axolotls which hinders cell migration. Dalton (1953) and Dalton and Krassner (1956, 1959) found evidence that there were differences in pituitary activity in black and white axolotls and that this affected pigment cell number and size, although probably not distribution. Landesman and Dalton (1964) continued this interesting line of research by examining the migration of propigment cells in balls of ectoderm and of ectoderm and somatopleure. The cells appeared to be able to migrate in both tissues and in endoderm as well. However, the postulated anti-migratory factor might not be formed or might be lost under the conditions of culture.

Twitty (1945) analysed the origin of pattern (melanophore only) on the flanks of *Tarichia torosa* (formerly *Triturus torosus*) and in axolotls (*Amblystoma mexicanum*). *T. torosa* has a horizontal band of melanophores on each flank whereas *A. mexicanum* has a few vertical bars of pigment on each flank in the black mutant. Twitty found that the propigment cells of *T. torosa* disperse at first over the whole of the flank of the larva but then reaggregate into bands. In culture the propigment cells at first emigrate from explants and then reaggregate into clumps. A similar type of behaviour was found in tissue culture of axolotl propigment cells. Although the explants composed a number of cell types these experiments demonstrate that a major factor responsible for pattern formation, namely aggregation, is probably inherent in the melanoblasts themselves.

Lehman (1950, 1953, 1957) and Lehman and Youngs (1959) carried out further studies on the development of pigment cell pattern in these two animals. In axolotls aggregates of xanthophores are found between the melanophore bands. Lehman (1951) demonstrated that once pigment synthesis is complete in the melanophores the cells no longer repel each other. Twitty (1945) had already shown that xanthoblasts reaggregate before the melanoblasts (confirmed by Lehman, 1957), although they emigrate from the crest after the melanoblasts. Lehman (1957) also made a most interesting observation in tissue culture, namely that isolated xanthoblasts or melanoblasts lose their adhesiveness to a substrate when surrounded by a number of cells of the opposite type. These experiments and observations suggest, as Lehman points out, that the banding in axolotls is produced by two factors: one, the aggregation of melano- and xanthoblasts and the other, the mutual interaction between them which results in the loss of adhesion (and presumably migration) of that cell type which is present at lower population density. Lehman and Youngs (1959) claimed that the underlying mesoderm played no part in the distribution of pigment cells in view of his experiments of grafts of *T. torosa* melanoblasts to larval axolotls.

However Finnegan (1955, 1958) carried out grafting experiments on *T. torosa* and *Amblystoma punctatum* (an unbanded species) in order to investigate more closely whether the mesoderm had any effect on pigment cell distribution. It is noticeable that the banding in *T. torosa* lies close to the boundary of the hypomere mesoderm. Finnegan found that grafts of neural crest, whether homo- or xenoplastic placed at or below the yolk border line (marking the edge of the hypomere) underwent an extensive dorsal migration and a very short ventral migration. Pigment cells tended to lie dorso-ventrally elongated. *Torosa* hosts allowed considerable dorsad migration of propigment cells from homoplastic grafts and greater migration from grafts of *A. punctatum*, but when *A. punctatum* was used as a host *torosa* propigment cells remained almost stationary in the graft, and *A. punctatum* melanoblast cells themselves underwent a limited migration, whereas guanoblasts migrated extensively. Finnegan suggests that these results show that the hypomeric mesoderm carries a factor inimical to the migration of melanoblasts in both species, though no test was made of the possibility of there being unpigmented "pigment" cells ventrally. However, I do not feel that these results definitely implicate the mesoderm, although they do demonstrate that there is apparently some barrier to ventral penetration of melanoblasts or is it in view of Lehman's results not a barrier to migration but merely that the ventral region contains a factor which prevents development of melanin? It can be appreciated that results which definitely establish one or other of these two theories have not yet been produced.

There is considerably less information about the behaviour of propigment cells and the establishment of pattern in other groups of vertebrates. There appears to be no information available on pigment cells behaviour in fish and reptiles. Rawles (1945), Foulks (1943) and Willier and Rawles (1940) showed that each generation of feathers requires a fresh supply of melanophores from the dermis. Consequently the avian dermis might become depleted of propigment cells unless they could in turn be recruited from other sources. Rawles showed that such cells can be recruited from the population in the peritoneal lining. By grafting it was possible to provide a peritoneum genetically different from the dermis so that after several moults the bird changed feather colour to the type characteristic of the peritoneum propigment cells (when in a feather germ).

Rawles (1948) also found that the extent of melanoblast migration in various breeds of chicks varies and that grafts of melanoblasts to a host of different breed may result in enhanced or diminished migration (see also Fox, 1949). The avian dermis probably effects a suppression of melanin formation so that the pigment cells cannot be detected until after they enter the epidermal primordia of the feathers. It has also been shown that even white breeds possess melanoblasts, but that these cells die in these breeds before they can deposit pigment. Grafts of melanoblast free skin transplanted into normal hosts are rapidly invaded by host melanoblasts, but grafts containing melanoblasts are not invaded by host propigment cells (Rawles, 1944, 1945). This result immediately suggests that a situation in which propigment cells repel each other may also exist in these birds. Although detailed study of the behaviour of propigment cells in different breeds of chicken has not been carried out it appears probable (Rawles, 1948) that the details of patterns of feather formed by the different pigmentation of different feathers is due to changes in the synthesis of melanin at different sites and times rather than to any feature of cell migration.

Watterson (1942) found that propigment cells can only enter the feather germ during a very limited period during its formation, though the block which is formed after entry to further melanoblast invasion can occasionally be broken by experimental means.

In mammals there is some evidence (Rawles, 1947) that the hair priordia attract melanoblasts. By this means their spread from the neural crest is impeded so that the belly tends to remain unpigmented.

Mesoderm cell movements

NOTOCHORD

The action of the notochord in elongating the embryo during neurulation has already been noted, but the mechanism by which the notochord extends itself will now be considered. Mookerjee et al. (1953) examined the adhesiveness of notochord cells of *Amblystoma tigrinum* and various newts by subjective methods and also described an increase in contact area between notochordal cells during development. At an early stage of development of the notochord its cells are very flattened and thin along the anterior–posterior axis but later they thicken along this axis so that the cells move mainly posteriorly, being held in place laterally by the chordal sheath. The cells appeared to develop increased adhesiveness during notochord development. This analysis appears to be one of the few pieces of work which has clearly revealed cell swelling as the main source of morphogenetic movement (passive).

SOMITE FORMATION

As so often before in this chapter it is incredible to note how little work has been carried out in this case on the morphogenetic movements which form the somites, though a considerable knowledge exists of the factors which will cause them to appear in unusual sites. As neurulation proceeds the amphibian or the avian embryo undergoes considerable lengthening, a process due to the presence of the notochord (Kitchin, 1949; C. O. Jacobson, 1962; Waddington et al. 1953). Probably the notochord pulls the rest of the embryo out as it elongates by a process of cell enlargement which is confined to its longitudinal axis. This elongation might be expected to have an effect on the structure of such tissues as the mesoderm, and this may explain the formation of somites. Waddington *et al.* (1953) suggested that an increase in adhesiveness of the pre-somite cells would, if it spread rearwards, lead to the formation of somites. Considerable changes in cell shape are visible during somite formation. Deuchar (1961) has described an initial change of the mesoderm cells from cuboid to a fusiform shape. These cells then become arranged in concentric layers, those belonging to a future somite all centering round a myocoel in a *Xenopus* embryo. Later these myoblasts and fibroblasts cells elongate, interdigitate and become oriented in an anterio-posterior direction. As a result the area of contact between cells which are going to lie in one somite tends to rise. Finally the cell contacts between one somite and the next are broken, possibly due to the extension described above, possibly due to the contraction of the cells in the somite. This description is of great interest and could only be bettered with quantitative data on the changes in shape of cells, their contact area and their adhesiveness. Deuchar investigated the adhesiveness of the cells in a non-quantitative way by observing their reaggregation after dispersal and found that ATP enhanced the rate of aggregation. Unfortunately no measure was made of the amount of ADP in the material used, since this may be the main factor which increases cell adhesiveness (see Chapter 4), on the other hand the ATP may have mainly acted on cell movement which is necessary for aggregation in these systems. The mesoderm cells are rich in ATPase. Frazer (1960) also examined the aggregation of presomite cells in the chick and found that they aggregated well and that aggregation was best in the presence of neural tissue. This is of interest in two ways: first, because Frazer presented other evidence that neural tissue in some way controls somite appearance in the chick mesoderm, and second, in the hope that further work will isolate this factor and reveal its effect on cell behaviour. As a result of the work of Deuchar and Frazer it is tempting to regard somite formation as being an aggregation. However, this idea raises further questions such as how is the centre of a somite determined so that aggregation occurs around it, and how the somites on either side are so precisely matched that they lie opposite each other. Kitchin showed that the matching of somites in register on either side of the axis is not controlled by the notochord. It is obvious that this small amount of work raises a number of points of great interest from which further research should be started.

THE CARDIAC PRIMORDIA

In vertebrates the mesoderm cells which give rise to the heart migrate *en masse* to their final sites (see description for amphibia by Wilens, 1955). De Haan (1958, 1963a,b,c, 1964) has investigated the movements of the precardiac cells in chick embryos. He found that the heart mesoderm forms aggregates which then migrate as independent and integrated units on the endodermal substrate to their final positions. Initially it appears that the clusters move randomly but finally take up a position in the crescentic cardiac region. Using time-lapse film methods (1963b) and operative techniques (1964) De Haan found that the clusters appear to follow oriented tracks towards the midline of the heart region. He postulated that the underlying endoderm cells may form these tracks. Other possibilities are that the movement is a chemotactic one or that the movement is more random than De Haan claims with trapping stopping the clusters in the right position. (De Haan carried out no statistical analysis of the randomness of this movement.) He also discovered (1963c) that the pulsation rate typical of each part of the heart is determined prior to this migration. This result suggests that the actual positioning of each cluster is accurately specified. In 1964 he published a paper in which he showed that if explants of endoderm and mesoderm were placed end to end so that the anterior ends were in contact (in other words each explant in opposite orientation, then no mesoderm migration took place whereas it did occur if both explants had the same orientation. De Haan suggests that this experiment indicated that a mechanism similar to contact guidance is responsible for this migration, but the "relay-race" type of chemotaxis found by Shaffer (see Chapter 5) would also account for the phenomenon. It should be realised that the endoderm also migrates, though more slowly, in the same direction as the mesoderm; the relevance of this movement to that of the mesoderm is obscure. Rosenquist and de Haan (1966) have published an excellent and detailed study of heart formation in the chick using autoradiographic methods. They confirmed much of de Haan's earlier work but found that the prospective cells of the epimyocardium, unlike those of the endocardium do not move randomly, but migrate and fold as a coherent sheet.

Cell migration and adhesiveness in other tissues

Bellairs (1953) provided a detailed description of cell migration in the development of the foregut in chick embryos. Cells migrate forward along the midline while other cells move backwards obliquely more laterally and downwards: as a result ridges of cell form laterally and finally they meet medially, thus forming the foregut by building a sheet of cells on the ventral side of the embryo separating the gut from the yolk sac.

The adrenal cortex has been suggested as the site of continuous cell migration, with the migrating cells replacing the inner side of the cortex. Although a considerable amount of work has been done on the question of whether or not this migration took place all work carried out without the use of autoradiographic labelling of cells will be ignored: indeed the very confused nature of the subject until recently particularly underlines the need for the use of cell labelling methods. Messier and Leblond (1960) used thymidine H^3 marking to study migration and proliferation in the rat and mouse and reported that the adrenal cortex probably was not a site of cell migration. Similarly, Walker and Rennels (1961) found no evidence that cells migrate from the periphery of the cortex in three weeks. Brenner (1963) examined the adrenal cortex in mice after stressing by injection of carbon tetrachloride, using tritiated thymidine labelling. He found that considerable centripetal cell migration took place in these stressed animals. Brenner suggested that the cells were not responding to carbon tetrachloride treatment and that migration took place normally.

Messier and Leblond (1960) also found evidence for cell migration at other sites in adult male rats, for example the lumenward movement of epithelial cells lining the stomach and intestine.

THE SKELETON AND ARMS IN ECHINODERMS

A very painstaking description of the formation of the skeleton in echinoderm embryos is due to Gustafson and Wolpert (1963). The ectoderm of the arms is passively extended by the growth of the skeleton: and a plug of primary mesenchyme lies on top of the skeleton inside each arm. These cells transmit the force of extension of skeleton to the arms. The skeleton is laid down by the primary mesenchyme cells. The primary mesenchyme cells are placed in a ring inside the ectoderm near the vegetal pole and have two ventrolateral "horns" soon after gastrulation starts. The pseudo-pods of these cells selectively explore the ectodermal wall attaching perhaps at their most adhesive sites. Initially the skeleton appears as two triangular crystals of triradiate structure which are randomly oriented and situated one in each junction of the ring with the ventro-lateral branches. Soon afterwards the skeletal rudiments are reoriented so that the three radii point in the correct directions. Since three cables of pseudopods meet where the crystal centre is, it seems likely that tensioning them appropriately orients the skeletal rudiment correctly. In turn the position and tension of these pseudo-pods will depend upon the siting of the mesenchyme cells on the inside of the ectoderm. Subsequently the skeleton follows the cable-like pseudopodal complexes. The pseudopod cables lead to the arms and the skeleton grows in these directions. Thus the skeletal pattern is controlled by the pseudopods of the primary mesenchyme and these in turn perhaps by the quantitative adhesive differences of different parts of the ectoderm inner surface.

PATTERNING OF CELL CONTACTS IN THE NERVOUS SYSTEM

The anatomy of the nervous system is such that sense and motor cells are connected by a number of neurons. As development pro-ceeds the neurons extend and develop fine processes which make contact with each other and with muscles and sense organs. These neurons form connections with other nerve cells, particularly in the central nervous system, such that very large numbers of them may transmit impulses when certain sense cells are stimulated, and these impulses may initiate a whole range of motor behaviour. The connections between axon and axon, or axon and sense organ or muscle show a variety of morphology (see Chapter 3). In many

cases a specific sensory stimulus produces a constant and specific type
of behaviour; similarly sensory data are usually brought into the
central nervous system in such a way that they accurately portray
the spatial patterning of stimulation, e.g. vision. Both these features
require that there is a mechanism whereby the sensory data are
accurately analysed and the appropriate motor action be correctly
synthesised every time the sensory data are presented. Since sense
data from different organs are frequently received by the central
nervous system by morphologically separate nerves entering the
central nervous system at different points, and since motor nerves
to different muscles may leave by different points, the nervous
system cannot be pictured as a system which receives sensory data in
an entirely random spatial pattern. One theory to account for the
non-random results of the working of the nervous system is that
there is an exact or nearly exact plan of connections between axons
such that there exists a very specific series of pathways for nerve
impulses to travel in. If this theory is correct it is of considerable
interest to discover whether it is due to processes like those which
establish the position of other cell types in embryos. However, the
nervous system might be organised in other ways. For example,
each sensory receptor may produce a certain temporal pattern of
impulses such that the pattern is different from those of other sensory
receptors. This is effectively the pattern theory of Weddell and
Sinclair (Sinclair, 1955). They maintain that there are no specific
fibres or endings and that the spatially and temporally dispersed
input to the brain is sufficient to account for the ability of the brain
to act accurately on sensory data. If this theory is correct then the
problem, from the point of view of this book, of establishing nerve
connections is much less difficult than under the previous theory
because no morphological specificity is required; it merely being
sufficient to somehow bring a sensory nerve from the brain (or in
the case of the eye and olfactory organs to the brain) to the sense
organ. This does not, of course, explain how motor innervation is
controlled but it might be postulated that given muscles respond to
given signals or that proprioception would provide a sensory feedback to control the action of specific muscles. But, as Melzack and
Wall (1962) point out, the pattern theory is somewhat vague and

inadequate because it does not specify what these patterns are or how they originate. In their review of a large amount of neuro-physiological work which is not appropriate to this book, they show that there is a large amount of evidence that temporal and spatial patterns provide the basis of sensory perception. They suggest analysis of these patterns may start by filtering presynaptically at the terminal arborisations, and that threshold, temporal and spatial summation and adaptation properties of central cells may also play their parts in analysing and ordering sensory data. It thus becomes clear that investigation of the specificity of the mechanisms whereby nerve connections are made is closely integrated with theories of nervous system operation, and that each field of study is important to the other. If one or other of the temporal pattern theories is correct it is not necessary to assume that highly specific connections must be made. A whole range of processes such as quantitative differences in adhesiveness, perhaps changing in a certain temporal pattern, specific adhesion, contact guidance and contact following, chemotaxis, and contact inhibition etc. may play their part in estab-lishing connections. Some of these processes might be inapplicable if it could be shown that nerve connections are formed with a very high degree of specificity.

Three of the theories which have been proposed to account for a fairly or very specific innervation are based almost entirely on specu-lation.

The first hypothesis is the so-called "learning" theory, namely that as the nervous system becomes electrically active (or in sections of the system which are regenerating) a random or semi-random activity is converted by use into one with ordered activity. Un-fortunately it is not clear what sort of mechanism might produce this result. Is it meant that actual morphological rearrangement of connections takes place or that functional blocking or opening of certain connections occurs? The chief evidence against this theory comes from the fact that amphibians which have developed in the presence of an anaesthetic from gastrulation show normal reflex behaviour very shortly after the anaesthetic is removed in late larval life (see Carmichael, 1927). However, such an experiment is not completely rigorous. In any case such a theory does not account for the major outgrowths of the nervous system and the fact that nerves reach or leave sensory organs and reach muscles.

The two rather similar theories of "neurobiotaxis" (Kappers, 1907) and "stimulogenous fibrillation" (Bok, 1915), proposed that the electrical activity of the nervous system affects the direction of outgrowth and the establishment of connections between axons. Both theories were proposed in the main in order to account for the development of tracts and nuclei within the central nervous system. Although Ingvar (1920) claimed to be able to obtain reaction of nerve cells in tissue culture to weak D.C. electric currents, subsequent workers were almost entirely unable to obtain any such effect (P. Weiss, 1934). However, in view of modern knowledge about electrophysiology, Weiss's experiments are not so conclusive as they once seemed because it may be that orientation of axon lengthening and the establishment of connections may require certain types of pattern of electrical impulse which hitherto have not been applied in these tissue culture tests. If cell surfaces are like polarisable electrodes (see Chapter 2) electrical impulses might alter cell adhesiveness.

A fourth theory, for which a good deal of experimental evidence has been put forward, has been proposed by Sperry (1942, 1951, 1958). He claims that there is a range of chemical specificities in the population of neurons or nerve endings such that connections will only develop between certain pairs of neurons or between given sense or motor cells and given neurons. It is supposed that these chemical specificities are arranged in such a way that each nerve or sense cell connects through a specified series of neurons to the central system, and that the specified pathways are just those which produce normal behaviour. This theory of course goes entirely against the evidence that temporal patterning is required so that the central system can "recognise" a given sense organ. In Sperry's earlier papers the exact nature of this specificity was not explained but more recently (1963) he has suggested that it is a specific chemotactic system. It should be realised that this hypothesis predicts the existence of a pair of unique chemicals or unique chemical systems for every functional neuronal connection in the animal body.

Owing to the extreme complexity of the nervous system it is at present very difficult to map the complete route and connections of any one axon in the central nervous system. Consequently it is impossible to discover to what extent morphologically, let alone functionally, specific connections are present. It may at first seem surprising that specific connections can be recognised simply by morphological means but randomness of connections is a measure of lack of specificity. If a neuron has connections with all of its nearest neighbours then there is strong reason for supposing that it is not connected specifically in a morphological sense. According to Sperry's hypothesis morphologically recognisable specificity of connections would be expected, whereas on the "learning" hypothesis the specificity might only be detectable at the functional level by recording from electrodes.

At first sight, it would seem that experimental interference with the establishment of connections during embryogenesis would be the most rewarding method of investigating these problems. However, in early embryonic stages when connections are first being established the system is capable of such extensive regulation after experimental interference that no overt evidence of any change caused by experiment emerges from most work. Consequently, studies on the regeneration of connections have been very widely carried out, for regulation of the nervous system after experimental interference does not usually take place later in development. One major disadvantage of regeneration studies is that connections have been previously established with the consequence that the central nervous system may have set up processes which specifically recognise impulses from sense organs so that these signals can easily be recognised even though they may reach the central nervous system by abnormal paths after regeneration. One of the main experimental designs with regeneration experiments has been to test whether groups of axons carrying some special function, e.g. the spatial pattern from the retina, will regenerate connections appropriate to that function even though the actual message is now meaningless to or contradictory to the central nervous system. If such regeneration occurs it suggests that regeneration of connections is not random and that it is not controlled by the central nervous system. An example of such an experiment is to cut the optic nerve, rotate the eye so that the retina now lies "upside-down" in relation to images, and to examine whether the animal behaves as though it sees objects upside down or not after regeneration. If it sees objects in their normal relationships after regeneration this would argue that the central nervous system or possibly a system in the retina had managed to learn or develop new connections unrelated to any former pattern. Mapping of the nerve connections might show that the connections were arranged in a largely random fashion. This would then suggest that the theory of specific connections was untrue.

Unfortunately the majority of work on the establishment or regeneration of neuron connections has used behavioural data as a test of whether or not specific connections have been established. This technique is very unsatisfactory because behavioural processes involve so much of the nervous system that it is hard to exclude the possibility of learning or analytical processes playing a part in the synthesis of apparently correct or incorrect behaviour. Electrophysiological recording of the nerves stimulated by sensory input provides a much better method of mapping the connections developed and of determining whether these are randomly arranged or positioned in an ordered and specific manner. However, the recording methods so far used in these studies have been insufficiently precise to exactly locate connections, all that can be done is to demonstrate the small area in which they probably lie. Furthermore insertion of electrodes may do considerable damage to unsuspected connections. Lastly, it should be remembered that morphological mapping is also required in any adequate study and that this is still very difficult to carry out in detail.

Work on the mechanism of the formation of neuron connections can be divided into (i) sensory (ii) motor and (iii) central nervous system connections. One class of sensory studies is almost unique: these are studies on the retina for in this instance the axons elongate back to make connections with the central system whereas nearly all other sensory and motor neurons emerge from the central nervous system and extend towards their peripheral targets. Since more research has been directed towards the establishment and re-establishment of connections between the neural retina and the optic tecta than on any other organ work on this subject will be considered first.

Connections between the retina and the central nervous system

If the optic nerve of an adult or larval amphibian is cut regeneration of the optic nerve to reconnect with the tecta takes place. No similar phenomenon is found in mammals after section of the optic nerve. Matthey (1926) first showed that this regeneration takes place in urodeles, and that as judged by behavioural tests normal or nearly normal vision is restored to the eye involved. These results were confirmed by Stone (1930) who showed that in addition eyes could be grafted between species with functional regeneration of the optic nerve of the grafted eye. Sperry (1943a,b; 1950b) rotated the eyes of adult salamanders and newts in their own orbits through 180° and combined this in a number of cases with section of the optic nerve. Regeneration took place in the latter cases and the animals as judged behaviourally, had vision which was reversed in both vertical and horizontal axes. The rotated eyes of course were reversed in a morphologically similar way. Sperry pointed out that animals whose eyes had not been rotated showed very precise perception and localisation of objects after regeneration of the optic nerve and that animals whose eyes had been rotated showed a worse than random localisation of objects. This suggested that possibly a remarkably accurate re-establishment of connections takes place so that when the eye is not reversed each retinal axon travels to the spatially correct site on the tectum and that they travel to the same sites even if the eye is reversed so that vision is now reversed. Sperry (1948a,b) repeated this type of experiment on various species of marine fish: on recovery of vision after regeneration of the optic nerve in those animals whose eyes had been turned through 180°, visuomotor responses were apparently inverted. Stone (1948) carried out an investigation of the histology of regeneration of the retina and optic nerve in urodeles and discovered that stage at which this type of regeneration develops. In adults section of the optic nerve is followed by the degeneration of all the neural retina with the exception of a peripheral ring of cells. The rest of the retina and the optic nerve regenerate from these cells. Stone (1948, 1953, 1960)

reimplanted eyes of *Amblystoma punctatum* with reversed orientation using embryonic, larval and adult stages. When stages were
used in which an optic nerve had been formed, this nerve was cut.
Later in larval or adult life the animals were tested for reversed
vision. It was found that using stages before the early optic cup stage
vision later was normal despite the rotation of the eye. At the early
optic cup stage the embryo underwent some change so that the
mechanism which controls visuomotor responses becomes fixed
whatever the orientation of the eye, so that a given response is due
to the stimulation of a given point of the eye. This mechanism is so
firmly established that if an eye (early optic cup stage) is transplanted to the body wall for about 32 days and then is re-implanted
in the orbit of a normal animal with reversed or normal orientation,
it still shows fixed visuomotor responses. Further evidence in favour
of the stability of this mechanism comes from the finding (Stone,
1948) that if an eye which has regenerated upside down is turned
back to its normal orientation and its optic nerve is cut again, its
second regeneration produces behaviourally correct vision. At the
early optic cup stage there is no neuronal connection between retina
and tectum. Unfortunately this experiment does not demonstrate
whether the process of fixation is located in the retina or the tectum,
because the normal animal would in any case have undergone this
fixation and would retain it centrally if it is located centrally. It
would have been interesting to have also made grafts to the orbits
of abnormal and blinded animals. Such an experiment might have
revealed whether the process is located centrally or peripherally.
De Long and Coulombre (1965) carried out an interesting histological examination of the establishment of retinotectal connections in
the chick during embryogenesis. Quadrants of the retina were
ablated on the 3rd, 4th and 5th day of development before connections are made. The projections from the eyes of 3-day embryos
were histologically normal, but those from 4- and 5-day embryos
lacked connections in that part of the tectum corresponding to the
ablated part of the retinae. This result suggests that a fairly specific
projection is established and that whatever process determines it is
set up between the 3rd and 4th day of development, before connections are made.

When an eye is rotated through 180° the dorso-ventral orientation and the horizontal orientation (nasotemporal) are reversed. Stone (1944, 1948) and Sperry (1945) obtained eyes with normal dorso-ventral orientation but reversed nasotemporal orientation by transplanting eyes from left to right orbits and vice versa. The behaviour of these animals was such that the reactions to lures moving in the horizontal plane was reversed but lures moving in the vertical plane were reacted to normally and correctly. Similarly by reversing a left eye through 180° and reimplanting it in a right orbit, the regeneration of its optic nerve results in an animal that reacts correctly to objects moving in the horizontal plane but not in the vertical plane. These results suggest that the mechanism which presents ordered information to the central nervous system does not treat each receptor on the retina as having some unique property, but rather does so by treating each receptor as having a vertical and an horizontal component, all vertical components being common in a horizontal traverse of the retina and vice versa. Sperry (1945) examined the regeneration of the optic nerve in various anurans after the optic chiasma had been excised and the nerve stumps so arranged that regeneration results in ipsilateral rather than the normal contra lateral regeneration. (However Gaze and Jacobson (1962) later showed that each eye has an ipsilateral as well as a contralateral projection (see later).) Behavioural criteria were used in assessing the results. It appeared that when the contra lateral projection was fed in ipsilaterally, vision was normal or nearly so, save that the animal appeared to treat information from its right eye as though it came from the left eye and vice versa. Similarly when right and left eyes were interchanged in another set of animals the same type of behaviour appeared although the contralateral projection was intact. These results suggest that there is some additional mechanism for recognising which eye perceives objects, and that the mechanisms which control the sensing of images on different parts of the retina are duplicated for each eye. Unfortunately we have no knowledge as to what part the ipsilateral projection may have in producing these results.

As far as I can judge these experiments demonstrate that the amphibian visual system possesses a mechanism for spatially interpreting retinal images, and that this mechanism once established is very stable; that in relation to retinal elements it recognises spatial position by vertical and horizontal components referred to the retina; and it is probably able to discriminate between left and right eye information. Sperry however interpreted these results in more detailed terms. He suggests that these results show that specific retinal neurons make specific connections with the tectal neurons, and second, that this specific connection is due to each neuron having a specific biochemical property which is matched by the appropriate tectal neuron. The first part of this suggestion is not unreasonable particularly if we regard this specificity as being a relationship of a point on the retina to a point on the tectum, in which the spatial relationship of two or more points on the retina is caused to appear on the tectum. Mapping work to be described and discussed shortly strongly suggests that it is correct. However, it seems to me that Sperry (1958) was over sanguine in dismissing the learning or neurobiotaxis theories on the grounds that functional and correct regeneration is established so quickly. We have almost no information on the rate at which connections can be made and broken (see however page 347). Again some of the mechanisms which have been proposed to explain segregation in aggregates might operate effectively enough to produce sufficiently specific connections.

Attardi and Sperry (1963) continued their work by testing whether there were destination and route preferences for retinal fibres in the goldfish, *Carassia auratus* (sic) and the cichlid fish *Astronotus ocellatus*. Before discussing their results it is worth following the details of the route which retinal fibres take into the tectum. An axon which is extending towards the tectum must first enter either the medial or lateral tract of the optic nerve. There then follow a series of alternatives at various positions of entering the parallel layer of the tectum or remaining in the optic tract. Within the tectum an axon may at any point either enter the plexiform layer of the tectum or remain in the parallel layer. Is the "choice" at any of these three points made randomly or not? Finally of course according to Sperry an exact selection of the terminal site of the axon and its connections must be made.

Attardi and Sperry removed a portion of the retina of these fish and simultaneously cut the optic nerve of that same eye. After regeneration had taken place a map was prepared of the connections made by the optic nerve to the tectum. If specific connections are made it would be expected that parts of the tectum might lack innervation because these regions correspond to the parts of the retina which have been removed. Either half the retina (ventral, dorsal, nasal or temporal) or rings of retina were removed in these experiments. The regenerated fibres appeared to be arranged in a random fashion in the optic nerve itself. They found that if a given half of the retina was removed then the corresponding part of the tectum lacked retinal fibres. In these experiments evidence that any part of the tectum corresponded to a given part of the visual field was based on the assumption that Gaze and Jacobson's (1962) electrophysiological mapping of the tectum for anurans would also apply to fish. Sperry (1963) claims that these results establish his theory that each axon is attracted to its correct site by some form of chemotactic stimulus. However it should be pointed out that although no figures for the number of retinal axons in these fish is given, Attardi and Sperry only show a small number of fibres in their maps, and it is of interest to enquire whether all fibres were located and traced. It does, on the other hand, seem reasonable to take these results as evidence in favour of the establishment of specific connections.

The use of behavioural tests to discover the mechanism whereby nerve connections are established or re-established has been super-seded by the use of electrophysiological methods. These were first used for investigating this type of problem by Gaze (1958) and Maturana et al. (1960). Gaze showed that there is a constant point-to-point representation of the amphibian retina on the contralateral tectum. In 1962 Gaze and Jacobson showed that this point-to-point projection is restored after regeneration of the optic nerve in Xenopus laevis. Maturana et al. (1960) confirmed Gaze's initial finding of a point-to-point projection in amphibia, and showed that it is in fact of remarkable complexity. They found that there are five types of ganglion cell, sustained edge detectors, convex edge detectors, contrast change detectors, dimming and darkness detectors. These results demonstrate that much of the analysis of retinal data is performed within the retina itself. Three of these types terminate in different layers of the tectum, while the darkness and changing contrast detectors terminate in a fourth layer. Each layer forms a map of the retina and each layer is in register. Consequently regeneration is ideally faced with the problem of attempting to produce a point-to-point positioning of retinal fibres with the projection of each type of fibre in register. Jacobson and Gaze (1964) carried out a similar study in the goldfish *Carassius auratus*, in which they showed that there was a point-to-point projection of the retina on the contralateral tectum (thereby justifying Attardi and Sperry's assumptions), and that the surface of the tectum contained three layers, the outer two layers having "on" and fast "on-off" units, and the third layer contained slowly adapting "on" and "off" units. A layer deep to these three layers contained all the above types of units.

With these techniques and information on the mapping of retino-tectal connections Gaze *et al.* (1963) reinvestigated the question of whether regeneration from the retina is always exact so that there is a point-to-point projection. Szekely (quoted by Gaze *et al.*, 1963) had found that newts with double temporal retinae (made by leaving the temporal half of say a right eye intact and transferring a temporal half from a left orbit to the nasal position) appeared to be blind after destruction of the rostral region of the contralateral tectum; similarly newts with double nasal retinae appear as though they are blind after destruction of the caudal part of the tectum. This result suggested that all neurons from the double temporal retinae project to one part (rostral) of the tectum and that the nasal neurons project to the caudal part. This result appears to be in agreement with the claim that specific connections occur in this system. When Gaze *et al.* (1963) carried out experiments using double nasal or temporal retinae whose connections had mostly reached the tecta, and which were nasotemporally specified, mapping of the retino-tectal projection showed that the projection from each half retina covered the whole tectum, and that it was arranged normally for the "correct" half retina and as a mirror image for the foreign half retina. No part of the tectum was removed in these experiments,

Assuming that all the tectum has developed properly (a point considered and adequately demonstrated by the same authors in 1965), this experiment shows that the specificity of the connections can be altered experimentally, and that the projection alters as a spatially coherent unit, either enlarging or reversing on the tectum. The fact that the "foreign" half retina projects in reverse on the tectum suggests that the retina at least in part specifies the main axes of the projection. As a result of these experiments Gaze et al. (1963) suggested that the place of termination of any axon on the retino-tectal projection is expressed by some type of gradient mechanism. On such a theory it would be expected that if half of the gradient is removed, the surviving half would then spread out over the whole tectum.

Gaze and Jacobson (1963) carried out an extensive series of electrophysiological mappings of the projection of regenerated optic nerves in *Rana temporaria* (45 animals). This paper is of considerable importance because it reports on the temporal and spatial sequence and functional details of optic nerve regeneration. Three stages of regeneration of connections to the tectum were found. In the first of these a disorganised projection reached a few parts of the tectum; this was succeeded by a partially organised projection related to only one axis (nasotemporal); and finally (normally) a point-to-point projection was re-established. However in some animals not only was there a normal projection to the contralateral tectum but that part (nasal) of the retina which normally projects only ipsilaterally was instead projected contralaterally with mirror image symmetry. One animal had a projection which combined elements of this abnormal projection together with the disorganised projection. Of 45 frogs three animals had projection patterns which were totally disarranged. It was found that there were two sites on each retina from which regenerating fibres first reach the tectum, one of these sites is the area centralis. The apparent specification of the two axes at different times agrees with Gaze's gradient hypothesis. The fact that initially the projection is disorganised, though functional, is considerable evidence against Sperry's hypothesis in which con-nections are made initially with the right specificity to ensure correct or nearly correct function. This fact rather encourages the idea that a learning process may take place during regeneration. Furthermore Gaze and Jacobson's demonstration that initially a few parts of the tectum bear a random projection of the retina suggests that the projection may operate only if different signals from parts of the retina can be compared: in other words not so much a point-to-point projection but a point referred to "adjacent point" to a similar point projection. This might indicate the operation of a central learning mechanism in the establishment of the projection.

Jacobson and Gaze (1965) reinvestigated the possible specificity of retino-tectal connections in the goldfish using electrophysiological mapping methods. After section of the temporal half of the optic nerve, there was no spread of the nasal projection to the whole of the tectum. Again after crushing of the optic nerve regeneration of the connections resulted in an orderly projection, and removal of half of the tectum did not result in a new siting of half the optic fibres. These results are interpreted by Jacobson and Gaze as demonstrating that a specific point-to-point projection exists in the goldfish, unlike the situation in *Xenopus laevis*. This apparent confirmation of Sperry's hypothesis for goldfish however lacks one element; Gaze and Jacobson's work on *Xenopus* involved the experimental alteration of retinae, and this type of operation produced those results which demonstrate that a rigid point-to-point specificity does not occur in the retino-tectal projections of these animals, whereas their work on the goldfish did not involve experimental interference with the retinae. Whether or not the retinal structure has an influence on the regeneration of the connections Gaze and Jacobson's work on the goldfish did not test this possibility.

Finally it should be mentioned that the tectum may play the major role in organising the retino-tectal projection and that we seriously lack experimental evidence about such a function for the tectum. Crelin (1952) removed or rotated one of the two tecta in *Amblystoma* larvae. He found that at early stages tecta can be regenerated and that rotation of the tectum starts to affect subsequent visuomotor co-ordination just before the stage at which the retina becomes polarised. This change in the tectum takes place before the retinal fibres reach it. Unfortunately these experiments still leave it unclear as to whether it is the retina, tectum or some other structure

which produces the change which results in the retinal projection
being moderately firmly specified.

Sensory connections in other organs

As with work on retino-tectal connections a great deal of
research on the establishment of sensory connections of other organs
has relied upon behavioural tests of the specificity of nerve connections. In a number of experiments it has been almost impossible to
distinguish between sensory and motor contributions to the observed
behaviour so that these pieces of work will be discussed in a later
section together with other work on motor connections.

Sperry (reviews 1951, 1958) has suggested that a strict specificity
determines the development of nerve connections of other sense
organs as well as in the eye. Sperry (1945) found that the vestibular
nerve appeared to reconnect to the various parts of the labyrinth
in fish in a specific manner during regeneration: a similar experiment
on the regeneration of the oculomotor nerve in fish is reported in
his (1947) paper. Sperry and Miner (1949) examined regeneration
of sensory connections of the trigeminal (V) root in *Triturus viridescens* and some anurans, using mainly larval stages. The localisation
of cutaneous stimuli was assessed by behavioural criteria. If the
branches are simply severed and not crossed behaviour in response
to cutaneous stimuli is normal after regeneration. When the trigeminal root was led through the root of cranial nerve VIII, regeneration was not so accurate. This second part of the experiment
was performed to discover whether pathways for the sensory
connections along the former routes of the trigeminal nerve might
not affect the accuracy of regeneration. In the experimental animals
the central part of the peripheral ophthalmic branch was surgically
united with the peripheral part of the mandibular branch in *T.
viridescens*, while in the anurans the peripheral ends of the peripheral
ophthalmic branch were united to their contralateral roots. The
trigeminal root was cross-united to root VIII in these experimental
animals, so that no route would be available to lead the regenerating
cells to the correct site. After regeneration in *T. viridescens* in which
the mandibular branch led impulses centrally into the ophthalmic
branch, cutaneous stimulation of the mandibular region resulted in

responses appropriate to the ophthalmic region. Again in the experiments in which contralateral ophthalmic branches were led into the ophthalmic branch of the other side, behaviour was such that stimuli appeared to be located on the wrong side of the head. These experiments appear to demonstrate that the behaviour and perhaps the connections are determined centrally and that the connections are not highly specific. Sperry and Miner thought that these results showed that these experiments demonstrate that specific central connections are specified peripherally by the sense cells but this seems to be a non-logical conclusion. It should be remembered that these nerves regenerate from the centre to the periphery.

A somewhat similar experimental design was used by Szekely (1959b), who replaced the primordium of the trigeminal ganglion by the vagus ganglion or vice versa. The skin of the fore part of the head is innervated by the trigeminal nerve whereas the vagus innervates the hind part of the head in these newts. In half of the animals with two vagus ganglia (on one side) stimulation of the cornea produced responses typical of vagal stimulation. This result suggests that perhaps the grafted vagus ganglion had probably made contact with the vagus nucleus. After metamorphosis normal corneal responses could be obtained, indeed on occasion when the normal vagus sensory area was stimulated. This result suggests that both graft and normal vagus ganglion have made contact with the abducens nucleus. In those animals with double trigeminal ganglia (on one side) about half the animals showed behaviour typical of stimulation of the vagal sensory region when the cornea was stimulated. The interpretation of these results is as Szekely points out rather difficult. At first sight they suggest that the vagal and trigeminal connections centrally are rather non-specific, and that perhaps the grafted ganglia are able to induce the motor system to make connections to the normal ganglia of their own type.

In adult amphibians stimulation of the skin of the head from the snout as far back as the level of the ear can produce a retraction of the eyeballs together with closure of the eyelids. This is termed the corneal reflex. The trigeminal nerve provides an afferent pathway and the abducens nerve a motor pathway. Weiss (1942) found that after grafts of an extra eye into the snout of urodeles, stimulation of the graft resulted in a corneal reflex in the host eye. Weiss suggested that nerves invading the grafted eye become altered ("modulated") by the cornea so that there are changes in central connections which bring about this form of reaction. Kollros (1943) found that the same reaction could be produced even if the eye was grafted in areas innervated by the vagus. This result apparently showed that the "modulation" postulated by Weiss might really take place. But Szekely (1959a) was able to evoke a corneal reflex from the regeneration blastema of a limb which had been amputated just before metamorphosis. After regeneration was complete the reflex disappeared, though it would appear again in the regeneration after a second or even a third amputation. This result casts considerable doubt on the "local sign-specificity" or "modulation" theory. Interestingly enough the morphological type of nerve ending is similar in the cornea and the blastema. Szekely (1966) suggests that a similar low threshold in both these regions, and a large number of nerve endings in both may play a role in producing the same response. Similarly Kornacker (1963) suggested that the corneal reflex might depend on the activity of a sufficient number of fibres. Kornacker (1963) carried out electrophysiological mapping of the projection on the brain of the afferent pathways from the skin of *Rana pipiens* involved in the corneal reflex. The sensory fibres terminate in the abducens nucleus. The fast conducting fibres projected medially in the nucleus, the slower ones laterally. The fast conducting fibres originated in the snout. He found that the corneal reflex could be easily stimulated by afferent impulses from the snout while the slower fibres originate from the region of the eye. These results show that the corneal reflex can be easily stimulated by adjacent sensation. This apparent lack of specificity in the origin of the corneal reflex may have a parallel in Szekely's work.

Miner (1951, quoted by Sperry, 1958) found that trunk sensory and motor fibres could be caused to innervate supernumerary limbs which were grafted to the flank of amphibia before the first nerve connections were formed. These limbs showed normal behaviour which suggests that specific connections do not form in this case. It is possible that a learning process took place or alternatively a process in which the peripheral sense and motor organs were in some way able to specify central connections.

A somewhat contrary experiment was also described by Miner at the same time. She rotated bands of skin on frog tadpoles by 180° so that in the band the adults had belly skin on the back and vice versa. The behaviour of these animals on mechanical stimulation of the rotated band of skin was very interesting. Stimulation of the belly skin lying dorsally resulted in the frog scratching its belly, and vice versa. Unlike the experiment with the supernumerary limb, in which abnormal nerves produced normal behaviour, here abnormal nerves produced abnormal behaviour which was typical only of the "type" of the sense organs. Of course in the first experiment the peripheral sense and motor organs were in their normal relative positions whereas in the second experiment the sense organs were in abnormal positions and the motor elements concerned were not interfered with experimentally. Again no evidence of the formation of specific connections arises from these experiments. Both experiments can be interpreted, as Miner did, as showing that local features of the sense organs imposes a "local sign specificity" which in some way determines the choice of central connections.

Motor nerve connections

In all experiments so far carried out on the mechanism of establishment of motor nerve connections, behavioural criteria have alone been used to test the specificity and other features of the process. Consequently sensory connections are involved, and part or all of the behaviour may be due to changes in the sensory system.

Sperry (1940; see also 1958) carried out a number of experiments in which flexor and extensor muscles were interchanged in rats and monkeys; all the animals failed to adapt their motor behaviour to correct for the reversal of muscles. This result suggested to Sperry that motor nerve connections are specific. Similarly Sperry (see review, 1958) transposed motor and also sensory nerves in the rat with results which again suggested to him that muscle innervation to produce correct behaviour was entirely specific in terms of the connections made between nerves and muscles. In these experiments nerve connections with the muscles were broken and regeneration was allowed to occur, so that incorrect nerves innervated given muscles. Again Sperry (1950a) cut their innervation to the pectoral fin in the fish *Sphaeroides spengleri* and disarranged the nerve ends so that regeneration might be incorrect. After regeneration the fin movements, which involve the action of two antagonist groups of muscles, appeared entirely normal. Sperry and Deupree (1956) repeated this experiment on the fish *Histrio histrio* which has remarkably complex pectoral fin behaviour, with similar results. They also found that pelvic fin nerves would not produce correct behaviour after being led to innervate the pectoral fin. These results can be equally well explained if central mechanisms emit coded messages appropriate to certain muscles and cannot readjust them to produce the right behaviour when proprioception and perhaps vision informs the central system that unsuitable behaviour has been produced.

On the other hand Weiss (1936) found that altering the innervation of limb muscles, either by transplanting limbs or muscles or by rerouting nerves, did not affect the motor behaviour of larval urodeles even though the muscles received incorrect innervation (see below). If supernumerary limbs were placed on the flank of a newt they showed exactly the same muscle contractions as the homologous limb of the host; the innervation in these cases was taken from the limb plexi, unlike Miner's work mentioned above. Weiss (1936) claimed that in larval urodeles with supernumerary limbs innervated from limb plexi, motor activity in these limbs was exactly synchronous with that in the homologous limb irrespective of whether the innervation belongs to hind or fore limb. There was no evidence that the host limbs in any way affected the behaviour of the supernumerary limb. This response of the supernumerary limb or even of individual muscles is in synchrony with the host's homologous limb and muscles and the homologous response even extends to the response of the graft being in synchrony with the muscle or muscles of the limb of the same side as the graft was taken from. Weiss termed this synchronous and homologous response the synonymous response. Weiss showed experimentally that this response is not controlled by the brain, and that proprioceptive stimuli from a supernumerary limb cause both it and the host's homologous limb to contract. However the response is not brought about by this proprioceptive information because it persists after the sensory connections have been cut. Weiss suggested that these results indicated that the peripheral part of the nervous system specifies the whole system in some way so that it transmits the appropriate message to the appropriate set of motor organs. He termed this process "specific modulation". Since cases had been found of a given nerve innervating two muscles which contract alternately, Weiss suggested that the specificity arose either in the end plate or in the muscles themselves. Little further work has been done to investigate Weiss' theory. It is however remarkable that Piatt (1952, 1957) found in larval *Amblystoma* that if limbs were grafted to abnormal sites (fore limb to hind limb or vice versa) or if nerve routes were altered so that normal limbs had abnormal innervation (hind limb innervation to forelimb), then these limbs did not show behaviour homologous with either the muscle or the nerve type, indeed their behaviour was very disorganised.

Miner (quoted by Sperry, 1958: see previous section) observed that a hind limb grafted into the trunk region of tadpoles which were innervated by thoracic nerves showed many of the features of the synonymous response. Szekely and Szentgothai (1962) implanted limb buds into the mid-thoracic region of 3-day chick embryos. After hatching these limbs were not used but on stimulation gave rise to various forms of behavioural response by the animal. All the animals responded (except one) to mechanical stimulation of the graft (wing or leg) by flexing both the wing or the leg. This unspecific type of behaviour is different from any previous finding and suggests that neither specific connections nor peripheral modulation occurs. In some of the animals a more complex type of behaviour was found in which the animal attempted to scratch its legs with its beak when the graft (wing or leg) was stimulated: once the bird caught sight of the slip on the graft used to stimulate it the bird localised the source of pain correctly. Interestingly the more complex form of behaviour only developed slowly after hatching. As Szekely (1966) points out it is almost impossible to explain these results in terms of the development of specific connections. There was some evidence that pressure stimulation produced the simple responses and pain stimulation produced the complex responses, again the more complex responses appeared to be associated with anatomically more complex nerve pathways. These results suggest that perhaps some complex analytical process is carried out with the sensory information from these pathways and that the appropriate motor messages are synthesised with respect to the frequency, intensity and number of activated fibres and perhaps with respect to coded sensory messages specific to each type of sensory receptor.

The majority of experiments which have been described in this section have shown that regenerating nerve fibres will establish functional connections as readily with one skeletal muscle as with another. Moreover these connections allow the muscles to contract in their normal manner. These results are taken by Weiss (1936, 1955) as evidence for his theory of "myotypic modulation". Many of Sperry's earlier results could be interpreted as examples of modulation: when he failed to obtain correct function this could be attributed to damage to the nerves or muscles during operation. He failed to carry out control operations on the correct nerves by routing them through unusual routes to their correct muscles. Sperry and Arora (1965) carried out an interesting set of experiments on regeneration of the oculomotor nerve in the cichlid fish *Astronotus ocellatus*. Behavioural criteria were used to judge the correctness or otherwise of regeneration. If the main trunk of the oculomotor nerve was cut the subsequent regeneration allowed normal eye movements. However if the individual branches to the various oculomotor muscles were cut and rerouted to the wrong muscle, for example the inferior oblique nerve to the superior rectus muscle or the superior oblique nerve into the medial rectus muscle, functional connections were re-established by the incorrect nerves, but the eye movements of such animals were abnormal. When a muscle was fed by both a normal and an abnormal nerve, the resulting innervation produced correct behaviour. These results strongly suggest that modulation does not occur, for if it did correct behaviour would be expected with incorrect innervation. However Sperry and Arora may have damaged the nerves during the process of rerouting them and no controls were carried out to guard against this possibility.

The central nervous system

Much of what has just been described has been explained by invoking all manner of changes in the central nervous system. But very little is known about the establishment of connections and nerve pattern in the central nervous system. The neural plate gives rise to the central nervous system, but nothing appears to be known about the stages of morphogenesis of the central nervous system immediately succeeding neurulation. At later stages we have little more information. Recently the application of autoradiographic methods has allowed the description of some remarkable events of cell behaviour during morphogenesis of the brain. DeLong and Sidman (1962) examined the formation of the colliculus in new-born mice which in part at least is built up by migration of cells from the ependyma to form the superior colliculus: interestingly enough removal of the eye affects the number of cells involved in the migration though not the extent of migration of each cell. Miale and Sidman (1961) also examined the formation of the cerebellum in the mouse, and found that the Purkinje cells and the neurons of the roof nuclei move outwards through the primitive ependyma to reach their final positions. The external granule cells migrate over the surface of the developing cerebellum till they reach positions (random or predetermined) where they sink into the cerebellum to become the granule cell neurons. Fujita (1964) examined cell migration in the central nervous system of the embryonic chick and found a rather similar development of the cerebellum to that described by Miale and Sidman. He also examined the formation of the tectum opticum (cortical layer) which is formed by migration, and showed that considerable lateral and ventral cell migration takes place within the spinal cord during its formation. The cerebral cortex of rats contains six layers, and Angevine and Sidman (1961) described their formation by waves of cell migration forming successive layers which lay superficial to the last layer formed. Berry and Rogers (1965) claimed that the layers were formed in the opposite order, the outermost one being formed first.

M

Remarkably little work has been carried out experimentally on the development of connections in the brain (see however Detwiler, 1936). Sperry (1948a, 1958) cut the brain stem between the optic lobe and the cerebellum in the adult newt *Triturus viridescens*. The effect of this might be to destroy visuomotor reactions, since regeneration of suitable pathways would be essential for normal reactions. If the regeneration were random with respect to whether junctions between left and right or left and left tracts on either side of the cut were made, then 50 per cent of the operated animals would be expected to have reversed vision (in the horizontal plane). Normal visuomotor reactions were found after regeneration which suggested to Sperry that the regeneration was specific not only to the visual field of the eye but also to which side of the body the eye was on. Later (1958) he reported the results of putting barriers such as mica plates into cuts in the brains of rats and found that the behaviour of the animals was unimpaired after a suitable time for regeneration had been allowed. These experiments involve the brain in massive and unknown damage so that few conclusions can be drawn from them, and alternative explanations are readily available to explain the results. Piatt (1948) reviewed a considerable amount of work on the development of the structure of the central nervous system but very little attention was paid to the possible role of cell migrations in this morphogenesis.

Stefanelli (1951) examined the development of the Mauthner neurons in the frog. Interestingly enough he foreshadowed some of Steinberg's experiments (1964) by showing that the Mauthner cell always took up an internal position in brain explants. Stefanelli removed the contralateral Mauthner cell and the homo- and contra-lateral fasiculi but this had no effect on the elongation of the fibres of the remaining cell. Rotation of the cell through 180° produced normal axon direction and transplantation to a new site produced axons oriented to the same site. Insertion of extra nervous material between the two Mauthner cells sometimes prevented the fibres crossing over from one side to the other as they normally do, but did not prevent their caudal extension. He argued that these experiments showed that nerve extension in this case was due to chemotaxis. But Piatt (1944) obtained Mauthner fibres of reversed orientation when the cells were transplanted to ectopic sites.

C. O. Jacobson (1964a,b) carried out a number of fascinating experiments on the mechanisms whereby fibre pattern is determined in the central nervous system. He excised bilaterally symmetrical parts of the neural plate in *Amblystoma mexicanum* embryos, and reversed their anterio-posterior positions. In some experiments only motor precursor regions components were reversed but in others both motor and sensory precursor parts of the brain were reversed. Many systematic changes were found in the brains of the "early feeding" stage larvae. Mauthner cells fibres sometimes grew out in the direction appropriate to the polarisation of the graft, a result always found by Stefanelli (1951), but by no means invariably so by Jacobson. But when the graft produced a reversed medulla oblongata without affecting the orientation of the Mauthner cell body (since it lay outside the reversed region) the fibres grew out from the cell body with reversed orientation (12 out of 13 cases). These results suggest that the direction that the Mauthner cell fibre takes is controlled by a substrate effect such as a gradient in adhesiveness. However Jacobson does not completely rule out the action of a chemotactic influence. Sensory roots were found to enter the brain at the point of emergence of a motor root irrespective of the appropriateness of the motor nerve root, which suggests either a chemotactic response or "contact guidance" behaviour. Similarly the course of the trigeminal root often is related to the orientation of the mesencephalon, whether normal or reversed. Jacobson (1964b) discusses Stefanelli's concept that chemotaxis leads the Mauthner fibres towards a definite position: he points out that one would expect according to this theory that fibres would tend to grow towards their normal position, even though their surroundings were experimentally disordered, but this did not occur frequently.

Again little is known about the morphogenesis of the spinal cord. Levi-Montalcini (1950) showed that the nuclei of Terni in the chick spinal cord arise by cell migration. Takeya and Watanabe (1961) suggested that the thickening of basal and alar plates of the spinal cord of the chick embryo is due to excessive proliferation in these regions of the cord. But Fujita (1964) found that neuroblasts migrate into these regions to enlarge their size. Wenger (1950) published a detailed analysis of the morphogenesis of the spinal cord in the chick: she removed either the right half, right dorsal quarter or the dorsal half of the cord at the brachial level. Very little disturbance of the existing nerve pattern in the intact parts of the cord resulted which suggests that even by the eighth day (youngest embryos) any cell migration which might occur has taken place and that further rearrangement is impossible. A number of experiments have been carried out on regeneration of the spinal cord after its section. Hooker (1925) cut the cord in tadpoles and found a functional regeneration from which he concluded that specific connections were re-established. Although a considerable number of papers have been published on either the gross anatomy of the regenerate or on the extent of restoration of function (nil in mammals), almost no attention has been paid to the role of cell migration in this process. However Butler and Ward (1965) have provided a detailed description of cell migration in the regeneration of the cord in larvae of *Amblystoma maculatum*: this regeneration gave fairly good recovery of behaviour. The first sign of regeneration is the migration of the ependymal cells followed by that of cells from the grey matter to form vesicles over each surface of the cut and to close off the central canal. The vesicles advance towards each other mainly (probably) as the result of cell migration into the vesicle walls. When they meet, nerve fibres from the white matter migrate across from one surface to the other, presumably using the vesicles as a substrate for cell attachment.

The development of nerve pattern

Up to this point I have dealt with the question of the establishment and re-establishment of correct function and whether or not this is brought about by specific or otherwise ordered cell movements, or by the purely random matching up of neurons. The second main feature of the development of the nervous system which is of interest here is how the actual pattern of the nerves is determined irrespective of the actual connections established by their component axons.

To a certain extent this question has been dealt with in Chapter 5 where forms of cell behaviour were discussed. In this section the applicability of these forms of cell behaviour to the question of the development of nerve pattern will be considered. But before doing this, work specifically on the behaviour of nerve cells will be considered. Speidel (1942, 1947, 1964) has reported a very interesting series of observations on the readjustment and regeneration of nerve connections. He observed (1942) the development of nerve connections in tadpoles of *Pseudacris feriarum*, *Hyla crucifer* and *Rana clamitans*. The cells were observed live and he noted that there was a continuous change in the number and position of connections, arborisations and end buds. Extension, retraction and arborisation of a nerve axon produced a continuous change in morphology. These observations give us an idea of the restless state of cell contacts that may obtain throughout the nervous system. In papers published in 1947 and 1964, he extended these painstaking observations of live material to the regeneration of the dorsal ramus of the vagus nerve in *Rana clamitans*. Cuts were made in the nerve and regeneration took place with a single nerve bridging the gap afterwards. The Schwann cells of the distal stump moved across the wound (fibrin etc.) to meet the nerve "sprouts" from the proximal stump. It appeared that these Schwann cells guided the axons across to the distal stump. The Schwann cells on the proximal side migrate only after the axons have entered the wound. But if two cuts are made near each other so that a small piece of nerve is left isolated Schwann cells migrate from either edge of this piece. This experiment suggests that Schwann cell migration is normally prevented by some feature of a nerve which has central connections. Interestingly enough the regenerations were obviously often grossly wrongly patterned, forming plexi with the central nervous system, bridges from side to side, and reversed innervation (i.e. regenerate from left proximal stump connects with right proximal stump). Abercrombie *et al.* (1949) investigated the relationship of rabbit Schwann cells and axons in culture: they found evidence that Schwann cells showed a particular adhesiveness to axons, possibly a specific adhesion.

The nature of contact guidance (Weiss 1934, 1945) and other features of nerve cell behaviour in culture have been discussed in Chapter 5. Weiss and his co-worker A. C. Taylor carried out a number of experiments to investigate whether or not contact guidance could explain the development of connections in the animal. Weiss and Taylor (1944) grew rat sciatic nerve in culture with fragments of other nerves nearby, the sciatic nerve showed no orientation towards the fragments and would grow into blind ends of pieces of aorta placed in the culture. These results argue against neurotropism, i.e. chemotaxis.

The actual detailed plan of parts of the nervous system has been particularly investigated by Piatt and by Taylor. Both have worked on the development of nerve pattern in the limbs of various amphibia. Taylor (1944) extirpated either dorsal or ventral spinal roots in the frog tadpole and examined the resulting innervations. He found that they were substantially correct in anatomical features, though some cutaneous sensory innervation developed additionally from the ventral roots when they alone were intact. He suggested that the anatomical correctness was due to the presence of pathways in the limb mesenchyme along which the nerve cells could advance by contact guidance. His results suggest that it is the routing mechanism which operates in the same way for either dorsal or ventral root nerves. Piatt (1957, 1958) found that the pattern of nerves in the forelimb of *Amblystoma larvae* was upset if the nerve supply did not come from the normal brachial plexus. This result would not be expected on Weiss' theory, for if tracts are laid down which will lead nerves along them by the action of contact guidance, there is no reason to suppose that different nerve supplies would produce different anatomical patterns of innervation. Weiss' theories are of course unsatisfactory in the respect, that even if true, they merely push the problem back to the question of how these tracts in the mesenchymes are formed.

It is clear that work in this field is in an unsatisfactory state and that no one has yet clearly demonstrated what mechanism is responsible for leading the nerves to their normal sites along normal pathways. Early work (see Detwiler, 1936) showed that organs attract nerves, but whether this is by chemotaxis, contact guidance, neurobiotaxis etc. is unknown. The few direct observational studies of nerve migration suggest that nerve pattern is not formed by the initial establishment of random connections which are then "tidied" up.

REFERENCES

ABERCROMBIE, M. (1950). The effects of antero-posterior reversal of lengths of the primitive streak in the chick. *Phil. Trans.*, **B.234**, 317–38.

ABERCROMBIE, M. (1957). The directed movement of fibroblasts: a discussion. *Proc. Zool. Soc. Calcutta.* Mookerjee Memorial Vol., 129–40.

ABERCROMBIE, M. (1961a). Behaviour of normal and malignant connective tissue cells *in vitro*. *Canad. Cancer Conf.*, **4**, 101–17.

ABERCROMBIE, M. (1961b). The bases of the locomotory behaviour of fibroblasts. *Exp. Cell. Res. Suppl.*, **8**, 188–98.

ABERCROMBIE, M. (1962). Contact-dependent behaviour of normal cells and the possible significance of surface changes in virus-induced transformation. *Cold Spring Harbor. Symp. Quant. Biol.*, **27**, 427–31.

ABERCROMBIE, M. (1964a). Cell contacts in morphogenesis. *Arch. Biol., Liège*, **75**, 351–67.

ABERCROMBIE, M. (1964b). The locomotory behaviour of cells. From *Biology of cells*, edited E. Willmer. Academic Press, New York, 177–202.

ABERCROMBIE, M. and AMBROSE, E. J. (1958). Interference microscope studies of cell contacts in tissue culture. *Exp. Cell Res.*, **15**, 332–45.

ABERCROMBIE, M. and AMBROSE, E. J. (1962). The surface properties of cancer cells: a review. *Cancer Res.*, **22**, 525–48.

ABERCROMBIE, M. and GITLIN, G. (1965). The locomotory behaviour of small groups of fibroblasts. *Proc. Roy. Soc.*, **B.162**, 289–302.

ABERCROMBIE, M. and HEAYSMAN, J. E. M. (1953). Observations on the social behaviour of cells in tissue culture—I. Speed of movement of chick heart fibroblasts in relation to their mutual contacts. *Exp. Cell Res.*, **5**, 111–31.

ABERCROMBIE, M. and HEAYSMAN, J. E. M. (1954). Observations on the social behaviour of cells in tissue culture—II. "Monolayering" of fibroblasts. *Exp. Cell Res.*, **13**, 276–91.

ABERCROMBIE, M. and HEAYSMAN, J. E. M. (1966). The directional movement of fibroblasts emigrating from cultured explants. *Ann. Med. Exp. Fenn.*, **44**, 161–65.

ABERCROMBIE, M., HEAYSMAN, J. E. M. and KARTHAUSER, H. M. (1957). Social behaviour of cells in tissue culture—III. Mutual influence of sarcoma cells and fibroblasts. *Exp. Cell Res.*, **13**, 276–91.

ABERCROMBIE, M., JOHNSON, M. L. and THOMAS, G. A. (1949). The influence of nerve fibres on Schwann cell migration investigated in tissue culture. *Proc. Roy. Soc.*, **B.136**, 448–60.

ABRAHAM, M. (1960). Processus de reorganisation dans les agregats formes partes cellules des gonades dissociees d'embryon de Poulet. *Arch. Anat. Micros.*, **49**, 333–44.

ADAM, N. K. (1941). *The physics and chemistry of surfaces*. 3rd edition. Oxford Univ. Press, London.

ADAMS, D. M. and RIDEAL, E. (1959). The surface behaviour of *Mycobacterium phlei*. *Trans. Faraday Soc.*, **85**, 185–9.

ALBERS, W. and OVERBEEK, J. Th. G. (1960). Stability of emulsions of water in oil—III. Flocculation and redispersion of water droplets covered by amphipolar monolayers. *J. Coll. Sci.*, **15**, 489–502.

ALESCIO, T. and CASSINI, A. (1962). Osservazioni preliminari sulla riassociazione in vitro dell'epitelio e del mesenchima del polmone embrionale di topo. *Monitore Zool. Ital.* Suppl. to Vol. 70: 97–103.

ALLEN, R. D. (1961). Structure and function in amoeboid movement. *Biol. Structure and Function.*, **2**, 549–56.

ALLEN, R. D., COOLEDGE, J. W. and HALL, P. J. (1960). Streaming in cytoplasm dissociated from the giant amoeba *Chaos chaos*. *Nature, Lond.*, **187**, 896–9.

ALLEN, R. D. (1964). Cytoplasmic streaming and locomotion in marine foramrifera. From *Primitive motile systems in cell biology*, edited R. D. Allen and N. Kamiya. Academic Press, New York, 407–31.

AMBELLAN, E. (1955). Effect of adenine mononucleotides on neural tube formation of frog embryo. *Proc. Nat. Acad. Sci. Wash.*, **41**, 428–32.

AMBELLAN, E. (1958). Comparative effects of mono-, di- and tri-phosphorylated nucleosides on amphibian morphogenesis. *J. Embryol. Exp. Morph.*, **6**, 86–93.

AMBROSE, E. J. (1956). A surface contact microscope for the study of cell movements. *Nature, Lond.*, **198**, 1194.

AMBROSE, E. J. (1961). The movements of fibrocytes. *Exp. Cell Res. Suppl.*, **8**, 54–73.

AMBROSE, E. J. (1962). Surface characteristics of neoplastic cells. From *Biological Interactions in normal and neoplastic growth*. Edited M. J. Brennan and W. L. Simpson. Little, Brown, Boston and New York.

AMBROSE, E. J. and EASTY, G. C. (1960) Membrane structure in relation to cellular motility. *Proc. R. Phys. Soc. Edin.*, **28**, 53–63.

AMBROSE, E. J., JAMES, A. M. and LOWICK, J. H. B. (1956). Differences between the electrical charge carried by normal and homologous tumour cells. *Nature, Lond.*, **177**, 576–7.

ANDERSON, E. and BEAMS, H. W. (1960). Cytological observations on the fine structure of the guinea-pig ovary with special reference to the oogonium, primary oocyte and associated follicle cells. *J. Ultrastr. Res.*, **3**, 432–46.

ANDERSON, N. G. (1953). The mass isolation of whole cells from rat liver. *Science*, **117**, 627–8.

ANDERSON, R. E. and WALFORD, R. L. (1960). Detection of leukocyte antibodies by means of I^{131}-labelled purified anti-human-globulin antibody. Problems of non-specific adsorption of globulin by leukocytes. *Blood*, **16**, 1523.

ANGEVINE, J. B. and SIDMAN, R. L. (1961). Autoradiographic study of cell migration during histogenesis of cerebral cortex in the mouse. *Nature, Lond.*, **192**, 766–8.

ANSEVIN, K. D. (1964). Aggregative and histoformative performance of adult frog liver cells *in vitro*. *J. Exp. Zool.*, **155**, 371–80.

ARMSTRONG, P. B. (1966). On the role of metal cations in cellular adhesion: effect on cell surface charge. *J. Exp. Zool.*, **163**, 99–109.

ASHWORTH, L. A. E. and GREEN, C. (1966). Plasma membranes: phospholipid and sterol content. *Science*, **151**, 210–1.

ATTARDI, P. and SPERRY, R. W. (1963). Preferential selection of central pathways by regenerating optic fibres. *Exp. Neurol.*, **7**, 46–64.

AUB, J. C., TIESLAU, C. and LANKESTER, A. (1963). Reactions of normal and tumour cell surfaces to enzymes—I. Wheat-germ lipase and associated mucopolysaccharides. *Proc. Nat. Acad. Sci., Wash.,* **50,** 613–19.

AUSTIN, C. R. (1956). Cortical granules in hamster eggs. *Exp. Cell Res.,* **10,** 533–40.

AUSTIN, C. R. (1961). *The Mammalian Egg.* Blackwell Scientific, Oxford.

BAINBRIDGE, D. R., BRENT, L. and GOWLAND, G. (1966). Distribution of allogenic ^{51}Cr-labelled lymph node cells in mice. *Transplant,* **4,** 138–53.

BALINSKY, B. I. (1959). An electron microscope investigation of the mechanism of adhesion of the cells in a sea-urchin blastula and gastrula. *Exp. Cell Res.,* **16,** 429–33.

BALINSKY, B. I. and WALTER, H. H. (1961). The immigration of presumptive mesoblast from the primitive streak in the chick as studied with the electron microscope. *Acta Embryol. Morph. Exp.,* **4,** 261–83.

BANG, F. B. (1955). Morphology of viruses. *Ann. Rev. Microbiol.,* **9,** 21–44.

BANGHAM, A. D., GLOVER, J. C., HOLLINGSHEAD, S. and PETHICA, B. A. (1962). The surface properties of some neoplastic cells. *Biochem. J.,* **84,** 513–17.

BANGHAM, A. D. and HORNE, R. W. (1962). Action of saponin on biological membranes. *Nature, Lond.,* **196,** 952–3.

BANGHAM, A. D. and PETHICA, B. A. (1958). The charged groups at the interface of some blood cells. *Biochem. J.,* **69,** 12–19.

BANGHAM, A. D. and PETHICA, B. A. (1961). The adhesiveness of cells and the nature of the chemical groups at their surfaces. *Proc. Roy. Soc. Edin.,* **28,** 43–52.

BARER, R. (1956). Phase contact and interference microscopy in cytology. From *Physical techniques in biological research,* edited G. Oster and A. W. Pollister. Academic Press, New York.

BARSKI, G. and BELEHRADEK, J. (1965). Etude microcinematographique du mecanisme d'invasion cancereuse en cultures de tissu normal associe aux cellules malignes. *Exp. Cell Res.,* **37,** 464–80.

BASCH, R. S. and STETSON, C. A. (1962). The relationship between hemagglutinogens and histocompatibility antigens in the mouse. *Ann. N.Y. Acad. Sci.,* **97,** 83–94.

BEAR, R. S. and SCHMITT, F. O. (1937). Optical properties of the axon sheaths of crustacean nerves. *J. Cell. Comp. Physiol.,* **9,** 275–88.

BECKER, F. S. (1960). Studies on the hemolytic properties of protamine. *J. Gen. Physiol.,* **44,** 433–42.

BELL, E. (1960). Some observations on the surface coat and intercellular matrix of the amphibian ectoderm. *Exp. Cell Res.,* **20,** 378–83.

BELL, L. (1961). Surface extension as the mechanism of cellular movement and cell division. *J. Theor. Biol.,* **1,** 104–6.

BELLAIRS, R. (1953). Studies on the development of the foregut in the chick blastoderm—2. The morphogenetic movements. *J. Embryol. Exp. Morph.,* **1,** 369–85.

BELLAIRS, R. (1965). The relationship between oocyte and follicle in the hen's ovary as shown by electron microscopy. *J. Embryol. Exp. Morph.,* **13,** 215–33.

BENNETT, H. S. (1956). The concepts of membrane flow and membrane vesiculation as mechanisms for active transport and ion pumping. *J. Biophys. Biochem. Cytol.,* **2,** Suppl., 99–103.

BENNETT, H. S. (1963). Morphological aspects of extracellular polysaccharides. *J. Histochem. Cytochem.,* **11,** 14–23.

BEN-OR, S. and DOLJANSKI, F. (1960). Single cell suspensions as tissue antigens. *Exp. Cell Res.*, **20**, 641–4.

BEN-OR, S., EISENBERG, S. and DOLJANSKI, F. (1960). Electrophoretic mobilities of normal and regenerating liver cells. *Nature, Lond.*, **188**, 1011–12.

BERMANN, F. (1960). Associations xenoplastiques d'organes embryonnaires: etude de quelques structures mixtes obtenues *in vitro* et *in vivo*. *C. R. Soc. Biol.*, **154**, 911–14.

BERRY, M. and ROGERS, A. W. (1965). The migration of neuroblasts in the developing cerebral cortex. *J. Anat.*, **99**, 691–709.

BERWALD, Y. and SACHS, L. (1963). *In vitro* cell transformation with chemical carcinogens. *Nature, Lond.*, **200**, 1182–4.

BERWICK, L. and COMAN, D. R. (1962). Some chemical factors in cellular adhesion and stickiness. *Cancer Res.*, **22**, 982–86.

BIERMAN, A. (1955). Electrostatic forces between non-identical colloidal particles. *J. Coll. Sci.*, **10**, 231–45.

BIKERMAN, J. J. (1957). Formation and rupture of adhesive joints. *Proc. 2nd Int. Cong. Surface Science*, **III**, 427–32.

BILLINGHAM, R. E. and MEDAWAR, P. B. (1951). The technique of free skin grafting in mammals. *J. Exp. Biol.*, **28**, 385–402.

BINGLEY, M. S. and THOMPSON, C. M. (1962). Bioelectric potentials in relation to movements in amoebae. *J. Theor. Biol.*, **2**, 16–32.

BIOT, M. A. (1957). Folding instability of a layered viscolastic medium under compression. *Proc. Roy. Soc.*, **A.242**, 444–54.

BIRKS, R., KATZ, B. and MILEDI, R. (1959). Dissociation of the "surface membrane complex" in atrophic muscle fibres. *Nature, Lond.*, **184**, 1507–8.

BLACK, W., DE JONGH, J. C. V., OVERBEEK, J. TH. G. and SPARNAAY, M. J. (1960). Measurement of retarded van der Waal's forces. *Trans. Farad. Soc.*, **56**, 1597–1608.

BLANCHETTE, E. (1961). A study of the fine structure of the rabbit primary oocyte. *J. Ultrastr. Res.*, **5**, 349–63.

BLASIE, J. K., DEWEY, M. M., BLAUROCK, A. E. and WORTHINGTON, C. R. (1965). Electron microscope and low-angle X-ray diffraction studies on outer segment membranes from the retina of the frog. *J. Mol. Biol.*, **14**, 143–52.

BLOUGH, H. A. (1963). The effect of vitamin A alcohol on the morphology of myxoviruses—I. The production and comparison of artificially produced virus. *Virology*, **19**, 349–58.

BOELL, E. J. and NACHMANSOHN, D. (1940). Localization of choline-esterase in nerve fibres. *Science*, **92**, 513–14.

BOEREMA, I. (1929). Die dynamik des Medullarrohrschlusses. *Arch. f. Entwickmech.*, **115**, 601–15.

BOK, S. T. (1961). Die entwicklung der hirnnerven und ihrer zentralen bahnen. Der stimulogene Fibrillation. *Folia Neurobiol.*, **9**, 475–565.

BONNER, J. T. (1947). Evidence for the formation of cell aggregates by chemotaxis in the development of the slime mould. *Dictyostelium discoideum*. *J. Exp. Zool.*, **106**, 1–26.

BONNER, J. T. (1959). Evidence for the sorting-out of cells in the development of the cellular slime moulds. *Proc. Nat. Acad. Sci. Wash.*, **45**, 379–84.

BONNER, J. T. and ADAMS, M. S. (1958). Cell mixtures of different species and strains of cellular slime moulds. *J. Embryol. Exp. Morph.*, **6**, 346–56.

BORSOS, T., RAPP, H. J. and MAYER, M. N. (1961). Studies on the second component of complement. I. *J. Immunol.*, **87**, 310–29.

BORSOS, T., DOURMASHKIN, R. R. and HUMPHREY, J. H. (1964). Lesions in erythrocyte membranes caused by immune haemolysis. *Nature, Lond.*, **202**, 251–2.

BOYSE, E. A. (1960). A method for the production of viable cell suspensions from solid tumors. *Transpl. Bull.*, **7**, 100–104.

BRAGG, A. N. (1938). The organization of the early embryo of *Bufo cognatus* as revealed especially by the mitotic index. *Z. Zellforsch. Mikros. Anat.*, **28**, 154–78.

BRAND, K. G. and SYVERTON, J. T. (1962). Results of species-specific hemagglutination tests on "transformed", non-transformed and primary cell cultures. *J. Nat. Cancer Inst.*, **28**, 147–57.

BRAUER, A. (1932). A topographical and cytological study of the sympathetic nervous components of the suprarenal of the chick embryo. *J. Morph.*, **53**, 277–325.

BRENNER, R. M. (1963). Radioautographic studies with tritiated thymidine of cell migration in the mouse adrenal cortex after a carbon tetrachloride stress. *Amer. J. Anat.*, **112**, 81–7.

BRESCH, D. (1955). Recherches preliminaires sur des associations d'organes embryonnaires de Poulet en culture *in vitro*. *Bull. Biologique*, **89**, 179–88.

BRIGHTMAN, M. W. and PALAY, S. I. (1963). The fine structure of ependyma in the brain of the rat. *J. Cell Biol.*, **19**, 415–39.

BRIGHTMAN, M. W. (1965). The distribution within the brain of ferritin injected into cerebrospinal fluid compartments—I. Ependymal distribution. *J. Cell Biol.*, **26**, 99–123.

BROWN, M. G., HAMBURGER, V. and SCHMITT, F. O. (1941). Density studies on amphibian embryos with special reference to the mechanism of organiser action. *J. Exp. Zool.*, **88**, 353–72.

BRUEMMER, N. and THOMAS, L. E. (1957). Cellular lipoproteins—II. A "ghost" cell preparation. *Exp. Cell Res.*, **13**, 103–8.

BUCK, R. C. and KRISHNAN, A. (1965). Site of membrane growth during cleavage of amphibian epithelial cells. *Exp. Cell Res.*, **38**, 426–8.

BUSCHKE, W. and WHITE, M. (1949). Studies on intercellular cohesion in corneal epithelium. *J. Cell. Comp. Physiol.*, **33**, 145–76.

BUTLER, E. G. and WARD, M. B. (1965). Reconstitution of the spinal cord following ablation in urodele larvae. *J. Exp. Zool.*, **160**, 47–66.

CARMICHAEL, L. (1927). A further study of the development of behaviour in vertebrates experimentally removed from the influence of external stimuli. *Psychol. Rev.*, **34**, 34–47.

CARREL, A. and EBELING, A. H. (1922). Pure cultures of large mononuclear leucocytes. *J. Exp. Med.*, **36**, 365–77.

CARTER, S. B. (1965). Principles of cell motility: the direction of cell movement and cancer invasion. *Nature, Lond.*, **206**, 1183–7.

CASIMIR, H. B. G. and POLDER, D. (1946). Influence of retardation on the London-van der Waals forces. *Nature, Lond.*, **168**, 787–8.

CATALANO, P., NOWELL, P., BERWICK, L. and KLEIN, G. (1960). Surface ultrastructure of tumor sublines differing in adhesiveness. *Exp. Cell Res.*, **20**, 633–5.

CAVANAUGH, D. J., BERNDT, W. O. and SMITH, T. E. (1963). Dissociation of heart cells by collagenase. *Nature, Lond.*, **200**, 261–2.

CHAMBERS, R. (1940). The relation of extraneous coats to the organisation and permeability of cellular membranes. *Cold Spring Harb. Rev. Quant. Biol.*, **8**, 144–53.

CHAMBERS, R. and ZWEIFACH, B. W. (1947). Intercellular cement and capillary permeability. *Physiol. Rev.*, **27**, 436–63.

CHAPMAN, D. (1966). Liquid crystals and cell membranes. *Ann. N.Y. Acad. Sci.*, **137**, 745–54.

CHIAKULAS, J. J. (1952). The role of tissue specificity in the healing of epithelial wounds. *J. Exp. Zool.*, **121**, 383–418.

CHILD, C. M. (1953). Exogastrulation and differential cell dissociation by sodium azide in *Dendraster excentricus* and *Patira miniata*. *Physiol. Zool.*, **26**, 28–58.

CHIQUOINE, A. D. (1960). The development of the zona pellucida of the mammalian ovum. *Amer. J. Anat.*, **10**, 149–55.

COHEN, S. (1959). Purification and metabolic effects of a nerve growth-promoting protein from snake venom. *J. Biol. Chem.*, **234**, 1129–37.

COLE, K. S. (1949a). Dynamic electrical characteristics of the squid axon membrane. *Arch. Sci. Physiol.*, **3**, 253–8.

COLE, K. S. (1949b). Some physical aspects of bioelectric phenomena. *Proc. Nat. Acad. Sci. Wash.*, **35**, 558–66.

COLE, K. S. (1962). The advance of electrical models for cells and axons. *Biophys. J. Suppl.*, **2**, 101–19.

COLE, K. S. and CURTIS, H. J. (1950). Bioelectricity and electric physiology. In *Medical Physics*, edited O. Glasser. Yearbook Publ., Chicago.

COLE, K. S. and GUTTMAN, R. M. (1942). Electric impedance of the frog egg. *J. Gen. Physiol.*, **25**, 765–75.

COLLINS, M. (1966a). Electrokinetic properties of dissociated chick embryo cells—I. pH-surface charge relationships and the effect of calcium ions. *J. Exp. Zool.*, **163**, 23–37.

COLLINS, M. (1966b). Electrokinetic properties of dissociated chick embryo cells—II. Calcium ion binding by neural retinal cells. *J. Exp. Zool.*, **163**, 39–47.

COLWIN, A. L. and COLWIN, L. (1963). Role of the gamete membranes in fertilization in *Saccoglossus kowalevski* (Enteropneusta)—I. The acrosomal region and its changes in early stages of fertilization. *J. Cell. Biol.*, **19**, 477–500; II. Zygote formation by gamete membrane formation. *J. Cell. Biol.*, **19**, 501–18.

COLWIN, A. L., COLWIN, L. and PHILPOTT, D. E. (1957). Electron microscope studies of early stages of sperm penetration in *Hydroides hexagonus* (*Annelida*) and *Saccoglossus kowalevski* (*Enteropneusta*). *J. Biophys. Biochem. Cytol.*, **3**, 489–501.

COMAN, D. R. (1944). Decreased mutual adhesiveness, a property of cells from squamous cell carcinomas. *Cancer Res.*, **4**, 625–9.

COMAN, D. R. (1953). Mechanisms responsible for the origin and distribution of blood-borne tumor metastases: a review. *Cancer Res.*, **13**, 397–404.

COMAN, D. R. (1954). Cellular adhesiveness in relation to the invasiveness of cancer: electron microscopy of liver perfused with a chelating agent. *Cancer Res.*, **14**, 519–21.

COMAN, D. R. (1960). Reduction in cellular adhesiveness upon contact with a carcinogen. *Cancer Res.*, **20**, 1202–4.

COMAN, D. R. (1961). Adhesiveness and stickiness: two independent properties of the cell surface. *Cancer Res.*, **21**, 1436–8.

COMAN, D. R. (1965). Directional movement of cells as affected by aggregation and dispersal. *Cancer Res.*, **25**, 870–1.

COMAN, D. R. and ANDERSON, T. F. A. (1955). A structural difference between the surfaces of normal and carcinomatous epidermal cells. *Cancer Res.*, **15**, 541–43.

COMAN, D. R. and DE LONG, R. P. (1951). The role of the vertebral venous system in metastasis of cancer to the spinal column. *Cancer*, **4**, 610–18.

COMAN, D. R., DE LONG, R. P. and McCUTCHEON, M. (1951). Studies on the mechanisms of metastasis. The distribution of tumors in various organs in relation to the distribution of arterial emboli. *Cancer Res.*, **11**, 648–51.

COOK, G. M. W., HEARD, D. H. and SEAMAN, G. V. F. (1960). A sialo-mucopeptide liberated by trypsin from the human erythrocyte. *Nature, Lond.* **188**, 1011–12.

COOK, G. M. W., HEARD, D. H. and SEAMAN, G. V. F. (1961). Sialic acids and the electrokinetic charge of the human erythrocyte. *Nature, Lond.*, **191**, 44–7.

COOK, G. M. W., HEARD, D. H. and SEAMAN, G. V. F. (1962). The electrokinetic characterisation of the Ehrlich Ascites carcinoma cell. *Exp. Cell Res.*, **28**, 27–39.

COOK, G. M. W., SEAMAN, G. V. F. and WEISS, L. (1963). Physicochemical differences between ascitic and solid forms of sarcoma 37 cells. *Cancer Res.*, **23**, 1813–18.

COOMBS, R. R. A., GLEESON-WHITE, M. M. and HALL, J. G. (1951). Factors influencing the agglutinability of red cells. *Brit. J. Exp. Path.*, **32**, 195–202.

COWDRY, E. V. (1955). *Cancer cells.* Saunders, Philadelphia.

CRAIN, S. M. (1956). Resting and action potentials of cultured chick embryo spinal ganglion cells. *J. Comp. Neurol.*, **104**, 285–330.

CRELIN, E. S. (1952). Excision and rotation of the developing *Amblystoma* optic tectum and subsequent tectal behaviour. *J. Exp. Zool.*, **120**, 547–77.

CRISP, D. J. (1958). Surface films of polymers. From *Surface phenomena in chemistry and biology*, edited J. F. Danielli, K. G. A. Pankhurst and A. C. Riddiford. Pergamon, London, 23–54.

CURTIS, A. S. (1957). The role of calcium in cell aggregation of *Xenopus* embryos. *Proc. R. Phys. Soc. Edin.*, **26**, 25–32.

CURTIS, A. S. G. (1958). A ribonucleoprotein from amphibian gastrulae. *Nature, Lond.*, **181**, 185.

CURTIS, A. S. G. (1960a). Cell contacts: some physical considerations. *Amer. Naturalist*, **94**, 37–56.

CURTIS, A. S. G. (1960b). Area and volume measurements by random sampling methods. *Med. and Biol. Illus.*, **10**, 261–6. (Last page contains serious misprints about values obtained in statistical tests).

CURTIS, A. S. G. (1961a). Control of some cell-contact reactions in tissue culture. *J. Nat. Cancer. Inst.*, **26**, 253–68.

CURTIS, A. S. G. (1961b). Timing mechanisms in the specific adhesion of cells. *Exp. Cell Res. Suppl.*, **8**, 107–22.

CURTIS, A. S. G. (1962a). Cell contact and adhesion. *Biol. Rev.*, **37**, 82–129.

CURTIS, A. S. G. (1962b). Pattern and mechanism in the reaggregation of sponges. *Nature, Lond.*, **196**, 245–8.

CURTIS, A. S. G. (1963). The effect of pH and temperature on cell re-aggregation. *Nature, Lond.*, **200**, 1235–6.

CURTIS, A. S. G. (1964). The adhesion of cells to glass. A study by interference reflection microscopy. *J. Cell Biol.*, **19**, 199–215.

CURTIS, A. S. G. (1965a).   Some interactions between cell and medium in culture. *Arch. Biol. Liège*, **76,** 209–15.

CURTIS, A. S. G. (1965b).   Cortical inheritance in the amphibian *Xenopus laevis*: preliminary results. *Arch. Biol.*, **76,** 523–46.

CURTIS, A. S. G. (1966).   Cell adhesion. *Sci. Progress*, **54,** 61–86.

CURTIS, A. S. G. (1967b).   The promotion of cell aggregation and adhesiveness by a pure serum protein (in press).

CURTIS, A. S. G. (1967b).   The measurement of cell adhesiveness in absolute values (in press).

CURTIS, A. S. G. and GREAVES, M. F. (1965).   The inhibition of cell aggregation by a pure serum protein. *J. Embryol. Exp. Morph.*, **13,** 309–26.

CURTIS, A. S. G. and VARDE, M. (1964).   Control of cell behaviour: topological factors. *J. Nat. Cancer Inst.*, **33,** 15–26.

DAGG, C. P. KARNOFSKY, D. A. and RODDY, J. (1956).   Growth of transplantable human tumours in the chick embryo and hatched chick. *Cancer Res.*, **16,** 589–94.

DALTON, H. C. (1950a).   Comparison of white and black axolotl chromatophores *in vitro. J. Exp. Zool.*, **115,** 17–35.

DALTON, H. C. (1950b).   Inhibition of chromatoblast migration as a factor in the development of genetic differences in pigmentation in white and black axolotls. *J. Exp. Zool.*, **115,** 151–70.

DALTON, H. C. (1953).   Relations between developing melanophores and embryonic tissues in the mexican axolotl. From *Pigment Cell Growth*, edited M. Gordon. Academic Press, New York, 17–25.

DALTON, H. C. and KRASSNER, Z. P. (1956).   Pituitary influence on pigment pattern development in the white axolotl. *J. Exp. Zool.*, **133,** 241–57.

DALTON, H. C. and KRASSNER, Z. P. (1959).   Role of genetic pituitary differences in larval axolotl pigment development. From *Pigment Cell Biology*, edited M. Gordon. Academic Press, New York, 51–60.

DAMERON, F. (1961).   L'Influence de divers mesenchymes sur la differenciation de l'epithelium pulmonaire de l'embryon de Poulet en culture *in vitro. J. Embryol. Exp. Morph.*, **9,** 628–33.

DAN, K. (1936).   Electrokinetic studies of marine ova—III. *Physiol. Zool.*, **9,** 43–57.

DAN, K. (1947a).   Behaviour of the cell surface during cleavage—VIII. On the cleavage of medusan eggs. *Biol. Bull.*, **93,** 163–88.

DAN, K. (1947b).   Electrokinetic studies of marine ova—V, VI, and VII. *Biol. Bull.*, **93,** 259–86.

DAN, K. (1952).   Cytoembryological studies of sea urchins—II. Blastula stages. *Biol. Bull.*, **102,** 74–89.

DAN, K. (1954).   Further study on the formation of "new membrane" in the eggs of the sea-urchin, *Hemicentrotus (Strongylocentrotus) pulcherrimus. Embryologia*, **2,** 99–114.

DAN, K. (1960).   Cytoembryology of echinoderms and amphibia. *Intern. Rev. Cytol.*, **9,** 321–67.

DAN, K. and DAN, J. C. (1940).   Behaviour of the cell surface during cleavage—III. On the formation of new surface in the eggs of *Strongylocentrotus pulcherrimus. Biol. Bull*, **78,** 486–501.

DAN, K. and DAN, J. C. (1942).   Behaviour of the cell surface during cleavage—IV.

Polar lobe formation and cleavage of the eggs of *Ilyanassa obsoleta*. Say. *Cytologia*, 12, 246–61.

DAN, K., DAN, J. C. and YANAGITA, T. (1938). Behaviour of the cell surface during cleavage. I.—*Protoplasma*, 28, 68–81.

DAN, K. and OKAZAKI, K. (1956). Cyto-embryological studies of sea urchins—III. Role of the secondary mesenchyme cells in the formation of the primitive gut in sea urchin larvae. *Biol. Bull.*, 110, 29–42.

DANIELLI, J. D. (1941). On the pH at the surface of ovalbumin molecules and the protein error with indicators. *Biochem. J.*, 35, 470–8.

DANIELLI, J. D. (1954). Morphological and molecular aspects of active transport. *Symp. Soc. Exp. Biol.*, 8, 502–16.

DANTCHAKOFF, V. (1935a). Sur les facteurs determinant l'emplacement des gonades chez le Poulet. *C. R. Acad. Sci.*, 200, 1495–7.

DANTCHAKOFF, V. (1935b). Sur l'equivalence des tissus somatique dans les gonades du Poulet. *C. R. Acad. Sci.*, 200, 1792–5.

DAVIES, D. A. L. and HUTCHISON, A. M. (1961). The serological determination of histocompatibility activity. *Brit. J. Exp. Path.*, 42, 587–91.

DAVIES, J. T. and RIDEAL, E. (1961). *Interfacial phenomena*. Academic Press, New York, xiv + 474.

DAVIES, J. T., HAYDON, D. A. and RIDEAL, E. (1956). Surface behaviour of *Bacterium coli*—I. The nature of the surface. *Proc. Roy. Soc.*, B.145, 375–83.

DE BRUYN, P. P. H. (1945). The motion of migrating cells in tissue culture of lymph nodes. *Anat. Rec.*, 93, 295–315.

DE BRUYN, P. P. H. (1946). The amoeboid movement of the mammalian leucocyte in tissue culture. *Anat. Rec.*, 95, 177–87.

DEFENDI, V. and GASIC, G. (1963). Surface mucopolysaccharides of polyoma virus transformed cells. *J. Cell. Comp. Physiol.*, 62, 23–31.

DE GIER, J. and VAN DEENEN, L. L. M. (1961). Some lipid characteristics of red cell membranes of various animal species. *Biochem. Biophys. Acta*, 49, 286–96.

DE HAAN, R. L. (1958). Modification of cell-migration patterns in the early chick embryo. *Proc. Nat. Acad. Sci., Wash.*, 44, 32–7.

DE HAAN, R. L. (1959). The effects of the chelating agent ethylene diamine tetra-acetic acid on cell adhesion in the slime mould *Dictyostelium discoideum*. *J. Embryol. Exp. Morph.*, 7, 335–43.

DE HAAN, R. L. (1963a). Migration patterns of the precardiac mesoderm in the early chick embryo. *Exp. Cell Res.*, 29, 544–60.

DE HAAN, R. L. (1963b). Regional organization of pre-pacemaker cells in the cardiac primordia of the early chick embryo. *J. Embryol. Exp. Morph.*, 11, 65–76.

DE HAAN, R. L. (1963c). Oriented cell movements in embryogenesis. From *Biological Organization*, 147–65.

DE HAAN, R. L. (1964). Cell interactions and oriented movements during development. *J. Exp. Zool.*, 157, 127–38.

DE LANNEY, L. E. (1941). The role of ectoderm in pigment production studied by transplantation and hybridisation. *J. Exp. Zool.*, 87, 323–45.

DE LAUBENFELS, M. V. (1928). Interspecific grafting using sponge cells. *J. Elisha Mitchell Sci. Soc.*, 44, 82–6.

DE LAUBENFELS, M. V. (1934). Physiology and morphology of Porifera exemplified by *Iotrochota birotulata* (Higgin). *Pap. Tortugas Lab.*, 27, 37–66.

358   THE CELL SURFACE

De Long, G. R. and Coulombre, A. J. (1965). Development of the retinotectal topographic projection in the chick embryo. *Exp. Neurol.*, **13**, 351–63.

De Long, G. R. and Sidman, R. L. (1962). Effects of eye removal at birth on histogenesis of the mouse superior colliculus: an autoradiographic analysis with tritiated thymidine. *J. Comp. Neurol.*, **118**, 205–24.

Derjaguin, B. V. (1954). A theory of the heterocoagulation, interaction and adhesion of dissimilar particles in solutions of electrolytes. *Disc. Faraday Soc.*, **18**, 85–98.

Derjaguin, B. V., Titjevskaia, A. S., Abricossova, I. I. and Malinka, A. D. (1954). Investigations of the forces of interaction of surfaces in different media and their application to the problem of colloid stability. *Disc. Faraday Soc.*, **18**, 24–41.

Derjaguin, B. V. and Landau, L. D. (1941). Theory of the stability of strongly charged lyophobic sols and of the adhesion of strongly charged particles in solutions of electrolytes. *Acta Physicochemica URSS*, **14**, 633–62.

De The, G., Heine, U., Somner, J. R., Arvy, L., Beard, D. and Beard, J. W. (1963). Multiplicity of cell response to the BAI strain A (myeloblastosis) avian tumor virus—IV. Ultrastructural characters of the thymus in myeloblastosis and of the adenosinetriphosphatase activity of thymic cells and associated virus. *J. Nat. Cancer Inst.*, **30**, 415–55.

Detwiler, S. R. (1936). Growth responses of spinal nerves to grafted brain tissue. *J. Exp. Zool.*, **74**, 477–95.

Deuchar, E. M. (1961). Enhancement of ATPase activity, somite segregation rate and aggregation of somite cells of Xenopus laevis embryos by treatment with ATP. *Exp. Cell Res.*, **23**, 21–8.

Devillers, C. (1949). Explantations en milieu synthetique de blastodermes de truite (*Salmo irideus*). *J. Cyto-embryol. Belgo-Neeland*, 67–73.

Devillers, C. (1951). La couche enveloppante du blastoderme de *Salmo*. Son rôle dans la mecanique embryonnaire. *C. R. Acad. Sci.*, **232**, 1599–1600.

Devillers, C. (1959). Co-ordination des forces epiboliques dans la gastrulation de *Salmo. Bull. Soc. Zool. Fr.*, **77**, 307–9.

Devillers, C. (1955). Adhesivite cellulaire et morphogenese. From *Problemes de structures, d'ultrastructures et de fonctions cellulaires*, edited J. A. Thomas. Masson, Paris, 139–166.

Devillers, C. and Rajchman, S. (1960). Structural and dynamic aspects of the development of the teleostean egg. From *Advances in Morphogenesis*, edited M. Abercrombie and J. Brachet. Academic Press, New York, 379–428.

Devis, R. and James, D. (1962). Electron microscopic appearance of close relationships between adult guinea-pig fibroblasts in tissue culture. *Nature, Lond.*, **194**, 695–6.

Devis, R. and James, D. J. (1964). Close association between adult guinea-pig fibroblasts in tissue culture studied with the electron microscope. *J. Anat.*, **98**, 63–8.

Dewey, M. M. and Barr, L. (1962). Intercellular connection between smooth muscle cells: the nexus. *Science*, **137**, 670–1.

Dickson, J. A. and Leslie, I. (1963). The reaggregation and metabolism of dissociated adult guinea-pig cells in filter well cultures. *J. Cell Biol.*, **19**, 449–51.

Dingle, J. T. and Lucy, J. A. (1962). Studies on the mode of action of excess of Vitamin A. *Biochem. J.*, **84**, 611–21.

Di Stefano, H. S. (1966). Substructure of membranes in cultivated chick embryo fibroblasts. *Arch. f. Zellforsch*, **70**, 322–33.

Dodd, M. C., Bigley, N. J. and Geyer, V. G. (1963). Further observations on the serological specificity of sialic acid for $Rh_0(D)$ antibody. *J. Immunol.*, **90**, 518–25.

Dodson, E. O. (1966). Aggregation of tumour cells. *Nature, Lond.*, **209**, 40–4.

Dollander, A. (1962). Conceptions actuelles et terminologie relatives à certains aspects de l'organisation corticale de l'oeuf d'amphibien. *Arch. Anat. Histol. Embryol.* Suppl. to vol. **44**, 93–103.

Douglas, H. W. and Parker, F. (1957). Electrophoretic studies on bacteria. Part 4. The cation charge reversal spectra of spores and cells of *B. megatherium, B. cereus* and *B. subtilis. Trans. Faraday Soc.*, **53**, 1494–9.

Douglas, H. W. and Shaw, D. J. (1958). Electrophoretic studies on model particles, Part 2. The mobility against *p*H behaviour and the cation charge reversal concentration of adsorbed carboxyl colloids. *Trans. Faraday Soc.*, **54**, 1748–53.

Dourmashkin, R. R., Dougherty, R. M. and Harris, R. J. C. (1962). Electron microscopic observations on Rous sarcoma virus and cell membranes. *Nature, Lond.*, **194**, 1116–9.

Dzyaloshinski, I. E., Lifshitz, E. M. and Pitaevski, L. P. (1959). ван-дер-вальcовы силы в жидких пленках. English translation in *J. Exp. Theor. Phys.*, **10**, 161–70.

Easton, J. M., Goldberg, B. and Green, H. (1962). Demonstration of surface antigens and pinocytosis in mammalian cells with ferritin-anti-body conjugates. *J. Cell Biol.*, **12**, 437–43.

Easty, G. C. and Easty, D. M. (1963). An organ culture system for the examination of tumour invasion. *Nature, Lond.*, **199**, 1104–5.

Easty, G. C., Easty, D. M. and Ambrose, E. J. (1960). Studies on cellular adhesiveness. *Exp. Cell. Res.*, **19**, 539–48.

Easty, G. C. and Mercer, E. H. (1960). An electron microscope study of the surfaces of normal and malignant cells in culture. *Cancer Res.*, **20**, 1608–13.

Easty, G. C. and Mercer, E. H. (1962). An electron microscope study of model tissues formed by the agglutination of erythrocytes. *Exp. Cell Res.*, **28**, 215–27.

Easty, G. C. and Mutolo, V. (1960). The nature of the intercellular material of adult mammalian tissues. *Exp. Cell Res.*, **21**, 374–85.

Edwards, G. A. and Fogh, J. (1959). Micromorphologic changes in human amnion cells during trypsinisation. *Cancer Res.*, **19**, 608–11.

Eisenberg, S., Ben-Or, S. and Doljanski, F. (1962). Electrokinetic properties of cells in growth processes—I. *Exp. Cell Res.*, **26**, 451–61.

Elbers, P. F. (1964). The cell membrane: image and interpretation. *Recent progress in surface science*, **2**, 443–503.

Eley, D. D. and Hedge, D. G. (1956). The interaction of protein monolayers with dissolved proteins—I. Fibrinogen, thrombin and plasma albumin. *Proc. Roy. Soc. B.*, **145**, 554–63.

Elton, G. A. H. (1948). Electroviscosity—I. The flow of liquids between surfaces in close proximity. *Proc. Roy. Soc. A.*, **194**, 259–74.

Elworthy, P. H. and McIntosh, D. S. (1964). The effect of solvent dielectric constant on micellisation by lecithin. *Kolloidz.*, **195**, 27–34.

Emmelot, P. and Bos, C. J. (1962). Adenosine triphosphatase in the cell membrane fraction from rat liver. *Biochem. Biophys. Acta.*, **58**, 374–5.

EMMELOT, P. and BOS, C. J. (1966). On the participation of neuraminidase-sensitive sialic acid in the K dependent phosphohydrolase of p-nitrophenyl phosphate by isolated rat liver plasma membrane. *Biochem. Biophys. Acta.*, **115**, 244–7.

ENDO, Y. (1961). Changes in the cortical layer of sea urchin eggs at fertilization as studied with the electron microscope—I. *Clypeaster japonicus. Exp. Cell Res.*, **25**, 383–97.

EPSTEIN, M. A. and HOLT, S. J. (1963). Electron microscope observations on the surface adenosine triphosphatase-like enzymes of HeLa cells infected with herpes virus. *J. Cell Biol.*, **19**, 325–47.

EPSTEIN, M. A., HUMMELER, K. and BERKALOFF, A. (1964). The entry and distribution of herpes virus and colloidal gold in HeLa cells after contact in suspension. *J. Exp. Med.*, **119**, 291–302.

ESSNER, E., SATO, H. and BELKIN, H. I. (1954). Experiments on ascites hepatoma—I. Enzymatic digestion and alkaline degradation of the cementing substances, and separation of cells in tumour islands. *Exp. Cell Res.*, **7**, 430–7.

FARQUHAR, M. G. and PALADE, G. (1963). Junctional complexes in various epithelia. *J. Cell. Biol.*, **17**, 375–412.

FAURÉ-FREMIET, E. (1925). Le mécanisme de la formation des complexes a partir de cellules d'éponges dissociees. *C. R. Soc. Biol.*, **93**, 618–20.

FAURÉ-FREMIET, E. (1932a). Morphogenese experimentale (reconstitution) chez *Ficulina ficus. Arch. Anat. Micros.*, **28**, 1–80.

FAURÉ-FREMIET, E. (1932b). Involution expérimentale et temoin de structure dans les cultures de *Ficulina ficus. Arch. Anat. Micros.*, **28**, 121–52.

FAWCETT, D. (1961). Intercellular bridges. *Exp. Cell Res.* Suppl., **8**, 174–87.

FELDMAN, M. (1955). Dissociation and reaggregation of embryonic cells of *Triturus alpestris. J. Embryol. Exp. Morph.*, **5**, 251–5.

FENN, W. O. (1922). The adhesiveness of leucocytes to solid surfaces. *J. Gen. Physiol.*, **5**, 169–79.

FERNANDEZ-MORAN, H. (1954). The submicroscopic structure of nerve fibres. *Prog. Biophys. Biochem. Cytol.*, **4**, 112–47.

FERNANDEZ-MORAN, H. (1957). Electron microscopy of nervous tissue. From *The metabolism of the nervous system*, edited D. Richter. Pergamon, London.

FERNANDEZ-MORAN, H. (1961). The fine structure of vertebrate and invertebrate photoreceptors as revealed by low-temperature electron microscopy. From *The structure of the eye*, edited G. K. Smelser. Academic Press, New York, 521–56.

FERNANDEZ-MORAN, H. and FINEAN, J. B. (1957). Electron microscope and low angle X-ray diffraction studies of the nerve myelin sheath. *J. Biophys. Biochem. Cytol.*, **3**, 725–48.

FINEAN, J. B. (1956). Recent ideas on the structure of myelin. From *Biochemical problems of lipids*, edited G. Popjak and E. Le Breton. Butterworths, London, 127–31.

FINEAN, J. B. (1957). The molecular structure of nerve myelin and its significance in relation to the nerve "membrane". From *The metabolism of the nervous system*, edited D. Richter. Pergamon, London.

FINEAN, J. B. (1959). Electron microscope and X-ray diffraction studies of a saturated synthetic phospholipide. *J. Biophys. Biochem. Cytol.*, **6**, 123.

FINEAN, J. B. (1963). The nature of the stability of the plasma membrane. *Circulation*, **28**, 1151–62.

FINEAN, J. B., HAWTHORNE, J. N. and PATTERSON, J. D. E. (1956). Structural and chemical differences between optic and sciatic nerve myelin. *J. Neurochem.*, **1**, 256–9.

FINGL, E., WOODBARG, L. A. and HECHT, H. H. (1952). Effects of innervation and drugs upon direct membrane potential of embryonic chick myocardium. *J. Pharmacol. and Exp. Therap.*, **104**, 103–14.

FINNEGAN, C. V. (1955). Ventral tissues and pigment pattern in salamander larvae. *J. Exp. Zool.*, **128**, 453–80.

FINNEGAN, C. V. (1958). The influence of the hypomere on chromatoblast migration in salamanders. *J. Exp. Zool.*, **138**, 453–91.

FIRKET, J. (1920). Recherches sur l'organogenèse des glandes sexuelles chez l'oiseaux. *Arch. Biol. Liège*, **30**, 393–516, Partie 2eme.

FISHER, H. W., PUCK, T. T. and SATO, G. (1958). Molecular growth requirements of single mammalian cells: the action of fetuin in promoting cell attachment to glass. *Proc. Nat. Acad. Sci. Wash.*, **44**, 4–10.

FISHER, H. W., PUCK, T. T. and SATO, G. (1959). Molecular growth requirements of single mammalian cells. *J. Exp. Med.*, **10**, 649–60.

FLICKINGER, R. A. (1954). Utilization of $C^{14}O_2$ by developing amphibian embryos with special reference to regional incorporation into individual embryos. *Exp. Cell Res.*, **6**, 172–80.

FORRESTER, J. A., AMBROSE, E. J. and MACPHERSON, J. A. (1962). Electrophoretic investigations of a clone of hamster fibroblasts and polyoma-transformed cells from the same population. *Nature, Lond.*, **196**, 1068–70.

FORRESTER, J. A., AMBROSE, E. J. and STOKER, M. G. P. (1964). Micro-electrophoresis of normal and transformed clones of hamster kidney fibroblasts. *Nature, Lond.*, **201**, 945–6.

FOULKS, J. G. (1943). An analysis of the source of melanophores in regenerating feathers. *Physiol. Zool.*, **16**, 351–80.

FOX, M. H. (1949). Analysis of some phases of melanoblast migration in the barred Plymouth Rock embryos. *Physiol. Zool.*, **22**, 1–22.

FRAZER, R. C. (1960). Somite genesis in the chick—III. The role of induction. *J. Exp. Zool.*, **145**, 151–63.

FUCHS, N. (1934). Uber die stabilitat und aufladung der aerosole. *Z. Physik.*, **89**, 736–43.

FUJITA, S. (1964). Analysis of neuron differentiation in the central nervous system by tritiated thymidine autoradiography. *J. Comp. Neurol.*, **122**, 311–27.

GALTSOFF, P. S. (1923). The amoeboid movement of dissociated sponge cells. *Biol. Bull.*, **45**, 153–61.

GALTSOFF, P. S. (1925a). Regeneration after dissociation. (An experimental study on sponges)—I. Behaviour of dissociated cells of *Microciona prolifera* under normal and altered conditions. *J. Exp. Zool.*, **42**, 183–222.

GALTSOFF, P. S. (1929). Heteroagglutination of dissociated sponge cells. *Biol. Bull.*, **57**, 250–60.

GANGULY, B. (1960). The differentiating capacity of dissociated sponge cells. *Arch. f. Entwickmech.*, **152**, 22–34.

GARBER, B. (1953). Quantitative studies on the dependence of cell morphology and motility upon the fine structure of the medium in tissue culture. *Exp. Cell Res.*, **5**, 132–46.

GARBER, B. (1963). Inhibition by glucosamine of aggregation of dissociated embryonic cells. *Devel. Biol.*, **7**, 630–41.

GARBER, B. and MOSCONA, A. A. (1964). Aggregation *in vitro* of dissociated cells—I. Reconstitution of skin in the chorio-allantoic membrane from suspensions of embryonic chick and mouse cells. *J. Exp. Zool.*, **155**, 179–202.

GAREN, A. and PUCK, T. T. (1951). The first two steps of the invasion of host cells by bacterial viruses. *J. Exp. Med.*, **94**, 177–89.

GARVIN, J. E. (1961). Factors affecting the adhesiveness of human leucocytes and platelets *in vitro*. *J. Exp. Med.*, **114**, 451–73.

GASIC, G. and BAYDAK, T. (1964). Adhesiveness of mucopolysaccharides to the surfaces of tumor cells and vascular endothelium. From *Biological interactions in normal and neoplastic growth*, edited M. J. Brennan and W. L. Simpson. Churchill, London, 709–13.

GASIC, G. and BERWICK, L. (1963). Hale stain for sialic acid-containing mucins. *J. Cell Biol.*, **19**, 223–8.

GASIC, G. J. and GALANTI, N. L. (1966). Proteins and disulphide groups in the aggregation of dissociated cells of sea sponges. *Science*, **151**, 203–5.

GASIC, G. and GASIC, T. (1962). Removal of sialic acid from the cell coat in tumour cells and vascular endothelium and its effects on metastases. *Proc. Nat. Acad. Sci. Wash.*, **48**, 1172–7.

GAZE, R. M. (1958). The representation of the retina on the optic lobe of the frog. *Q.J. Exp. Physiol.*, **43**, 209–14.

GAZE, R. M. and JACOBSON, M. (1962). The projection of the binocular visual field on the optic tecta of the frog. *Q.J. Exp. Physiol.*, **47**, 273–80.

GAZE, R. M. and JACOBSON, M. (1963). A study of the retinotectal projection during regeneration of the optic nerve in the frog. *Proc. Roy. Soc.* **B.**, **157**, 470–548.

GAZE, R. M., JACOBSON, M. and SZEKELY, G. (1963). The retino-tectal projection in *Xenopus* with compound eyes. *J. Physiol.*, **165**, 484–99.

GAZE, R. M., JACOBSON, M. and SZEKELY, G. (1965). On the formation of connexions by compound eyes in *Xenopus*. *J. Physiol.*, **176**, 409–17.

GENT, W. L. G., GREGSON, N. A., GAMMACK, D. B. and RAPER, J. H. (1964). The lipid-protein unit in myelin. *Nature, Lond.*, **204**, 553–5.

GEREN, B. B. (1954). The formation from the Schwann cell surface of myelin in the peripheral nerves of chick embryos. *Exp. Cell Res.*, **7**, 558–62.

GERISCH, G. (1960). Zellfunktionen und zellfunktionswechsel in der entwicklung von *Dictyostelium discoideum*—I. *Arch. entwicklung*, **152**, 632–54.

GERISCH, G. (1961). Zellfunktionen und zellfunktionswechsel in der entwicklung von *Dictyostelium discoideum*—II. Aggregation homogener zellpopulationen und zentrenbildung. *Devel. Biol.*, **3**, 685–724.

GESSNER, B. M. and GINSBURG, V. (1964). Effect of glycosidases on the fate of transfused lymphocytes. *Proc. Nat. Acad. Sci. Wash.*, **52**, 750–5.

GEY, G. O., SHAPRAS, P. and BORYSKO, E. (1954). Activities and responses of living cells and their components as recorded by cinephase microscopy and electron microscopy. *Ann. N.Y. Acad. Sci.*, **58**, 1089–109.

GIERSBURG, H. (1929). Beitrage zur entwicklungsphysiologie amphibien—II. Neurulation bei *Rana* und *Triton*. *Arch. Anat. Embryol.*, **103**, 387–424.

GILLESPIE, T. and RIDEAL, E. K. (1956). The coalescence of drops at an oil–water interface. *Trans. Faraday Soc.*, **52**, 173–83.

GILLETTE, R. (1944). Cell number and cell size in the ectoderm during neurulation. (*Amblystoma maculatum*). *J. Exp. Zool.*, **96**, 201–22.

GIUDICE, G. (1962). Reconstitution of whole larvae from disaggregated cells of sea urchin embryos. *Devel. Biol.*, **5**, 402–11.

GLAESER, R. M. and MEL, H. C. (1964). The electrophoretic behaviour of osmium tetroxide-fixed and potassium permanganate-fixed rat erythrocytes. *Biochem. Biophys. Acta.*, **79**, 606–17.

GLASER, O. C. (1914). On the mechanism of morphological differentiation in the nervous system. *Anat. Rec.*, **8**, 525–51.

GLASER, O. C. (1916). The theory of autonomous folding in embryogenesis. *Science*, **44**, 505–9.

GLAUERT, A. M., DINGLE, J. T. and LUCY, J. A. (1962). Action of saponin on biological cell membranes. *Nature, Lond.*, **196**, 953–5.

GLAUERT, A. M., DANIEL, M. R., LUCY, J. A. and DINGLE, J. T. (1963). Studies on the mode of action of excess of Vitamin A—VII. Changes in the fine structure of erythrocytes during haemolysis by Vitamin A. *J. Cell Biol.*, **17**, 111–21.

GLYNN, L. E. and HOLBOROW, E. J. (1959). Distribution of blood-group substances in human tissues. *Brit. Med. Bull.*, **15**, 150–3.

GOLDACRE, R. F. (1952). The folding and unfolding of protein molecules as a basis of osmotic work. *Int. Rev. Cytol.*, **1**, 135–64.

GOLDACRE, R. F. (1961). The role of the cell membrane in the locomotion of amoebae, and the source of the motive force and its control by feedback. *Exp. Cell Res. Suppl.*, **8**, 1–16.

GOLDACRE, R. J. (1964). On the mechanism and control of amoeboid movement. From *Primitive motile systems in cell biology*, edited R. D. Allen and N. Kamiya. Academic Press, New York, 237–53.

GOLDBERG, B. and GREEN, H. (1959–60). The cytotoxic action of immune gamma globulin and complement on Krebs ascites tumor cells—I. Ultrastructural studies. *J. Exp. Med.*, **109**, 505–10.

GOLDSMITH, J. B. (1928). The history of the germ cells in the domestic fowl. *J. Morphol. Physiol.*, **46**, 275–315.

GOLOSOW, N. and GROBSTEIN, C. (1962). Epitheliomesenchymal interaction in pancreatic morphogenesis. *Devel. Biol.*, **4**, 242–55.

GORTER, E. and GRENDEL, R. (1925). On bimolecular layers of lipoids on the chromocytes of blood. *J. Exp. Med.*, **41**, 439–43.

GOWANS, J. L. and KNIGHT, E. J. (1964). The route of recirculation of lymphocytes in the rat. *Proc. Roy. Soc. B.*, **159**, 257–82.

GRANGER, G. A. and WEISER, R. S. (1964). Homograft target cells: specific destruction *in vitro* by contact interaction with immune macrophages. *Science*, **145**, 1427–9.

GRAPER, L. (1929). Die primitiventwicklung des Huhnchens nach stereokinematographischen untersuchungen, kontrolliert durch vitale farbmarkierung und verglichen mit entwicklung ander Wirbeltiere. *Arch. f. Entwickmech.* **116**, 382–429.

GREEN, H. and LOVINCZ, A. L. (1957). The role of a natural antibody in the rejection of Krebs 2 mouse tumor cells by the chick embryo. *J. Exp. Med.*, **106**, 111–26.

GREGG, G. J. H. (1956). Serological investigations of cell adhesion in the slime moulds, *Dictyostelium discoideum*, *Dictyostelium purpureum*, and *Polysphondylium violaceum*. *J. Gen. Physiol.*, **39**, 813–20.

GREGG, G. J. H. (1960). Surface antigen dynamics in the slime mould, *Dictyostelium discoideum*. *Biol. Bull.*, **118**, 70–8.

GREGG, G. J. H. and TRYGSTAD, C. (1958). Surface antigen defects contributing to developmental failure in aggregating variants of the slime mould, *Dictyostelium discoideum*. *Exp. Cell Res.*, **15**, 358–69.

GRIFFIN, J. L. and ALLEN, R. D. (1960). The movement of particles attached to the surface of amoebae in relation to current theories of amoeboid movement. *Exp. Cell Res.*, **20**, 619–22.

GROBSTEIN, C. (1953). Analysis *in vitro* of the early organisation of the rudiment of the mouse submandibular gland. *J. Morph.*, **93**, 19–144.

GROBSTEIN, C. (1955). Tissue disaggregation in relation to determination and stability of cell type. *Ann. N.Y. Acad. Sci.*, **60**, 965–1160.

GROBSTEIN, C. (1961). Cell contact in relation to embryonic induction. *Exp. Cell Res. Suppl.*, **8**, 234–45.

GROVER, J. (1961). The relation between the embryonic age of dissociated chick lung cells and their capacity for reaggregation and histogenesis *in vitro*. *Exp. Cell. Res.*, **24**, 171–3.

GRUBB, R. (1955). An estimate of the number of Rh receptors on a single red cell. *Acta Genet. Statist. Med.*, **5**, 377–80.

GRUNDFEST, H. (1961). Ionic mechanisms in electrogenesis. *Ann. N.Y. Acad. Sci.*, **94**, 405–57.

GRUNDFEST, H. (1964). Impulse-conducting properties of cells. From *The general physiology of cell specialization*, edited D. Mazia and A. Tyler. McGraw-Hill, New York, 277–322.

GUSTAFSON, T. and KINNANDER H. (1960). Cellular mechanisms in morphogenesis of the sea urchin gastrula. *Exp. Cell Res.*, **21**, 361–73.

GUSTAFSON, T. and WOLPERT, L. (1961). Studies on the cellular basis of morphogenesis in the sea urchin embryo. Directed movements of primary mesenchyme cells in normal and vegetalized larvae. *Exp. Cell Res.*, **24**, 64–79.

GUSTAFSON, T. and WOLPERT, L. (1962). Cellular mechanisms in the morphogenesis of the sea urchin larva. Change in cell sheets. *Exp. Cell Res.*, **27**, 260–79.

GUSTAFSON, T. and WOLPERT, L. (1963). The cellular basis of morphogenesis and sea urchin development. *Int. Rev. Cytol.*, **15**, 139–214.

GWATKIN, R. B. L. and THOMSON, J. L. (1964). "Pronase"—A new method for dispersing the cells of mammalian tissues. *Nature, Lond.*, **201**, 1242–3.

HACHISU, S. and FURUSAWA, K. (1963). On the long range attraction force between colloid particles of tungstic acid—I. *Science of Light*, **12**, 1–8.

HADEK, R. (1963). Submicroscopic changes in the penetrating spermatozoon of the rabbit. *J. Ultrastr. Res.*, **8**, 161–9.

HADEK, R. (1964). Submicroscopic studies on the cortical villi in the rabbit ovum. *J. Ultrastr. Res.*, **10**, 58–65.

HADORN, E., ANDERS, G. and URSPRUNG, H. (1959). Kombinate aus teilweise dissoziierten imaginalscheiben verschiedener mutanten und arten von Drosophila. *J. Exp. Zool.*, **142**, 159–75.

HALPERN, B., PEJSACHOWICZ, B., FEBVRE, H. L. and BARSKI, G. (1966). Differences in patterns of aggregation of malignant and non-malignant mammalian cells. *Nature, Lond.*, **209**, 157–61.

HAMA, K. (1960). The structure of the desmosomes in frog mesothelium. *J. Biophys. Biochem. Cytol.*, **7**, 575–7.

HAMAKER, H. C. (1937). The London–Van der Waals attraction between spherical particles. *Physica.*, **4**, 1058–72.

HAMBURGER, R. N., PIOUS, D. A. and MILLS, S. E. (1963). Antigenic specificities acquired from the growth medium by cells in tissue culture. *J. Immunol.*, **6**, 439–49.

HAMPTON, J. C. (1958). An electron microscope study of the hepatic uptake and accretion of submicroscopic particles injected into the blood stream and into the bile duct. *Acta Anatomica.*, **32**, 262–91.

HARBER, D. S., STERN, B. K. and REILLY, R. W. (1964). Removal and dissociation of epithelial cells from the rodent gastro-intestinal tract. *Nature, Lond.*, **203**, 319–20.

HARRIS, H. (1961). Chemotaxis. *Exp. Cell Res.* Suppl., **8**, 199–208.

HARRIS, J. E. and FORD, C. E. (1963a). Cellular traffic of the thymus; experiments with chromosome markers. *Nature, Lond.*, **201**, 884–5.

HARRIS, J. E. and FORD, C. E. (1963b). The role of the thymus: immigration of cells from thymic grafts to lymph nodes in mice. *Lancet*, **7277**, 389–90.

HARRISON, R. G. (1914). The reaction of embryonic cells to solid structures. *J. Exp. Zool.*, **17**, 521–44.

HARVEY, E. N. (1954). Tension at the cell surface. *Protoplasmatologia*, **5**.

HARVEY, E. N. and DANIELLI, J. F. (1938). Properties of the cell surface. *Biol. Rev.*, **13**, 319–41.

HAYDON, D. A. (1961). The surface charge of cells and some other small particles as indicated by electrophoresis—I. The zeta-potential surface charge relationships. *Biochem. Biophys. Acta.*, **50**, 430–57.

HAYDON, D. A. and SEAMAN, G. V. F. (1962). An estimation of the surface ionogenic groups of the human erythrocyte and of *Escherichia coli*. *Proc. Roy. Soc. B.*, **156**, 533–49.

HAYES, R. L. (1965). An *in vitro* technique for reaggregation of dissociated tissue in a centrifugal field. *Exp. Cell Res.*, **37**, 1–11.

HAYES, T. L., LINDGREN, F. I. and GOFMAN, J. W. (1963). A quantitative determination of the osmium tetroxide-lipoprotein interaction. *J. Cell Biol.*, **19**, 251–5.

HAYRY, P., MYLLYLA, G., SAXEN, E. and PENTINNEN, K. (1966). The inhibitory mechanism of serum on the attachment of HeLa cells on glass. *Ann. Med. Exp. Fenn.*, **44**, 166–70.

HAYRY, P., PENTINNEN, K. and SAXEN, E. (1965). The different effects of some methods of disaggregation on the electrophoretic mobility of the HeLa cell. *Ann. Med. Exp. Fenn.*, **43**, 91–4.

HEALY, T. W. (1961). Flocculation-dispersion behaviour of quartz in the presence of a polyacrylamide flocculant. *J. Coll. Sci.*, **16**, 609–17.

HEALY, T. W. and LA MER, V. K. (1964). The energetics of flocculation and redispersion by polymers. *J. Coll. Sci.*, **19**, 323–32.

HEARD, D. and SEAMAN, G. V. F. (1960). The influence of pH and ionic strength on the electrokinetic stability of the human erythrocyte membrane. *J. Gen. Physiol.*, **43**, 635–54.

HEARD, D. and SEAMAN, G. V. F. (1961). The action of lower aldehydes on the human erythrocyte. *Biochem. Biophys. Acta.*, **53**, 366–74.

HEARD, D., SEAMAN, G. V. F. and SIMON-Reuss, I. (1961). Electrophoretic mobility of cultured mesodermal tissue cells. *Nature, Lond.*, **190**, 1009.

HEBB, C. R. and CHU, M-Y. W. (1960). Reversible injury of L-strain mouse cells by trypsin. *Exp. Cell Res.*, **20**, 453–7.

HEIDENHAIN, M. (1907). From *Plasma und Zelle*, G. Fischer, Jena, 51.

HERBST, C. (1900). Uber das auseinandergehen im Furchung's und Gewebe-zellen in kalkfreiem medium. *Arch. f. Entwickmech*, **9**, 424–63.

HERZENBERG, L. A. and HERZENBERG, L. A. (1961). Association of H-2 antigens with the cell membrane fraction of mouse liver. *Proc. Nat. Acad. Sci. Wash.*, **47**, 762–7.

HILLIER, J. and HOFFMAN, J. F. (1953). On the ultrastructure of the plasma membrane as determined by the electron microscope. *J. Cell. Comp. Physiol.*, **42**, 203–48.

HIRAMOTO, R., GOLDSTEIN, M. N. and PRESSMAN, D. (1960). Limited fixation of antibody by viable cells. *J. Nat. Cancer Inst.*, **24**, 255–65.

HIS, W. (1874). Unsere korperform und das physiologische problem ihrer enstehung. F. Vogel, Leipzig.

HODES, M. E., PALMER, C. G. and LIVENGOOD, D. (1961). The action of synthetic surfactants in membranes of tumour cells. *Exp .Cell Res.*, **24**, 298–310.

HOKIN, L. E. and HOKIN, M. R. (1960). The role of phosphatidic acid and phosphoinositide in transmembrane transport elicited by acetylcholine and other humoral agents. *Int. Rev. Neurobiol.*, **2**, 99–136.

HOLTFRETER, J. (1939). Gewebeaffinitat, ein Mittel der embryonalen formbildung. *Arch. Exp. Zellf.*, **23**, 169–209.

HOLTFRETER, J. (1943a). Properties and functions of the surface coat in amphibian embryos. *J. Exp. Zool.*, **93**, 251–323.

HOLTFRETER, J. (1943b). A study of the mechanics of gastrulation. Part I. *J. Exp. Zool.*, **94**, 261–318.

HOLTFRETER, J. (1944). A study of the mechanics of gastrulation; Part II. *J. Exp. Zool.*, **95**, 171–212.

HOLTFRETER, J. (1947a). Observations on the migration, aggregation and phagocytosis of embryonic cells. *J. Morph.*, **80**, 25–56.

HOLTFRETER, J. (1947b). Significance of the cell membrane in embryonic processes. *Ann. N.Y. Acad. Sci.*, **49**, 709–60.

HOOKER, D. (1925). Studies on regeneration in the spinal cord—III. Re-establishment of anatomical and physiological continuity after transection in frog tadpoles. *J. Comp. Neurol.*, **38**, 315–47.

HORSTADIUS, S. (1950). *The neural crest.* Oxford Univ. Press, Oxford.

HUANG, C., WHEELDON, L. and THOMPSON, T. E. (1964). The properties of lipid bilayer membranes separating two aqueous phases: formation of a membrane of simple composition. *J. Mol. Biol.*, **8**, 148–60.

HUBBARD, M. J. and ROTHSCHILD, V. (1939). Spontaneous rhythmic impedance changes in the trout's egg. *Proc. Roy. Soc.* **B.**, **127**, 510–26.

HUMPHREYS, T. (1961). The fate of Ehrlich ascites cells injected intravenously into chick embryos. *Transpl. Bull.*, **26**, 118–20.

HUMPHREYS, T. (1963). Chemical dissolution and *in vitro* reconstruction of sponge cell adhesions—I. Isolation and functional demonstration of the components involved. *Devel. Biol.*, **8**, 27–47.

HUMPHREYS, T. (1965a). Aggregation of chemically dissociated sponge cells in the absence of protein synthesis. *J. Exp. Zool.*, **160**, 235–9.

HUMPHREYS, T. (1965b). Cell surface components participating in aggregation: evidence for a new cell particulate. *Exp. Cell Res.*, **40**, 539–43.

HUMPHREYS, W. J. (1964). Electron microscope studies of the fertilized egg and the two-cell stage of *Mytilus edulis*. *J. Ultrastr. Res.*, **10**, 244–62.

HUNT, (1937). The development of gut and its derivatives from the mesectoderm and mesentoderm of early chick blastoderms. *Arch. f. Entwickmech.*, **68**, 349–70.

HUNTER, R. J. (1960). Constancy of the surface charge density of human erythrocytes at different ionic strengths. *Arch. Biochem. Biophys.*, **88**, 308–12 (see also inserted correction).

HUXLEY, J. (1921). Differences in viability in different types of regenerates from dissociated sponges, with a note on the entry of somatic cells by spermatozoa. *Biol. Bull.*, **40**, 127–9.

HYMAN, L. B. (1940). *The Invertebrata*, Vol. 1. McGraw-Hill, New York.

IKUSHIMA, N. (1959). Development of the organized embryo from the temporarily disaggregated gastrula in amphibia. *Mem. Coll. Sci. Kyoto*, **26**, 241–8.

IMMERS, J. (1956). Cytological features of the development of the eggs of *Paracentrotus lividus* reared in artificial sea water devoid of sulphate ions. *Exp. Cell Res.*, **10**, 546–8.

INGVAR, S. (1920). Reaction of cells to the galvanic current in tissue culture. *Proc. Soc. Exp. Biol. Med.*, **17**, 198–9.

IZQUIERDO, L. and VIAL, J. D. (1962). Electron microscope observations on the early development of the rat. *Z. f. Zellforsch.*, **56**, 157–79.

JACOBSON, C. O. (1962). Cell migration in the neural plate and the process of neurulation in the axolotl larva. *Zool. Bid. Uppsala.*, **35**, 433–49.

JACOBSON, C. O. (1964a). Motor nuclei, cranial nerve roots and fibre pattern in the medulla oblongata after reversal experiments on the neural plate of axolotl larvae—I. Bilateral operations. *Zool. Bid. Uppsala.*, **36**, 73–158.

JACOBSON, C. O. (1964b). Experimental investigations on the development of the urodele brain. *Acta. Univ. Uppsala.*, **36**, 1–15.

JACOBSON, M. (1962). The representation of the retina on the optic tectum of the frog. Correlation between retinotectal magnification factor and the retinal ganglion cell count. *Q.J. Exp. Physiol.*, **47**, 170–8.

JACOBSON, M. and GAZE, R. M. (1964). Types of visual response from single units in the optic tectum and optic nerve of the goldfish. *Q.J. Exp. Physiol.*, **49**, 199–209.

JACOBSON, M. and GAZE, R. M. (1965). Selection of appropriate tectal connections by regenerating optic nerve fibres in adult goldfish. *Exp. Neurol.*, **13**, 418–30.

JAKUS, M. A. (1956). Studies on the cornea—II. The fine structure of Descemets membrane. *J. Biophys. Biochem. Cytol.*, **2**, suppl., 243–52.

JAMES, A. M., LOVEDAY, D. E. and PLUMMER, D. T. (1964). Some physical investigations of the behaviour of bacterial surfaces. *Biochem. Biophys. Acta.*, **79**, 351–63.

JEHLE, H. (1963). Intermolecular forces and biological specificity. *Proc. Nat. Acad. Sci. Wash.*, **50**, 516–24.

JEHLE, H., PARKE, W. and SALYERS, A. (1965). Charge fluctuation interactions and molecular biophysics. *Biophysics*, **9**, 433–47.

JELINEK, R. and FRIEBOVA, Z. (1966). Influence of mitotic activity on neurulation movements. *Nature, Lond.*, **209**, 822–3.

JEON, K. W. and BELL, L. G. E. (1965). Chemotaxis in large free-living amoebae. *Exp. Cell Res.*, **38**, 536–55.

JOHNSSON, R. and HEGYELI, A. (1965). Induced orientation of the growth of malignant cells *in vitro. Proc. Nat. Acad. Sci. Wash.*, **54**, 1375–8.

JOLY, M. (1953). La rheologie superficielle: un modele pour les processus d'ecoulement des systemes colloidaux. *Kolloid. Z.*, **139**, 43–51.

JOLY, M. (1956). Non-Newtonian surface viscosity. *J. Coll. Sci.*, **11**, 519–31.

JONES, B. M. (1960). Some studies relating to the problem of cellular adhesion. *Proc. R. Phys. Soc. Edin.*, **28**, 35–42.

JONES, B. M. (1965). Inhibitory effect of p-benzoquinone on the aggregation behaviour of embryo-chick fibroblast cells. *Nature, Lond.*, **205**, 1280–2.

JONES, B. M. (1966). A unifying hypothesis of cell adhesion. *Nature, Lond.*, **212**, 362–5.

JONES, P. C. T. (1966). A contractile protein model for cell adhesion. *Nature, Lond.*, **212**, 365–9.

JOUWERSMA, C. (1960). On the theory of peeling. *Polymer Sci.*, **45**, 253–55.

JOY, R. T. and FINEAN, J. B. (1963). A comparison of the effects of freezing and of treatment with hypertonic solutions on the structure of nerve myelin. *J. Ultrastr. Res.*, **8**, 264–82.

JUST, E. E. (1919). The fertilization reaction in *Echinarachnius parva* (Parts I, II and III). *Biol. Bull.*, **30**, 1–70.

KABAT, E. A. and MAYER, M. M. (1958). *Experimental immunochemistry.* 2nd edition. Thomas, Springfield, xii + 905.

KAMAT, V. B. and WALLACH, D. F. H. (1965). Separation and partial purification of plasma-membrane fragments from Ehrlich ascites carcinoma microsomes. *Science*, **148**, 1343–5.

KANNO, Y. and LOWENSTEIN, W. L. (1963). A study of the nucleus and cell membrane of oocytes with an intra-cellular electrode. *Exp. Cell Res.*, **31**, 149–66.

KAPPERS, C. U. A. (1917). Further contributions on neurobiotaxis—IX. *J. Comp. Neurol.*, **27**, 261–98.

KARLSSON, U. and SCHULTZ, R. (1964). Plasma membrane apposition in the central nervous system after aldehyde perfusion. *Nature, Lond.*, **201**, 1230–1.

KARLSSON, U. and SCHULTZ, R. (1965). Fixation of the central nervous system for electron microscopy by aldehyde perfusion—I. Preservation with aldehyde perfusates v. direct perfusion with osmium tetroxide with special reference to membranes and the extra cellular space. *J. Ultrastr. Res.*, **12**, 160–86.

KARRER, H. E. (1960a). Cell interconnections in normal human cervical epithelium. *J. Biophys. Biochem. Cytol.*, **7**, 181–3.

KARRER, H. E. (1960b). Electron microscopic study of the phagocytosis process in lung. *J. Biophys. Biochem. Cytol.*, **7**, 357–66.

KATCHALSKY, A. (1961). Membrane permeability and the thermodynamics of irreversible processes. From *Membrane transport and metabolism*, edited A. Kleinzeller and A. Kotyk. Academic Press, New York, 69–86.

KATCHALSKY, A. (1964). Polyelectrolytes and their biological interactions. *Biophys. J.*, **4**, 9–42.

KATCHALSKY, A., DANON, D., NEVO, A. and DE VRIES (1959). Interactions of basic polyelectrolytes with the red blood cell—II. Agglutination of red blood cells by polymeric bases. *Biochem. Biophys. Acta.*, **33**, 120–38.

KATZMAN, R. and WILSON, C. E. (1961). Extraction of lipid and lipid cation from frozen brain tissue. *J. Neurochem.*, **7**, 113–27.

KAVANAU, J. L. (1963). Structure and functions of biological membranes. *Nature, Lond.*, **198**, 525–30.

KELUS, A., GURNER, B. W. and COOMBS, R. R. A. (1959). Blood group antigens in HeLa cells shown by mixed agglutination. *Immunol.*, **2**, 262–7.

KEMP, N. E. (1956). Electron microscopy of growing oocytes of *Rana pipiens*. *J. Biophys. Biochem. Cytol.*, **2**, 281–92.

KESSEL, R. G. (1960). The role of cell division in gastrulation of *Fundulus heteroclitus*. *Exp. Cell. Res.*, **20**, 277–82.

KIRKWOOD, J. G. (1955). The influence of fluctuations in protein charge and charge configuration on the rate of enzymatic reactions. *Disc. Faraday Soc.*, **20**, 78–82.

KIRKWOOD, J. G. and SCHUMAKER, J. B. (1952). Forces between protein molecules in solution arising from fluctuations in proton charge and configuration. *Proc. Nat. Acad. Sci. Wash.*, **38**, 863–71.

KITCHENER, J. A. and PROSSER, A. P. (1957). Direct measurement of the long-range van der Waals forces. *Proc. Roy. Soc. A.*, **242**, 403–9.

KITCHIN, I. C. (1949). The effects of notochordectomy in *Amblystoma mexicanum*. *J. Exp. Zool.*, **112**, 393–415.

KITE, J. H. and MERCHANT, D. J. (1961). Studies of some antigens of the L strain mouse fibroblast. *J. Nat. Cancer Inst.*, **26**, 419–34.

KLENK, E. and UHLENBRUCK, G. (1960). Uber neuraminsaurehaltige mucoide aus menschenerythrocyten-stroma, ein beitrag zur chemie der agglutinogene. *Z. f. Physiol. Chem.*, **319**, 151–60.

KLUG, A. and BERGER, J. E. An optical method for the analysis of periodicities in electron micrographs, and some observations on the mechanism of negative staining. *J. Mol. Biol.*, **10**, 565–9.

KODANI, M. (1962). *In vitro* alteration of blood group phenotypes of human epithelial cells exposed to heterologous blood group substances. *Proc. Soc. Exp. Biol. Med.*, **109**, 252–9.

KOJIMA, K. and SAKAI, I. (1964). On the role of stickiness of tumor cells in the formation of metastases. *Cancer Res.*, **24**, 1887–91.

KOLLROS, J. J. (1943). Experimental studies on the development of the corneal reflex in Amphibia—III. The influence of the periphery upon the reflex center. *J. Exp. Zool.*, **92**, 121–42.

KONO, T. and COLOWICK, S. P. (1961). Isolation of skeletal muscle cell membrane and some of its properties. *Arch. Biochem. Biophys.*, **93**, 520–33.

KORN, E. D. (1966). Structure of biological membranes. *Science*, **153**, 1491–8.

KORNAKER, K. (1963). Some properties of the afferent pathway in the frog corneal reflex. *Exp. Neurol.*, **7**, 224–39.

KRAGH, A. M. and LANGSTON, W. B. (1962). The flocculation of quartz and other suspensions with gelatine. *J. Coll. Sci.*, **17**, 101–23.

KRAEMER, P. M. (1966). Sialic acid of mammalian cell lines. *J. Cell Physiol.*, **67**, 23–34.

KREDEL, F. E. (1927). The physical relation of cells in tissue culture. *Johns Hopkins Hosp. Bull.*, **40**, 216–27.

KUCHLER, R. J., MARLOWE, M. L. and MERCHANT, J. (1960). The mechanism of cell binding and cell-sheet formation in L strain fibroblasts. *Exp. Cell Res.*, **20**, 428–37.

KUHL, W. (1937). Untersuchungen uber das verhalten kunstlich getrennter furchungszellen und zellaggregate einiger amphibienarten mit hilfe des zeitrafferfilmes (Laufbild und Teilbild analyse). *Arch. f. Entwickmech.*, **136**, 593–671.

KURODA, Y. (1964). Studies on cartilage cells *in vitro*—II. Changes in aggregation and in cartilage-forming activity of cells maintained in monolayer cultures. *Exp. Cell Res.*, **35**, 337–41.

LA MER, V. K. (1964). Introduction. (This short paper is a discussion of the proper use of the terms *flocculation* and *coagulation*.) *J. Coll. Sci.*, **19**, 291–3.

LANDESMAN, R. and DALTON, H. C. (1964). Tissue environment and morphogenesis of axolotl melanophores. *J. Exp. Zool.*, **114**, 255–61.

LASANSKY, A. and DE ROBERTIS, E. (1960). Electron microscopy of retinal photoreceptors. *J. Biophys. Biochem. Cytol.*, **7**, 493–8.

LAWS, J. O. and STRICKLAND, L. H. (1961). The adhesion of liver cells. *Exp. Cell Res.*, **24**, 240–54.

LE BARON, F. N. and FOLCH, J. (1956). The isolation from brain tissue of a trypsin-resistant fraction containing combined inositol, and its relation to neurokeratin. *J. Neurochem.*, **1**, 101–8.

LEESON, T. S. and KALANT, H. (1961). Effects of *in vivo* decalcification on ultrastructure of adult rat liver. *J. Biophys. Biochem. Cytol.*, **10**, 95–104.

LEFFORD, F. (1964). The effect of donor age on the emigration of cells from chick embryo explants *in vitro*. *Exp. Cell Res.*, **35**, 557–71.

LEHMAN, H. E. (1950). The suppression of melanophore differentiation in salamander larvae following orthotopic exchanges of neural folds between species of *Amblystoma* and *Triturus*. *J. Exp. Zool.*, **114**, 435–64.

LEHMAN, H. E. (1951). An analysis of the dynamic factors responsible for the phenomenon of pigment suppression in salamander larvae. *Biol. Bull.*, **100**, 127–52.

LEHMAN, H. E. (1953). Analysis of the development of pigment patterns in larval salamanders with special reference to the influence of epidermis and mesoderm. *J. Exp. Zool.*, **124**, 571–620.

LEHMAN, H. E. (1957). The developmental mechanics of pigment pattern formation in the black axolotl, *Amblystoma mexicanum*—I. *J. Exp. Zool.*, **135**, 355–86.

LEHMAN, H. E. and YOUNGS, L. M. (1959). Extrinsic and intrinsic factors influencing amphibian pigment pattern formation. From *Pigment cell biology*, edited M. Gordon. Academic Press, New York, 1–36.

LEIGHTON, J., KLINE, I., BELKIN, M. and TETENBAUM, Z. (1956). Studies on human cancer using sponge-matrix tissue culture—III. The invasive properties of a carcinoma (Strain HeLa) as influenced by temperature variations, by conditioned media and in contact with rapidly growing chick embryonic tissue. *J. Nat. Cancer Inst.*, **16**, 1353–73.

LESSEPS, R. J. (1963). Cell surface projections: their role in the aggregation of embryonic chick cells as revealed by electron microscopy. *J. Exp. Zool.*, **153**, 171–82.

LESTER, G. R. (1961). Contact angles of liquids at deformable solid surfaces. *J. Coll. Sci.*, **16**, 315–26.

LEVI-MONTALCINI, R. (1950). The origin and development of the visceral system in the spinal cord of the chick embryo. *J. Morph.*, **86**, 253–77.

LEVI-MONTALCINI, R. (1965). Morphological and metabolic effects of the nerve growth factor. *Arch. Biol., Liège.*, **76**, 387–417.

LEVIN, E. and THOMAS, L. E. (1961). Cellular lipoproteins—III. The insoluble lipoprotein of rat liver fraction. *Exp. Cell Res.*, **22**, 363–9.

LEVINE, S. (1946). On the interaction of colloidal particles—II. Discussion of basis of theory. *Trans. Faraday Soc.*, **44**, 833–43.

LEWIS, W. H. (1947). Mechanics of invagination. *Anat. Rec.*, **97**, 139–56.

LEWIS, W. H. (1949). Gel layers of cells and eggs and their role in early development. *Lecture Ser. Jackson Memorial Laboratory*, 59–77.

LEWIS, W. H. and LEWIS, M. R. (1912). The cultivation of sympathetic nerves from the intestine of chick embryos in saline solution. *Anat. Rec.*, **6**, 7–31.

LEWIS, W. H. and WEBSTER, L. T. (1921). Migration of lymphocytes in plasma cultures of human lymph nodes. *J. Exp. Med.*, **22**, 261–9.

LIEBERMAN, I. and OVE, P. (1957). Purification of a serum protein required by a mammalian cell in tissue culture. *Biochem. Biophys. Acta.*, **25**, 449–50.

LIEBERMAN, I. and OVE, P. (1958). A protein growth factor for mammalian cells in culture. *J. Biol. Chem.*, **233**, 637–42.

LIFSHITZ, E. M. (1956). теория молекчлярных сил притяжения между твердыми телами. (The theory of molecular attractive forces between solids.) *J. Exp. Physics.*, **2**, 73–83.

LING, G. N. (1962). *A physical theory of the living state.* Blaisdell, New York.

LISTGARTEN, M. A. (1964). The ultrastructure of human gingival epithelium. *Amer. J. Anat.*, **114**, 49–69.

LOEB, L. (1921). Amoeboid movement, tissue formation and consistency of protoplasm. *Amer. J. Physiol.*, **56**, 140–67.

LOEWENSTEIN, W. R. (1966). Permeability of membrane junctions. *Ann. N.Y. Acad. Sci.*, **137**, 441–72.

LOVTRUP, S. (1962). On the surface coat in the amphibian embryo. *J. Exp. Zool.*, **150**, 197–206.

LOVTRUP, S. (1965). Morphogenesis in the amphibian embryo. Gastrulation and neurulation. *Zool. Gothoburgensia*, **I**, 1–139.

LOWICK, J. H. B., PURDOM, L., JAMES, A. M. and AMBROSE, E. J. (1961). Some microelectrophoretic studies of normal and tumour cells. *J. Roy. Micros. Soc.*, **80**, 47–57.

LUBINSKA, L. (1961). Sedentary and migratory states of Schwann cells. *Exp. Cell Res.* Suppl., **8**, 74–90.

LUCEY, E. C. A. and CURTIS, A. S. G. (1959). Time-lapse film study of cell reaggregation. *Med. Biol. Illus.*, **9**, 86–93.

LUCY, J. A. (1964). Globular lipid micelles and cell membranes. *J. Theor. Biol.*, **7**, 360–73.

LUCY, J. A. and GLAUERT, A. M. (1964). Structure and assembly of macromolecular lipid complexes composed of globular micelles. *J. Mol. Biol.*, **8**, 727–48.

LUNDBERG, A. (1955). Microelectrode experiments on unfertilised sea urchin eggs. *Exp. Cell Res.*, **9**, 393–8.

LUZZATI, V. and HUSSON, F. (1962). The structure of the liquid–crystalline phases of lipid–water systems. *J. Cell Biol.*, **12**, 207–19.

LYKLEMA, J. and OVERBEEK, J. TH. G. (1961).   On the interpretation of electrokinetic potentials. *J. Coll. Sci.*, **16**, 501–12.

MACPHERSON, I. and STOKER, M. (1962).   Polyoma transformation of hamster cell clones—an investigation of genetic factors affecting cell competence. *Virology*, **16**, 147–51.

MADDEN, R. E. and BURK, D. (1962).   Production of viable single cell suspensions from solid tumors. *J. Nat. Cancer Inst.*, **27**, 841–61.

MADDY, A. H. and MALCOLM, B. R. (1965).   Protein conformation in the plasma membrane. *Science*, **150**, 1616–8.

MALENKOV, A. G., VASILIEV, E. M., MODYANOVA, E. A., POZHKOVA, Z. A. and SHTAMM, E. V. (1963).   о природе сцепления клеток печни. *Biofizika*, **8**, 354–60.

MARCHESI, V. T. and GOWANS, J. L. (1964).   The migration of lymphocytes through the endothelium of venules in lymphocytes: an electron microscope study. *Proc. Roy. Soc.* **B.**, **159**, 283–90.

MARCHESI, V. T., SEARS, M. L. and BARNETT, R. J. (1964).   Electron microscopic studies of nucleoside phosphatase activity on blood vessels and glia of the retina. *Investig. Ophthal.*, **3**, 1–21.

MARGOLIASH, E., SCHENK, J. R., MARGIE, M. P., BUROKAS, S., RICHTER, W. R., BARLOW, G. M. and MOSCONA, A. A. (1965).   Characterisation of specific cell aggregating materials from sponge cells. *Biochem. Biophys. Res. Comms.*, **20**, 383–8.

MARSLAND, D. (1956).   Protoplasmic contractility in relation to gel structure: temperature-pressure experiments on cytokinesis and amoeboid movement. *Int. Rev. Cytol.*, **5**, 199–227.

MATALON, R. and SCHULMAN, J. H. (1949).   Mechanisms of adsorption, solution and penetration. *Disc. Faraday Soc.*, **6**, 27–39.

MATEYKO, G. M. and KOPAC, M. J. (1963).   Cytophysical studies on living normal and neoplastic cells. *Ann. N.Y. Acad. Sci.*, **105**, 183–286.

MATTHEW, R. (1926).   Recuperation de la vue après greffe de l'oeil chez le Triton adulte. *C.R. Soc. Biol.*, **94**, 4–5.

MATURANA, H. R., LETTVIN, J. Y., McCULLOCH, W. S. and PITTS, W. H. (1960).   Anatomy and physiology of vision in the frog (*Rana pipiens*). *J. Gen. Physiol.* Suppl. 2, **43**, 129–75.

MAYHEW, E. (1966).   Cellular electrophoretic mobility and the mitotic cycle. *J. Cell. Comp. Physiol.*, **49**, 717–25.

MAYHEW, E. and O'GRADY, E. A. (1965).   Electrophoretic mobilities of tissue culture cells in exponential and parasynchronous growth. *Nature, Lond.*, **207**, 86–7.

MAYHEW, E. and NORDLING, S. (1966).   The electrophoretic mobility of fixed tissue cells: studies with formaldehyde and osmium tetroxide. *Exp. Cell Res.*, **43**, 72–6.

McCOLLESTER, D. L. (1962).   A method for isolating skeletal-muscle cell membrane components. *Biochem. Biophys. Acta.*, **57**, 427–37.

McLOUGHLIN, C. B. (1961a).   The importance of mesenchymal factors in the differentiation of chick epidermis—I. *J. Embryol. Exp. Morph.*, **9**, 370–84.

McLOUGHLIN, C. B. (1961b).   The importance of mesenchymal factors in the differentiation of chick epidermis—II. *J. Embryol. Exp. Morph.*, **9**, 385–409.

McLOUGHLIN, B. C. B. (1963).   Mesenchymal influences on epithelial differentiation. *Symp. Soc. Exp. Biol.*, **17**, 359–88.

MEHRISHI, J. N. and SEAMAN, G. V. F. (1966).   Temperature dependence of the electrophoretic mobility of cells and quartz particles. *Biochem. Biophys. Acta.*, **112**, 154–9.

MELZACK, R. and WALL, P. D. (1962).   On the nature of cutaneous sensory mechanisms. *Brain*, **85**, 331–56.

MERCER, E. (1957).   Electron micrographs of cross-sections of protein monolayers. *Nature, Lond.*, **180**, 87.

MERCER, E. and WOLPERT, L. (1962).   An electron microscope study of the cortex of the sea urchin (*Psammechinus miliaris*) egg. *Exp. Cell. Res.*, **27**, 1–13.

MERCHANT, D. J. and KAHN, R. H. (1958).   Fiber formation in suspension culture of L strain fibroblasts. *Proc. Soc. Exp. Biol. Med.*, **97**, 359–62.

MESSIER, B. and LEBLOND, C. P. (1960).   Cell proliferation and migration as revealed by radioautography after injection of thymidine-H^3 into male rats and mice. *Amer. J. Anat.*, **106**, 247–65.

MEYER, D. B. (1964).   The migration of primordial germ cells in the chick embryo. *Dev. Biol.*, **10**, 154–90.

MIALE, I. L. and SIDMAN, R. L. (1961).   An autoradiographic analysis of histogenesis in the mouse cerebellum. *Exp. Neurol.*, **4**, 277–96.

MICHL, J. (1961).   Metabolism of cells in tissue culture *in vitro*—I. The influence of serum protein fractions on the growth of normal and neoplastic cells. *Exp. Cell Res.*, **23**, 324–34.

MICHL, J. (1965).   Carbamyl phosphate as an essential component of the flattening factor for cells in culture. *Nature, Lond.*, **207**, 412.

MILLER, D. and CRANE, R. K. (1961).   The digestive function of the epithelium of the small intestine. *Biochem. Biophys. Acta.*, **52**, 293–8.

MINTZ, B. (1962).   Experimental study of the developing mammalian egg: removal of the zona pellucida. *Science*, **138**, 594–5.

MINTZ, B. (1964).   Gene expression in the morula stage of mouse embryos, as observed during development of t^{12}/t^{12} lethal mutants *in vitro*. *J. Exp. Zool.*, **157**, 267–72.

MITCHELL, P. S. (1956).   Hypothetical thermokinetic and electrokinetic mechanisms of locomotion in micro-organisms. *Proc. Roy. Phys. Soc. Edin.*, **25**, 32–4.

MITCHISON, J. M. (1952).   Cell membranes and cell division. *Symp. Soc. Exp. Biol.*, **6**, 105–27.

MITCHISON, J. M. and SWANN, M. M. (1952).   Optical changes in the membranes of the sea-urchin egg at fertilization, mitosis and cleavage. *J. Exp. Biol.*, **29**, 357–62.

MITCHISON, J. M. and SWANN, M. M. (1953).   Measurements of sea-urchin eggs with an interference microscope. *Q.J. Micros. Sci.*, **94**, 381–9.

MITCHISON, J. M. and SWANN, M. M. (1954a).   The mechanical properties of the cell surface—I. The cell elastimeter. *J. Exp. Biol.*, **31**, 443–60.

MITCHISON, J. M. and SWANN, M. M. (1954b).   The mechanical properties of the cell surface—II. The unfertilized sea-urchin egg. *J. Exp. Biol.*, **31**, 461–72.

MITCHISON, J. M. and SWANN, M. M. (1955).   The mechanical properties of the cell surface—III. The sea-urchin egg from fertilization to cleavage. *J. Exp. Biol.*, **32**, 734–50.

MOELWYN-HUGHES, E. A. (1964).   *Physical Chemistry*. Oxford Univ. Press, Oxford, vii + 1334.

MONNÉ, L. and HARDE, S. (1951).   On the cortical granules of the sea-urchin egg. *Arkiv. f. Zool.*, **1** (series 2), 487–98.

MOOKERJEE, S., DEUCHAR, E. M. and WADDINGTON, C. H. (1953). The morphogenesis of the notochord in amphibia. *J. Embryol. Exp. Morph.*, **1**, 399–409.

MÖLLER, G. (1961). Demonstration of mouse isoantigens at the cellular level by the fluorescent antibody technique. *J. Exp. Med.*, **114**, 415–34.

MÖLLER, G. (1963). Phenotypic expressions of isoantigens of the H-2 system in embryonic and newborn mice. *J. Immunol.*, **90**, 271–9.

MÖLLER, G. and MÖLLER, E. (1962). Phenotypic expression of the mouse isoantigens. *J. Cell. Comp. Physiol.*, **60**, Suppl., 107–28.

MOORE, A. R. (1940). Osmotic and structural properties of the blastular wall in *Dendraster excentricus*. *J. Exp. Zool.*, **84**, 73–83.

MOORE, A. R. (1941). Osmotic block of gastrulation in *Dendraster*. *J. Exp. Zool.*, **87**, 101–11.

MOORE, A. R. and BURT, A. S. (1939). On the locus of the forces causing gastrulation in the embryo of *Dendraster excentricus*. *J. Exp. Zool.*, **82**, 159–71.

MORGAN, C., HSU, K. C., RIFKIND, R. A., KNOX, A W. and ROSE, H. M. (1961). The application of ferritin-conjugated antibody to electron microscope studies of influenza virus in infected cells—I. The cellular surface. *J. Exp. Med.*, **114**, 825–32.

MORGAN, T. H. (1895). The formation of the fish embryo. *J. Morph..*, **10**, 419–72.

MORGAN, W. T. J. (1963). Some observations on the carbohydrate-containing components of human ovarian cyst mucin. *Ann. N.Y. Acad. Sci.*, **106**, 177–90.

MORGAN, W. T. J. and WATKINS, W. M. (1959). Some aspects of the biochemistry of the human blood group substances. *Brit. Med. Bull.*, **15**, 109–12.

MOSCONA, A. (1957a). The development *in vitro* of chimeric aggregates of dissociated embryonic chick and mouse cells. *Proc. Nat. Acad. Sci.*, *Wash.*, **43**, 184–94.

MOSCONA, A. (1957b). Formation of lentoids by dissociated retinal cells of the chick embryo. *Science*, **125**, 598–9.

MOSCONA, A. (1960). Patterns and mechanisms of tissue reconstruction from dissociated cells. From *Developing cell systems and their controls*, edited D. Rudnick. Ronald Press, New York.

MOSCONA, A. (1961a). Effect of temperature on adhesion to glass and histogenetic cohesion of dissociated cells. *Nature, Lond.*, **190**, 408–9.

MOSCONA, A. (1961b). Rotation mediated histogenetic aggregation of dissociated cells. *Exp. Cell Res.*, **22**, 455–75.

MOSCONA, A. A. (1962). Analysis of cell recombinations in experimental synthesis of tissues *in vitro*. *J. Cell. Comp. Physiol.*, **60**, Suppl. 1, 65–80.

MOSCONA, A. A. (1963a). Studies on cell aggregation: demonstration of materials with a selective cell-binding activity. *Proc. Nat. Acad. Sci. Wash.*, **49**, 742–7.

MOSCONA, A. A. (1963b). Inhibition by trypsin inhibitors of dissociation of embryonic tissue by trypsin. *Nature, Lond.*, **199**, 379–80.

MOSCONA, A. and MOSCONA, H. (1952). The dissociation and aggregation of cells from organ rudiments of the early chick embryo. *J. Anat.*, **86**, 287–301.

MOSCONA, A. A. and MOSCONA, M. H. (1966). Aggregation of embryonic cells in a serum-free medium and its inhibition at suboptimal temperatures. *Exp. Cell Res.*, **41**, 697–702.

MOSCONA, M. H. and MOSCONA, A. A. (1963). Inhibition of adhesiveness and aggregation of dissociated cells by inhibitors of protein and RNA synthesis. *Science*, **142**, 1070–1.

Moscona, M. H. and Moscona, A. A. (1966). Inhibition of cell aggregation *in vitro* by puromycin. *Exp. Cell Res.*, **41**, 703–6.

Moskowitz, M. (1963). Aggregation of cultured mammalian cells. *Nature, Lond.*, **200**, 854–6.

Moskowitz, M. and Amborski, F. (1964). Nutritional and physical factors affecting the aggregation of cultured mammalian cells. *Nature, Lond.*, **203**, 1236–7.

Mudd, S., McCutcheon, M. and Lucke, B. (1934). Phagocytosis. *Physiol. Revs.*, **14**, 210–75.

Mueller, P., Rudin, D. O., Tien, H. T. and Wescott, W. C. (1962). Reconstitution of excitable cell membrane structure *in vitro*. *Circulation*, **26**, 1167–70.

Muir, A. R. and Peters, A. (1962). Quintuple-layered membrane junctions at terminal bars between endothelial cells. *J. Cell Biol.*, **12**, 443–6.

Mullins, L. J. (1961). The macromolecular properties of excitable membrane. *Ann. N. Y. Acad. Sci.*, **94**, 390–404.

Myllylä, G., Häyry, P., Penttinen, K. and Saxen, E. (1966). Serum lipoproteins in primary cell attachment and growth behaviour of cells on glass. *Ann. Med. Exp. Fenn.*, **44**, 171–6.

Nace, G. W. (1963). A developmentally significant antigen on the surface of frog embryo cells. *Devel. Biol.*, **7**, 280–4.

Nakai, J. (1956). Dissociated dorsal root ganglia in tissue culture. *Amer. J. Anat.*, **99**, 81–129.

Nakai, J. (1960). Studies on the mechanism determining the course of nerve fibers in tissue culture—II. The mechanism of fasciculation. *Z. f. Zellforsch.*, **52**, 427–49.

Nakajima, S. (1965). Selectivity in fasciculation of nerve fibers *in vitro*. *J. Comp. Neurol.*, **125**, 193–203.

Nakanishi, Y. H., Katao, H. and Iwasaki, T. (1963). Inhibitory effects of chloramphenicol on the histogenetic aggregation of dissociated cells. *Jap. J. Genet.*, **38**, 257–60.

Nawar, G. M. (1955). Descriptive and experimental studies on the origin of the autonomic ganglia in the chick embryo. *Diss. Abstr.*, **15**, 1467–8.

Neifakh, S. A., Avramov, J. A., Gaitskhoki, V. S., Kazakova, T. B., Monakhov, N. K., Repin, V. S., Turovski, V. S. and Vassiletz, I. M. (1965). Mechanism of the controlling function of mitochondria. *Biochem. Biophys. Acta.*, **100**, 329–43.

Nelson, A. (1963). Immune adherence. *Adv. Immunol.*, **3**, 131–80.

Neter, E. (1963). Attachment of antibodies and antigens onto cell surfaces. From *Cell-bound antibodies*, edited B. Amos and H. Koprowski, Wista Inst., Philadelphia, 35–43.

Neville, D. M. (1960). The isolation of a cell membrane fraction from rat liver. *J. Biophys. Biochem. Cytol.*, **8**, 413–22.

New, D. A. T. (1955). A new technique for the cultivation of the chick embryo *in vitro*. *J. Embryol. Exp. Morph.*, **3**, 320–31.

New, D. A. T. (1959). The adhesive properties and expansion of the chick blastoderm. *J. Embryol. Exp. Morph.*, **7**, 146–64.

Nieuwkoop, P. D. and Florschutz, P. A. (1950). Some especial features of the gastrulation and neurulation in the egg of *Xenopus laevis*, Daud., and some other anurans. *Arch. Biol., Liège*, **61**, 113–50.

N

NORDLING, S. and MAYHEW, E. (1966). Surface area changes and the electro-phoretic mobility of tissue culture cells. *Exp. Cell Res.*, **41**, 77–83.

NORDLING, S., PENTINNEN, K. and SAXEN, E. (1965). The effects of different methods of washing, drying and sterilizing glass surfaces on cell attachment and growth behaviour. *Exp. Cell Res.*, **37**, 161–8.

NORDLING, S., VAHERI, A., SAXEN, E. and PENTINNEN, K. (1965). The effects of anionic polymers on cell attachment and growth behaviour, with a note on a similarity in the effect of fresh human serum. *Exp. Cell Res.*, **37**, 406–19.

NOSSAL, G. J. V. (1959). Antibody production by single cells—III. The histology of antibody production. *Brit. J. Exp. Path.*, **40**, 301–11.

NOWELL, P. C. and BERWICK, L. (1958). The surface ultrastructure of normal and leukemic rat lymphocytes. *Cancer Res.*, **18**, 1067–9.

NUSSBAUM, J. L., BIETH, R. and MANDEL, P. (1963). Phosphatides in myelin sheath and repetition of sphingomyelin in the brain. *Nature, Lond.*, **198**, 586–7.

O'BRIEN, J. S. and SAMPSON, E. L. (1965). Myelin membrane: a molecular abnorm-ality. *Science*, **150**, 1613–4.

ODA, M. and PUCK, T. T. (1961). The interaction of mammalian cells with anti-bodies. *J. Exp. Med.*, **113**, 599–610.

ODLAND, G. (1958). The fine structure of the interrelationships of cells in the human epidermis. *J. Biophys. Biochem. Cytol.*, **4**, 529–38.

ODOR, D. L. (1960). Electron microscope studies on ovarian oocytes and un-fertilised tubal ova in the rat. *J. Biophys. Biochem. Cytol.*, **7**, 567–74.

OKADA, T. S. (1959). Regeneration of cartilaginous matrix from the dissociated chondrocytes *in vitro*. *Exp. Cell Res.*, **16**, 437–40.

OKADA, T. S. (1965). Immunohistological studies on the reconstitution of nephric tubules from dissociated cells. *J. Embryol. Exp. Morph.*, **13**, 299–307.

OLDFIELD, F. (1963). Orientation behaviour of chick leucocytes in tissue culture and their interactions with fibroblasts. *Exp. Cell Res.*, **30**, 125–38.

O'NEILL, C. H. (1964). Preparation and properties of isolated cell surface mem-branes. *Rec. Progr. Surface Sci.*, **2**, 427–42.

O'NEILL, C. H. and WOLPERT, L. (1961). Isolation of the cell membrane of *Amoeba proteus*. *Exp. Cell Res.*, **24**, 592–5.

OPPENHEIMER, J. (1936). The development of isolated blastoderms of Fundulus heteroclitus. *J. Exp. Zool.*, **72**, 247–69.

OVERBEEK, J. TH. G. (1952). From *Colloid Science*, Vol. 1: *Irreversible systems*, chapters, 2, 4, 5, 6, 7 and 8. Edited H. R. Kruyt. Elsevier, Amsterdam.

OVERBEEK, J. TH. G. and SPARNAAY, M. J. (1954). London-van der Waals attraction between macroscopic objects. *Disc. Faraday Soc.*, **18**, 12–24.

OVERTON, J. (1962). Desmosome development in normal and reassociating cells in the early chick blastoderm. *Devel. Biol.*, **4**, 532–48.

OVERTON, J. (1963). Intercellular connections in the outgrowing stolon of Cordylophora. *J. Cell Biol.*, **17**, 661–71.

OVERTON, J. and SHOUP, J. (1964). Fine structure of cell surface specializations in the maturing duodenal mucosa of the chick. *J. Cell Biol.*, **21**, 75–85.

OWEN, R. D. (1962). Earlier studies of blood groups in the rat. *Ann. N.Y. Acad. Sci.*, **97**, 37–42.

PALAY, S. L. and KARLIN, L. J. (1959). An electron microscope study of the intestinal villus—I. The fasting animal. *J. Biophys. Biochem. Cytol.*, **5**, 363–71.

PANKHURST, K. G. A. (1958). Monolayer studies of tanning reactions. From *Surface phenomena in chemistry and biology*, edited J. F. Danielli, K. G. A. Pankhurst and A. C. Riddiford. Pergamon, London, 100–16.

PASTEELS, J. (1935). Sur les mouvements morphogénétiques de gastrulation suscitant l'apparition de la ligne primitive chez les oiseaux. *C.R. Soc. Biol.*, **120**, 1362–7.

PASTEELS, J. (1940). Un aperçu comparatif de la gastrulation chez les Chordés. *Biol. Revs.*, **15**, 59–106.

PASTEELS, J. (1942). Sur l'existence eventuelle d'une croissance au cours de la gastrulation des Vertébrés. *Acta. Biol. Belg.*, **2**, 130–3.

PATINKIN, D. and DOLJANSKI, F. (1965). The effect of volume expansion on the electrophoretic mobility of cells. *J. Cell. Comp. Physiol.*, **66**, 343–50.

PATLAK, C. S. (1953). A mathematical contribution to the study of orientation of organisms. *Bull. Math. Biophys.*, **15**, 431–76.

PENTTINEN, K., SAXEN, E., SAXEN, L., TOIVONEN, S. and VAINIO, T. (1958). Importance of the growth medium on the morphological behaviour, inductive action and immunology of cells. *Ann. Med. Exp. Fenn.*, **36**, 27–46.

PEREIRA, H. G. and DE FIQUEIRDO, M. V. T. (1962). Mechanism of hemagglutination by adenovirus types 1, 2, 4, 5, and 6. *Virology*, **18**, 1–8.

PERLMANN, P. (1959). Immunochemical analysis of the surface of the sea-urchin egg—an approach to the study of fertilisation. *Experientia*, **15**, 41–52.

PETER, K. (1939). Untersuchungen uber die entwicklung des dotterentoderms—IV. Das schicksal des Dotterentoderms beim Huhnchen. *Z. Mikr. Anat. Forsch.*, **46**, 627–39.

PETHICA, B. A. (1961). The physical chemistry of cell adhesion. *Exp. Cell Res.*, suppl., **8**, 123–40.

PETHICA, B. A. and ANDERSON, P. J. (1953). The role of the stroma phospholipids and cholinesterase in the haemolytic reaction. *Internat. Conf. on biochemical problems of lipids*, 129–35.

PETRY, G., QVERBECK, L. and VOGELL, W. (1961). Sind desmosomen statische oder temporare zellverbindungen? *Naturwiss*, 166–7.

PIATT, J. (1944). Experiments on the decussation and course of Mauthner's fibres in *Amblystoma punctatum. J. Comp. Neurol.*, **80**, 335–50.

PIATT, J. (1948). Form and causality in neurogenesis. *Biol. Revs.*, **23**, 1–45.

PIATT, J. (1952). Transplantation of aneurogenic forelimbs in place of the hindlimb in *Amblystoma. J. Exp. Zool.*, **120**, 247–85.

PIATT, J. (1957). Studies on the problem of nerve pattern—II. Innervation of the intact forelimb by different parts of the central nervous system in *Amblystoma. J. Exp. Zool.*, **134**, 103–25.

PIATT, J. (1958). Transplantation of aneurogenic forelimbs in place of the hindlimb in *Amblystoma. J. Exp. Zool.*, **120**, 247–86.

PIDOT, A. L. and DIAMOND, J. M. (1964). Streaming potential in a biological membrane. *Nature, Lond.*, **201**, 701–2.

POLET, H. (1966). The effect of hydrocortisone on the membranes of primary human amnion cells *in vitro. Exp. Cell Res.*, **41**, 316–23.

PONDER, E. (1949). Lipid and protein components in the surface ultrastructure of the erythrocyte. *Disc. Faraday Soc.*, **6**, 152–60.

PONDER, E. (1955). Red cell structure and its breakdown. *Protoplasmatologia*, **10**, Part 2.

PONDER, E. and PONDER, R. V. (1959). A zone phenomenon and a progressive reaction occurring with a radioactive hemolysin, sodium erucate-I[131]. *J. Gen. Physiol.*, **41**, 651–5.

PORTER, K. R. (1956). Observations on the fine structure of animal epidermis. From *Proc. 3rd Internat. Conf. Electron Microscopy, London*, 1954, 539–46.

POTTER, D. D., FURSHPAN, E. J. and LENNOX, E. S. (1966). Connections between cells of the developing squid as revealed by electrophysiological methods. *Proc. Nat. Acad. Sci., Wash.*, **55**, 328–36.

POULIK, M. D. and LAUF, P. K. (1965). Heterogenity of water-soluble structural components of human red cell membrane. *Nature, Lond.*, **208**, 874–6.

PUCK, T. and SAGIK, B. (1953). Virus and cell interactions with ion exchangers. *J. Exp. Med.*, **97**, 807–20.

PULVERTAFT, R. J. V. and WEISS, L. (1963). Some effects of micro-electrophoresis suspending fluids on tissue cells. *J. Path. Bact.*, **85**, 473–9.

PURDOM, L., AMBROSE, E. J. and KLEIN, G. (1958). A correlation between surface electrical charge and some biological characteristics during the stepwise progression of a mouse sarcoma. *Nature, Lond.*, **181**, 1586–7.

RANCK, J. B. (1964). Synaptic "learning" due to electro-osmosis: a theory. *Science*, **144**, 187–9.

RAND, R. P. and BURTON, A. C. (1964). Mechanical properties of the red cell membrane—I. Membrane stiffness and intracellular pressure. *Biophys. J.*, **4**, 115–35 (see also *ibid*, **4**, 491).

RAPPAPORT, C. (1966a). Effect of temperature on dissociation of adult mouse liver with sodium tetraphenylboron. *Proc. Soc. Exp. Biol. Med.*, **121**, 1022–5.

RAPPAPORT, C. (1966b). Role on "intercellular" matrix in aggregation of mammalian cells. *Proc. Soc. Exp. Biol. Med.*, **121**, 1025–8.

RAPPAPORT, C. and HOWZE, G. B. (1966). Dissociation of adult mouse liver by sodium tetraphenylboron, a potassium complexing agent. *Proc. Soc. Exp. Biol. Med.*, **121**, 1010–6.

RAPPAPORT, C. and HOWZE, G. B. (1966). Further studies on the dissociation of adult mouse tissue. *Proc. Soc. Exp. Biol. Med.*, **121**, 1016–21.

RAPPAPORT, C., POOLE, J. P. and RAPPAPORT, H. P. (1960). Studies on properties of surfaces required for growth of mammalian cells in synthetic medium. *Exp. Cell Res.*, **20**, 465–510.

RAWLES, M. (1944). The migration of melanoblasts after hatching into pigment-free skin grafts of the common fowl. *Physiol. Zool.*, **17**, 167–83.

RAWLES, M. (1945). Behaviour of melanoblasts derived from the celomic lining in interbreed grafts of wing skin. *Physiol. Zool.*, **18**, 1–16.

RAWLES, M. (1947). Origin of pigment cells from the neural crest in the mouse embryo. *Physiol. Zool.*, **20**, 248–66.

RAWLES, M. (1948). Origin of melanophores and their role in development of colour patterns in vertebrates. *Physiol. Revs.*, **28**, 383–408.

REBHUN, L. I. (1962). Electron microscope studies on the vitelline membrane of the surf clam, *Spisula solidissima*. *J. Ultrastr. Res.*, **6**, 107–22.

REROLLE, M. (1959). Affrontements cutanés *in vitro*: le comportement épidermique. *Arch. Anat. Micros.*, **48**, 325–43.

RHUMBLER, L. (1902). Zur mechanik des gastrulationvorganges insbesondere der invagination. Eine entwicklungsmechanische studie. *Arch. f. Entwickmech.*, **14**, 401–76.

RICHARDSON, S. H., HULTIN, H. O. and GREEN, D. E. (1963). Structural proteins of membrane systems. *Proc. Nat. Acad. Sci., Wash.*, **50**, 821–7.

RIEMERSMA, J. C. and BOOIJ, H. L. (1962). The reaction of osmium tetroxide with lecithin: application of staining procedures. *J. Histochem. Cytochem.*, **10**, 89–95.

RINALDINI, L. M. (1958). The isolation of living cells from animal tissues. *Int. Rev. Cytol.*, **7**, 587–647.

RIS, H. (1941). An experimental study of the origin of melanophores in birds. *Physiol. Zool.*, **14**, 48–66.

RISKI, M. T. M. (1961). The influence of glucosamine hydrochloride on cellular adhesiveness in *Drosophila melanogaster*. *Exp. Cell Res.*, **24**, 111–9.

ROBERTSON, J. D. (1958a). Structural alterations in nerve fibres produced by hypotonic and hypertonic solutions. *J. Biophys. Biochem. Cytol.*, **4**, 349–64.

ROBERTSON, J. D. (1958b). A molecular theory of cell membrane structure. *4th Intern. Conf. on Electron Microscopy*, **2**, 159–71.

ROBERTSON, J. D. (1959). The ultrastructure of cell membranes and their derivatives. *Biochem. Soc. Symp.*, **16**, 3–43.

ROBERTSON, J. D. (1960a). The molecular biology of cell membranes. From *Molecular Biology*. Academic Press, New York, 87–151.

ROBERTSON, J. D. (1960b). The molecular structure and contact relationships of cell membranes. *Prog. Biophys. Chem.*, **11**, 344–418.

ROBERTSON, J. D. (1963a). Unit membranes: a review with recent studies of experimental alterations and a new subunit structure in synaptic membranes. From *Cellular membranes in development*, edited M. Locke. Academic Press, New York, 1–81.

ROBERTSON, J. D. (1963b). The occurrence of a subunit pattern in the unit membranes of club endings in Mauthner cell synapses in goldfish brains. *J. Cell Biol.*, **19**, 201–21.

ROBERTSON, J. D. (1966). Granulo-fibrillar and globular substructure of unit membranes. *Ann. N.Y. Acad. Sci.*, **137**, 421–40.

ROOTS, B. I. and JOHNSTON, P. V. (1964). Neurons of ox brain nuclei: their isolation and appearance by light and electron microscopy. *J. Ultrastr. Res.*, **10**, 350–61.

ROSE, W. and HEIMS, R. W. (1962). Moving interfaces and contact angle rate-dependency. *J. Coll. Sci.*, **17**, 39–48.

ROSENBERG, M. D. (1960). Microexudates from cells grown in tissue culture. *Biophys. J.* **1**, 137–59.

ROSENBERG, M. D. (1962). Long-range interactions between cell and substratum. *Proc. Nat. Acad. Sci., Wash.*, **48**, 1342–9.

ROSENBERG, M. D. (1963a). Cell guidance by alterations in monomolecular films. *Science*, **139**, 411–2.

ROSENBERG, M. D. (1963b). The relative extensibility of cell surfaces. *J. Cell Biol.*, **17**, 289–97.

ROSENQUIST, G. C. (1966). A radioautographic study of labeled grafts in the chick blastoderm. Development from primitive-streak stages to stage 12. *Carnegie Inst., Wash. Publ.*, 625: *Contrib. Embryology*, **38**, 71–110.

ROSENQUIST, C. C. and DE HAAN, R. L. (1966). Migration of precardiac cells in the chick embryo: a radioautographic study. *Carnegie Inst., Wash. Publ.*, 625: *Contrib. Embryology*, **38**, 111–21.

ROSIN, S. (1943). Experiments zur entwicklungsphysiologie der pigmentierung bei amphiben. *Rev. Suisse Zool.*, **50,** 485–578.

ROTHSCHILD, V. (1938). The biophysics of the egg surface of *Echinus esculentus* during fertilisation and cytolysis. *J. Exp. Biol.*, **15,** 209–16.

ROTHSCHILD, V. (1956). *Fertilisation.* Methuen, London, ix + 170.

ROTHSCHILD, V. (1961). Sperm energetics. From *The cell and the organism*, edited J. A. Ramsay and V. B. Wigglesworth. Cambridge Univ. Press, Cambridge, 9–21.

ROTHSCHILD, V. and SWANN, M. M. (1952). The fertilisation reaction in the sea-urchin. The block to polyspermy. *J. Exp. Biol.*, **29,** 469–83.

ROTHSTEIN, A. (1954). The enzymology of the cell surface. *Protoplasmatologia*, II.E.4.

ROUS, P. and JONES, F. S. (1916). A method for obtaining suspensions of living cells from the fixed tissues, and for the plating out of individual cells. *J. Exp. Med.*, **23,** 549–55.

ROUX, W. (1894). Uber den "Cytotropismus" der Furchungszellen des grasfrosches. (*Rana fusca*). *Arch. f. Entwickmech.*, **1,** 43–68.

ROUX, W. (1896). Uber die selbstordnung (Cytotaxis) sich "beruhrender" Furchungszellen des Froscheies durch zellenzusammen-fugung, zellentrennung und zellengleiten. *Arch. f. Entwickmech.*, **3,** 381–468.

RUHENSTROTH-BAUER, G., FUHRMAN, G. F., KUBLER, W., RUEFF, F. and MUNK, K. (1962). Zur bedeutung der neuraminsauren in der zellmembran fur das wachstum maligner zellen. *Z. f. Krebsforsch.*, **65,** 37–43.

RUNNSTROM, J. (1928). Die veranderungen der plasmakolloide beider entwicklungserregung des seeigeleies. *Protoplasma*, **4,** 388–514.

RUNNSTROM, J. (1952). The cell surface in relation to fertilisation. *Symp. Soc. Exp. Biol.*, **6,** 39–88.

RUNNSTROM, J. (1957). Cellular structure and behaviour under the influence of animal and vegetal factors in sea-urchin development. *Arkiv. f. Zool.*, **13,** 523–37.

RUNNSTROM, J. and KRIZAT, G. (1961). The sperm reception and the activating system in the surface of the unfertilised sea-urchin egg. *Arkiv. f. Zool.*, **13,** 95–112.

ST. AMAND, G. S. and TIPTON, S. R. (1954). The separation of neuroblasts and other cells from grasshopper embryos. *Science*, **119,** 93–4.

SACHTLEBEN, P. and RUHENSTROTH-BAUER, G. (1962). Die Anderung der elektrischen oberflachenladung von erythrozyten durch agglutinierende und sensibilisierende substanzen. *Med. Exp.*, **6,** Part I, 183–92; Part II, 226–36.

SAISON, R. and INGRAM, D. E. (1962). A report on blood groups in pigs. *Ann. N.Y. Acad. Sci.*, **97,** 226–32.

SALPETER, M. and SINGER, M. (1960). Differentiation of the submicroscopic adepidermal membrane during limb regeneration in adult *Triturus*, including a note on the use of the term, Basement membrane. *Anat. Rec.*, **136,** 27–40.

SAMUEL, E. W. (1961). Orientation and rate of locomotion of individual amebas in the life cycle of the cellular slime mould *Dictyostelium mucoroides*. *Dev. Biol.*, **3,** 317–35.

SAUNDERS, B. G. and WRIGHT, J. E. (1962). Immunogenetic studies in two trout species of the genus *Salmo*. *Ann. N.Y. Acad. Sci.*, **97,** 111–5.

SANYAL, S. and MOOKERJEE, S. (1960). Experimental dissociation of cells from *Hydra*. *Arch. f. Entwickmech.*, **152,** 131–6.

SARA, M. (1956). Esperienze di aggregazione cellulare mista dopi dissociate nelle calispongi. *Bull. Zool.*, **23**, 113–9.

SAWYER, P. N., BRATTAIN, W. H. and BODDY, P. J. (1964). Electrochemical precipitation of human blood cells and its possible relation to intravascular thrombosis. *Proc. Nat. Acad. Sci., Wash.*, **51**, 428–32.

SAXEN, E. and PENTTINEN, K. (1961). Host factors in cell culture: further studies on the growth-controlling action of fresh human serums. *J. Nat. Cancer Inst.*, **26**, 1367–80.

SCHAFFER, J. (1927). From *Handbuch der mikroskopische anatomie des menschen*, Volume 2, edited W. von Mollendorf. Springer, Berlin, 35.

SCHECHTMAN, A. M. (1942). The mechanism of amphibian gastrulation. *Univ. Calif. Publ. Zool.*, **51**, 1–39.

SCHELUDKO, A. (1962). Sur certaines particularités des lames mouseuse—I, II and III. *Proc. K. Acad. Wetenschappen, Amsterdam*. Series B, **65B**, 76–108.

SCHENKEL, J. H. and KITCHENER, J. (1961). A test of the Derjaguin-Verwey-Overbeek theory with a colloidal suspension. *Trans. Faraday Soc.*, **56**, 161–73.

SCHIEFFERDECKER, P. (1886). Methode zur Isolirung von Epithelzellen. *Z. f. Mikros*, **3**, 483.

SCHMITT, F. O., BEAR, R. S. and PONDER, E. (1938). The red cell envelope considered as a Wiener mixed body. *J. Cell. Comp. Physiol.*, **11**, 309–13.

SCHMITT, F. O., BEAR, R. S. and PALMER, K. J. (1941). X-ray diffraction studies on the structure of the nerve myelin sheath. *J. Cell. Comp. Physiol.*, **18**, 31–42.

SCHNEIDER, N. (1958). Sur les possibilities de propagation d'un sarcome de souris sur de organes embryonnaires de poulet a differents stades du developpement. *Arch. Anat. Micros.*, **47**, 573–604.

SCHWAN, H. P. and MOROWITZ, H. J. (1962). Electrical properties of the pleuropneumonia-like organism A 5959. *Biophys. J.*, **2**, 395–407.

SCOTT, A. (1960). Surface changes during cell division. *Biol. Bull.*, **119**, 260–72.

SCOTT, R. W. (1959). Tissue affinity in *Amaroecium*—I. Aggregation of dissociated fragments and their integration into one fragment. *Acta Embryol. Morph. Exp.*, **2**, 209–26.

SEAMAN, G. V. F. and HEARD, D. (1960). The surface of the washed human erythrocyte as a polyanion. *J. Gen. Physiol.*, **44**, 251–63.

SEAMAN, G. V. F., KOK, D'A, and HEARD, D. (1962). The electrophoretic mobility of human red cells re-suspended in their native serum. *Clin. Sci.*, **23**, 115–23.

SEAMAN, G. V. F. and UHLENBRUCK, G. (1962). The action of proteolytic enzymes on the red cells of some animal species. *Biochem. Biophys. Acta.*, **64**, 570–2.

SEDAR, A. W. and FORTE, J. G. (1964). Effects of calcium depletion on the junctional complex between oxyntic cells of gastric glands. *J. Cell Biol.*, **22**, 173–88.

SELMAN, G. G. (1958). The forces producing neural closure in amphibia. *J. Embryol. Exp. Morph.*, **6**, 448–65.

SELMAN, G. G. and WADDINGTON, C. H. (1955). The mechanism of cell division in the cleavage of the newt's egg. *J. Exp. Biol.*, **32**, 700–33.

SETALA, K., AYRAPAA, O., NYHOLM, M. M., STJERNVALL, L. and AHO, Y. (1960). Das gegenseitige verhalten von "artifiziellen zellmembranen" und synthetischen tumorauslosersubstanzen tensionaktiver natur, elektronenoptisch untersucht. *4th Intern. Conf. on Electron Microscopy*.

SHAFFER, B. M. (1957a). Aspects of aggregation in cellular slime moulds—I. Orientation and chemotaxis. *Amer. Nat.*, **91**, 19–35.

SHAFFER, B. M. (1957b). Properties of slime-mould amoebae of significance for aggregation. *Q. J. Micros. Sci.*, **98**, 377–92.

SHAFFER, B. M. (1957c). Variability of behaviour of aggregating cellular slime moulds. *Q. J. Micros. Sci.*, **98**, 393–405.

SHAFFER, B.M. (1958). Integration in aggregating cellular slime moulds. *Q.J. Micros. Sci.*, **99**, 103–21.

SHAFFER, B. M. (1961a). The cells founding aggregation centres in the slime mould *Polysphondylium violaceum*. *J. Exp. Biol.*, **38**, 833–49.

SHAFFER, B. M. (1961b). Differentiation in the lower fungi. Species differences in the aggregation of the Acrasieae. *Recent Adv. in Botany*, 294–8.

SHAFFER, B. M. (1962). The Acrasina. *Adv. in Morphogenesis*. Academic Press, New York, **2**, 109–82.

SHAFFER, B. M. (1964). Intracellular movement and locomotion of cellular slime-mould amoebae. From *Primitive motile systems in cell biology*, edited R. D. Allen and N. Kamiya. Princeton Univ. Press, Princeton.

SHAPIRO, B. E. (1958). Influences of the salinity and *p*H of blastocoelic perfusates on the initiation of amphibian gastrulation. *J. Exp. Zool.*, **139**, 381–402.

SHARP, J. A. and BURWELL, R. G. (1960). Interaction ("Peripolesis") of macrophages and lymphocytes after skin homografting or challenge with soluble antigens. *Nature, Lond.*, **188**, 474–5.

SHOGER, R. D. (1960). The regulative capacity of the node region. *J. Exp. Zool.*, **143**, 221–38.

SINGER, J. M., ORESKES, I. and ALTMAN, G. (1962). The mechanism of particulate carrier reactions—IV. Adsorption of human γ-globulin to tanned sheep erythrocytes and their sensitisation for agglutination with rheumatoid arthritis serum. *J. Immunol.*, **89**, 227–33.

SIMON, D. (1957a). Sur la localisation de cellules germinales primordiales chez l'embryon de Poulet et sur leur mode de migration vers les ébauches gonadiques. *C.R. Acad. Sci.*, **244**, 1541–3.

SIMON, D. (1957b). La localisation primaire des cellules germinales dans l'embryon de Poulet; preuves experimentales. *C.R. Soc. Biol.*, **151**, 1010–2.

SIMON, D. (1960). Contribution a l'etude de la circulation et du transport des gonocytes primaires dans les blastodermes d'oiseau cultives *in vitro*. *Arch. Anat. Micros.*, Suppl., **49**, 93–176.

SIMON-REUSS, I., COOK, G. M. W., SEAMAN, G. V. F. and HEARD, D. H. (1964), Electrophoretic studies of some types of mammalian tissue cell. *Cancer Res.*, **24**. 2038–43.

SINCLAIR, D. C. (1955). Cutaneous sensation and the doctrine of specific energy. *Brain*, **78**, 584–614.

SJORSTRAND, F. S. (1962). Critical evaluation of ultrastructural patterns with respect to fixation. From *The Interpretation of Ultrastructure*, Volume 1, edited R. J. C. Harris. Academic Press, New York, 47–68.

SJORSTRAND, F. S. (1963a). A new repeat structural element of mitochondrial and certain cytoplasmic membranes. *Nature, Lond.*, **199**, 1262–4.

SJORSTRAND, F. S. (1963b). The ultrastructure of the plasma membrane of columnar epithelium cells of the mouse intestine. *J. Ultrastr. Res.*, **8**, 517–41.

SJORSTRAND, F. S. (1963c). A new ultrastructural element of the membrane in mitochondria and of some cytoplasmic membranes. *J. Ultrastr. Res.*, **9**, 340–61.

SJORSTRAND, F. S. and ELFVIN, L. G. (1962). The layered asymmetric structure of the plasma membrane in the exocrine pancreas cells of the cat. *J. Ultrastr. Res.*, **7**, 504–34.

SJORSTRAND, F. S. and ELFVIN, L. G. (1964). The granular structure of mitochondrial membranes and of cytomembranes as demonstrated in frozen-dried tissue. *J. Ultrastr. Res.*, **10**, 263–92.

SNEATH, J. S. and SNEATH, P. H. A. (1959). Adsorption of blood-group substances from serum on to red cells. *Brit. Med. Bull.*, **15**, 154–64.

SOBEL, H. (1958). The behaviour *in vitro* of dissociated embryonic pituitary tissue. *J. Embryol. Exp. Morph.*, **6**, 518–26.

SOBOTKA, H. (1956). Penetrability of molecular layers. *J. Coll. Sci.*, **11**, 435–44.

SOLOMON, A. K. (1961). Measurement of the equivalent pore radius in cell membranes. From *Membrane transport and metabolism*, edited A. Kleinzeller and A. Kotyak. Academic Press, New York, 94–9.

SOTELO, J. R. and PORTER, K. R. (1959). An electron microscope study of the rat ovum. *J. Biochem. Biophys. Cytol.*, **5**, 327–34.

SPEIDEL, C. C. (1942). Studies of living nerves—VII. Growth adjustments of cutaneous terminal arborisations. *J. Comp. Neurol.*, **76**, 57–74.

SPEIDEL, C. C. (1947). Correlated studies of sense organs and nerves of the lateral-line in living frog tadpoles—I. Regeneration of denervated organs. *J. Comp. Neurol.*, **87**, 29–55.

SPEIDEL, C. C. (1964). Correlated studies of sense organs and nerves of the lateral-line in living frog tadpoles—IV. Patterns of vagus nerve regeneration after single and multiple operations. *Amer. J. Anat.*, **114**, 133–60.

SPERRY, R. W. (1940). The functional results of muscle transposition in the hind limb of the rat. *J. Comp. Neurol.*, **73**, 379–404.

SPERRY, R. W. (1942). Re-establishment of visu-motor co-ordination by optic nerve regeneration. *Anat. Rec.*, **84**, 20.

SPERRY, R. W. (1943a). Visuomotor co-ordination in the newt (*Triturus viridescens*) after regeneration of the optic nerve. *J. Comp. Neurol.*, **79**, 33–55.

SPERRY, R. W. (1943b). Effect of 180 degree rotation of the retinal field on visuo-motor co-ordination. *J. Exp. Zool.*, **92**, 263–79.

SPERRY, R. W. (1945). Restoration of vision after crossing of optic nerves and after contralateral transplantation of eye. *J. Neurophysiol.*, **8**, 15–28.

SPERRY, R. W. (1947). Nature of functional recovery following regeneration of the oculomotor nerve in amphibians. *Anat. Rec.*, **97**, 293–316.

SPERRY, R. W. (1948a). Patterning of central synapses in regeneration of the optic nerves in teleosts. *Physiol. Zool.*, **21**, 351–61.

SPERRY, R. W. (1948b). Orderly patterning of synaptic associations in regeneration of intracentral fiber tracts mediating visuomotor co-ordination. *Anat. Rec.*, **102**, 63–75.

SPERRY, R. W. (1950a). Myotopic specificity in teleost neurons. *J. Comp. Neurol.*, **93**, 277–87.

SPERRY, R. W. (1950b). Neural basis of the spontaneous optokinetic response produced by visual inversion. *J. Cell. Comp. Physiol.*, **43**, 482–9.

N*

SPERRY, R. W. (1951). Mechanisms of neural maturation. From *Handbook of experimental psychology*. John Wiley, New York, 236–80.

SPERRY, R. W. (1958). Physiological plasticity and brain circuit theory. From *Biological and biochemical bases of behaviour*, edited H. F. Harlow and C. N. Woolsey. Univ. Wisconsin Press.

SPERRY, R. W. (1963). Chemoaffinity in the orderly growth of nerve fiber patterns and connections. *Proc. Nat. Acad. Sci., Wash.*, **50**, 703–10.

SPERRY, R. W. and ARORA, H. L. (1965). Selectivity in regeneration of the oculo-motor nerve in the cichlid fish, *Astronotus ocellatus*. *J. Embryol. Exp. Morph.*, **14**, 307–17.

SPERRY, R. W. and DEUPREE, N. (1956). Functional recovery following alterations in nerve-muscle connections of fishes. *J. Comp. Neurol.*, **106**, 143–62.

SPERRY, R. W. and MINER, N. (1949). Formation within sensory nucleus V of synaptic associations mediating cutaneous localization. *J. Comp. Neurol.*, **90**, 403–23.

SPIEGEL, M. (1954). The role of specific antigens in cell adhesion. *Biol. Bull.* **107**, 130–55.

SPIEGEL, M. (1955). The reaggregation of dissociated sponge cells. *Ann. N.Y. Acad. Sci.*, **60**, 1056–78.

SPRATT, N. T. (1946). Formation of the primitive streak in the explanted chick blastoderm marked with carbon particles. *J. Exp. Zool.*, **103**, 259–304.

SPRATT, N. T. (1947). Regression and shortening of the primitive streak in the explanted chick blastoderm. *J. Exp. Zool.*, **104**, 69–100.

SPRATT, N. T. (1955). Analysis of the organizer center in the early chick embryo—I. Localization of prospective notochord and somite cells. *J. Exp. Zool.*, **128**, 121–64.

SPRATT, N. T. (1957a). Analysis of the organizer center in the early chick embryo—II. Studies of the mechanics of notochord elongation and somite formation. *J. Exp. Zool.*, **134**, 577–612.

SPRATT, N. T. (1957b). Analysis of the organizer center in the early chick embryo—III. Regulative properties of the chorda and somite centers. *J. Exp. Zool.*, **135**, 319–54.

SPRATT, N. T. (1963). Role of the substratum, supracellular continuity and differential growth in morphogenetic cell movements. *Dev. Biol.*, **7**, 51–63.

SPRATT, N. T. and HAAS, H. (1961a). Integrative mechanisms in development of the early chick blastoderm—II. Role of morphogenetic movements and regenerative growth in synthetic and topographically disarranged blastoderms. *J. Exp. Zool.*, **147**, 57–93.

SPRATT, N. T. and HAAS, H. (1961b). Integrative mechanisms in development of the early chick blastoderm—III. Role of cell population size and growth potentiality in synthetic systems larger than normal. *J. Exp. Zool.*, **147**, 271–93.

SPRATT, N. T. and HAAS, H. (1965). Germ layer formation and the role of the primitive streak in the chick—I. Basic architecture and morphogenetic tissue movements. *J. Exp. Zool.*, **158**, 9–38.

SRIVASTAVA, S. N. and HAYDON, D. A. (1964). Estimate of the Hamaker constant for paraffinic hydrocarbons in aqueous suspensions. *Trans. Faraday Soc.*, **60**, 971–8.

STABLEFORD, L. T. (1949). The blastocoel fluid in amphibian gastrulation. *J. Exp. Zool.*, **112**, 529–46.

STEFANELLI, A. (1951). The Mauthnerian apparatus in the *Ichthyopsida*; its nature and function and correlated problems of neurohistogenesis. *Q. Rev. Biol.*, **26**, 17–34.

STEFANELLI, A. and ZACCHEI, A. M. (1958). Sulle modalita di aggregazione di cellule embrionali di pollo disgregate con tripsina. *ActaEmbryol.Morph.Exp.*, **2**, 1–12.

STEINBERG, M. S. (1958). On the chemical bonds between animal cells: a mechanism for type-specific association. *Amer. Nat.*, **92**, 65–82.

STEINBERG, M. S. (1962a). The role of temperature in the control of aggregation of dissociated embryonic cells. *Exp. Cell Res.*, **28**, 1–10.

STEINBERG, M. S. (1962b). Calcium complexing by embryonic cell surfaces: relation to intercellular adhesiveness. From *Biological interactions in normal and neoplastic growth*, edited M. J. Brennan and W. L. Simpson. Little, Brown, Boston, 127–40.

STEINBERG, M. S. (1962c). On the mechanism of tissue reconstruction by dissociated cells—I. Population kinetics, differential adhesiveness and the absence of directed migration. *Proc. Nat. Acad. Sci.*, *Wash.*, **48**, 1577–82.

STEINBERG, M. S. (1963a). "ECM"; its nature, origin and function in cell aggregation. *Exp. Cell Res.*, **30**, 257–79.

STEINBERG, M. S. (1963b). On the mechanism of tissue reconstruction by dissociated cells—II. Time course of events. *Science*, **137**, 762–3.

STEINBERG, M. S. (1964a). On the mechanism of tissue reconstruction by dissociated cells—III. Free energy relations and the reorganisation of fused, heteronomic tissue fragments. *Proc. Nat. Acad. Sci. Wash.*, **48**, 1769–76.

STEINBERG, M. S. (1964b). The problem of adhesive selectivity in cellular interactions. From *Cellular membranes in development*, edited M. Locke. Academic Press, New York, 321–66.

STEINBERG, M. S. and ROTH, S. A. (1964). Phases in cell aggregation and tissue reconstruction. An approach to the kinetics of cell aggregation. *J. Exp. Zool.*, **157**, 327–38.

STOCKENIUS, W. (1959). An electron microscope study of myelin figures. *J. Biophys. Biochem. Cytol.*, **5**, 491–500.

STOCKENIUS, W. (1960). Osmic tetroxide fixation of lipids. *4th Int. Conf. on Electron Microscopy*, 716–25.

STOCKENIUS, W. (1962). Some electron microscopical observations on liquid–crystalline phases in lipid–water systems. *J. Cell Biol.*, **12**, 221–9.

STOCKENIUS, W. (1963). The molecular structure of lipid–water systems and cell membrane models studied with the electron microscope. From *The interpretation of Ultrastructure*, edited R. J. C. Harris. Academic Press, New York, 349–66.

STOKER, M. (1964). Regulation of growth and orientation in hamster cells transformed by polyoma virus. *Virol.*, **24**, 165–74.

STONE, L. S. (1930). Heteroplastic transplantation of eyes between the larvae of two species of *Amblystoma. J. Exp. Zool.*, **55**, 193–261.

STONE, L. S. (1944). Functional polarization in retinal development and its reestablishment in regenerating retinae of grafted salamander eyes. *Proc. Soc. Exp. Biol. Med.*, **57**, 13–14.

STONE, L. S. (1948). Functional polarization in developing and regenerating retinae of transplanted eyes. *Ann. N.Y. Acad. Sci.*, **49**, 856–65.

STONE, L. S. (1953). Normal and reversed vision in transplanted eyes. *A.M.A. Arch. Ophthalmol.*, **49**, 28 35.

STONE, L. S. (1960). Polarization of the retina and development of vision. *J. Exp. Zool.*, **145**, 85–95.

STRAUMFJORD, J. V. and HUMMEL, J. P. (1959). Anionic polymers—IV. Micro-electrophoresis of ascites tumor cells and the effect of polyxenylphosphate. *Cancer Res.*, **19**, 913–7.

SUTER, E. (1956). Interaction between phagocytes and pathogenic micro-organisms. *Bact. Revs.*, **20**, 94–132.

SWIFT, C. H. (1914). The origin and early history of the primordial germ cells in the chick. *Amer. J. Anat.*, **15**, 483–516.

SZEKELY, G. (1959a). The apparent "corneal specificity" of sensory neurons. *J. Embryol. Exp. Morph.*, **7**, 375–9.

SZEKELY, G. (1959b). Functional specificity of cranial sensory neuroblasts in Urodela. *Acta Biol. Acad. Sci., Hung.*, **10**, 107–16.

SZEKELY, G. (1966). Embryonic determination of neural connections. *Advances in Morphogenesis*, **5**, 181–219.

SZEKELY, G. and SZENTGOTHAI, J. (1962). Reflex and behaviour patterns elicited from implanted supernumerary limbs in the chick. *J.Embryol. Exp. Morph.*, **10**, 140–51.

SZULMAN, A. E. (1964). The histological distribution of the blood group sub-stances in man as disclosed by immune fluorescence—III. The A, B and H antigens in embryos and fetuses from 18 mm. in length. *J. Exp. Med.*, **119**, 503–16.

TACHIBANA, T. and OKUDA, M. (1960). The effect of ripple on monolayers of high polymers. *Kolloid Z.*, **171**, 15–20.

TAIT, J. (1918). Capillary phenomena observed in blood cells: thigmocytes, phagocytosis, amoeboid movement, differential adhesiveness of corpuscles, emigration of leucocytes. *Q.J. Exp. Physiol.*, **12**, 1–33.

TAKEYA, H. and WATANABE, T. (1961). Differential proliferation of the ependyma in the developing neural tube of amphibian embryo. *Embryologia.*, **6**, 169–76.

TAMARIN, A. and SREEBENY, L. M. (1963). An analysis of desmosome shape, size and orientation by the use of histometric and densitometric methods with electron microscopy. *J. Cell Biol.*, **18**, 125–34.

TAYLOR, A. C. (1944). Selectivity of nerve fibers from the dorsal and ventral roots in the development of the frog limb. *J. Exp. Zool.*, **96**, 159–85.

TAYLOR, A. C. (1961). Attachment and spreading of cells in culture. *Exp. Cell Res.* Suppl., **8**, 154–73.

TAYLOR, A. C. (1962). Responses of cells to pH changes in the medium. *J. Cell Biol.*, **15**, 201–9.

TAYLOR, A. C. and ROBBINS, E. (1963). Observations on microextensions from the surface of isolated vertebrate cells. *Devel. Biol.*, **7**, 660–73.

TERAYAMA, H. (1962). Surface electric charge of ascites hepatomas and the dis-sociation of islands of tumor cells. *Exp. Cell Res.*, **28**, 113–9.

THOMAS, L. E. and LEVIN, E. O. (1962). Cellular lipoproteins—IV. The insoluble lipoproteins of various tumors. *Exp. Cell Res.*, **28**, 265–9.

THOMPSON, C. M. and WOLPERT, L. (1963). The isolation of motile cytoplasm from *Amoeba proteus. Exp. Cell Res.*, **32**, 156–60.

THOMPSON, T. E. (1964). The properties of bimolecular phospholipid membranes. From *Cellular membranes in development*, edited M. Locke. Academic Press, New York, 83–96.

TILL, J. E. and McCULLOCK, E. A. (1961). A direct measurement of the radiation sensitivity of normal mouse bone marrow cells. *Radiat. Res.*, **14**, 213–22.

TIMOSHENKO, S. (1940). *Theory of plates and shells*. McGraw-Hill, New York.

TODARO, G. J., LAZAR, G. K. and GREEN, H. (1965). The initiation of cell division in a contact-inhibited mammalian cell line. *J. Cell. Comp. Physiol.*, **66**, 325–34.

TOWNES, P. L. and HOLTFRETER, J. (1955). Directed movements and selective adhesion of embryonic amphibian cells. *J. Exp. Zool.*, **128**, 53–120.

TREVAN, D. J. and ROBERTS, D. C. (1960). Sheet formation by cells of an ascites tumour *in vitro*. *Brit. J. Cancer*, **114**, 724–9.

TRINKAUS, J. P. (1951). A study of the mechanism of epiboly in the egg of *Fundulus heteroclitus*. *J. Exp. Zool.*, **118**, 269–319.

TRINKAUS, J. P. (1961). Affinity relationships in heterotypic cell aggregates. *Colloques. Internat. C.N.R.S.*, **101**, 209–25.

TRINKAUS, J. P. (1963a). Behaviour of dissociated retinal pigment cells in heterotypic cell aggregates. *Ann. N.Y. Acad. Sci.*, **100**, 413–34.

TRINKAUS, J. P. (1963b). The cellular basis of *Fundulus* epiboly: adhesivity of blastula and gastrula cells in culture. *Devel. Biol.*, **7**, 513–32.

TRINKAUS, J. P. and GROSS, M. C. (1961). The use of tritiated thymidine for marking migratory cells. *Exp. Cell. Res.*, **24**, 52–7.

TRINKAUS, J. P. and GROVES, P. W. (1955). Differentiation in culture of mixed aggregates of dissociated tissue cells. *Proc. Nat. Acad. Sci. Wash.*, **41**, 787–95.

TRINKAUS, J. P. and LENTZ, J. P. (1964). Direct observation of type-specific segregation in mixed cell aggregates. *Devel. Biol.*, **9**, 115–36.

TRIPLETT, E. L. (1958). The development of the sympathetic ganglia, sheath cells and meninges in amphibians. *J. Exp. Zool.*, **138**, 283–312.

TRURNITH, J. and SCHIDLOVSKY, G. (1960). Thin cross-sections of artificial stacks of monomolecular films. *Proc. European Reg. Conf. Electron Microscopy, Delft.*, 721–5.

TSUBO, I. and BRANDT, P. W. (1962). An electron microscopic study of the malpighian tubules of the grasshopper, *Dissosteira carolina*. *J. Ultrastr. Res.*, **6**, 28–35.

TUFT, P. (1957). Changes in the osmotic activity of the blastocoel and archenteron contents during the early development of *Xenopus laevis*. *Proc. Roy. Phys. Soc. Edin.*, **26**, 42–8.

TUFT, P. (1961a). A morphogenetic effect of beta-mercaptoethanol. *Nature, Lond.*, **191**, 1072–3.

TUFT, P. (1961b). Role of water regulating mechanisms in Amphibian morphogenesis: a quantitative hypothesis. *Nature, Lond.*, **192**, 1049–51.

TUFT, P. (1962). The uptake and distribution of water in the embryo of *Xenopus laevis* (Daudin). *J. Exp. Biol.*, **39**, 1–19.

TURKINGTON, R. W. (1962). Thyrotropin-stimulated ATPase in isolated thyroid cell membranes. *Biochem. Biophys. Acta*, **65**, 386–8.

TWITTY, V. C. (1945). The developmental analysis of specific pigment patterns. *J. Exp. Zool.*, **100**, 141–8.

TWITTY, V. C. and NIU, M. C. (1948). Causal analysis of chromatophore migration. *J. Exp. Zool.*, **108**, 405–37.

TWITTY, V. C. and NIU, M. C. (1954). The motivation of cell migration, studied by isolation of embryonic cells singly and in small groups *in vitro*. *J. Exp. Zool.*, **125**, 541–73.

TYLER, A. (1946). An auto-antibody concept of cell structure, growth and differentiation. *Growth*, **10**, Symposium 6, 7–19.

TYLER, A. (1948). Fertilization and immunity. *Physiol. Revs.*, **28**, 180–219.

UMANSKY, R. (1966). The effect of cell population density on the developmental fate of reaggregating mouse limb bud mesenchyme. *Devel. Biol.*, **13**, 31–56.

VAKAET, L. (1956). Etude expérimentale par la méthode des cultures d'organes des réactions de l'épithélium bronchial de l'embryon de poulet. *Arch. Anat. Micros.*, **45**, 48–64.

VAKAET, L. (1960). Quelques précisions sur la cinématique de la ligne primitive chez le Poulet. *J. Embryol. Exp. Morph.*, **8**, 321–6.

VAKAET, L. (1964). Diversité fonctionelle de la ligne primitive du blastoderme de poulet. *C.R. Soc. Biol.*, **158**, 1964–6.

VAN BREEMAN, V. L., REGER, J. F. and COOPER, W. G. (1956). Observations on the basement membranes in rat kidney. *J. Biochem. Biophys. Cytol.*, **2**, suppl. 283–6.

VAN DEENEN, L. L. M., HOUTSMULLER, V. M. T., DE HAAS, G. H. and MULDER, E. (1962). Monomolecular layers of synthetic phosphatides. *J. Pharm. Pharmacol.*, **14**, 429–44.

VAN DEN TEMPEL, M. (1958). Distance between emulsified oil globules upon coalescence. *J. Coll. Sci.*, **13**, 125–33.

VANDENHEUVEL, F. A. (1963). Study of biological structure at the molecular level with stereomodel projections—I. The lipids in the myelin sheath of nerve. *J. Amer. Oil Chemists Soc.*, **40**, 455–71.

VANDENHEUVEL, F. A. (1963). Structural studies of biological membranes: the structure of myelin. *Ann. N.Y. Acad. Sci.*, **122**, 57–76.

VAN HARREVELD, A., CROWELL, J. and MALHOBA, S. K. (1965). A study of extra-cellular space in central nervous tissue by freeze-substitution. *J. Cell Biol.*, **25**, 117–38.

VAN LIMBORDH, J. (1957). *The development of gonadal asymmetry on the duck embryo.* Oosterbaan and Le Covistre, Gres.

VASSAR, P. S. (1963). The electric charge density of human tumor cell surfaces. *Laborat. Investig.*, **12**, 1072–7.

VASSAR, P. S. and CULLING, C. F. A. (1964). Cell surface effects of phytohaemag-glutinin. *Nature, Lond.*, **202**, 610–11.

VERWEY, E. J. W. and OVERBEEK, J. TH. G. (1948). *Theory of the stability of lyophobic colloids.* Elsevier, Amsterdam.

VOGT, W. (1925). Gestaltungsanalyse am amphibienkeim mit ortlicher vitalfarbung —II. Gastrulation und mesodermbildung bei urodelen und anuren. *Arch. f. Entwickmech.*, **120**, 385–706.

VOIGTLANDER, G. (1932). Neue untersuchungen uber den "Cytotropismus" der Furchungszellen. *Arch. f. Entwickmech.*, **127**, 151–215.

VOLD, M. J. (1961). The effect of adsorption on the van der Waals interaction of spherical colloidal particles. *J. Coll. Sci.*, **16**, 1–12.

VOROPAEVA, T. N. (1963), DERJAGUIN, B. V. and KABANOV, E. N. (1963). From *Research in surface forces*, edited, B. V. Derjaguin. In translation. Consultants Bureau, New York, 116.

WADDINGTON, C. H. (1932). Experiments on the development of chick and duck embryos cultivated *in vitro*. *Phil. Trans. Roy. Soc.* **B.**, **221**, 179–230.

WADDINGTON, C. H. (1939). Order of magnitude of morphogenetic forces. *Nature, Lond.*, **144**, 637.

WADDINGTON, C. H. (1942). Observations on the forces of morphogenesis in the amphibian embryo. *J. Exp. Biol.*, **19**, 284–93.

WADDINGTON, C. H. (1952). The epigenetics of birds. Cambridge Univ. Press, Cambridge, xvii + 272.

WADDINGTON, C. H., PERRY, M. M. and OKADA, E. (1961). "Membrane knotting" between blastomeres of *Limnaea*. *Exp. Cell Res.*, **23**, 631–3.

WALKER, B. E. and RENNELS, E. G. (1961). Adrenal cortical cell replacement in the mouse. *Endocrinal.*, **68**, 365–74.

WALLACH, D. F. H. and EYLAR, E. M. (1961). Sialic acid in the cellular membranes of Ehrlich ascites carcinoma cells. *Biochem. Biophys. Acta.*, **52**, 594–6.

WALLACH, D. F. H. and ULLREY, D. (1962). Studies on the surface and cytoplasmic membranes of Ehrlich ascites carcinoma cells—I. The hydrolysis of ATP and related nucleotides by microsomal membranes. *Biochem. Biophys. Acta.*, **64**, 526–39.

WARTENBERG, H. and SCHMIDT, W. (1961). Elektronenmikroskopische untersuchungen der strukturellen veranderungen im rindenbereich des Amphibieneies im ovar und nach der befruchtung. *Z. f. Zellforsch*, **54**, 118–46.

WARTENBERG, H. and STEGNER, H. E. (1960). Uber die elektronmikroskopische feinstructur des menschlichen ovarialeies. *Z. f. Zellforsch.*, **52**, 450–74.

WATTERSON, R. L. (1942). The morphogenesis of down feathers with special reference to the developmental history of melanophores. *Physiol. Zool.*, **15**, 234–59.

WAUGH, D. F. and SCHMITT, F. O. (1940). Investigation of the thickness and ultrastructure of cellular membranes by the analytical leptoscope. *Cold Spring Harbor Symp. Quant. Biol.*, **8**, 233–41.

WEISBERGER, A. S., GUYTON, R. A., HEINLE, R. W. and STORAASLI, J. P. (1951). The role of the lungs in the removal of transfused lymphocytes. *Blood*, **6**, 916–25.

WEISS, L. (1958). The effects of trypsin on the size, viability and dry mass of sarcoma 37 cells. *Exp. Cell Res.*, **14**, 80–3.

WEISS, L. (1959a). Studies on cellular adhesion in tissue culture—I. The effect of serum. *Exp. Cell Res.*, **17**, 499–507.

WEISS, L. (1959b). Studies on cellular adhesion in tissue culture—II. The adhesion of cells to gel surfaces. *Exp. Cell Res.*, **17**, 508–15.

WEISS, L. (1960a). The adhesion of cells. *Internat. Rev. Cytol.*, **9**, 187–225.

WEISS, L. (1960b). Studies on cellular adhesion in tissue culture—III. Some effects of calcium. *Exp. Cell Res.*, **21**, 71–7.

WEISS, L. (1961a). The measurement of cell adhesion. *Exp. Cell Res.* Suppl., **8**, 141–53.

WEISS, L. (1961b). Studies on cellular adhesion in tissue culture—IV. The alteration of substrata by cell surfaces. *Exp. Cell Res.*, **25**, 504–17.

WEISS, L. (1962a). Cell movement and cell surfaces: a working hypothesis. *J. Theor. Biol.*, **2**, 236–50.

WEISS, L. (1962b). The mammalian tissue cell surface. *Biochem. Soc. Symp.*, 232–54.

WEISS, L. (1963a). The pH value at the surface of *Bacillus subtilis*. *J. Gen. Microbiol.*, **32**, 331–40.

WEISS, L. (1963b). Studies on cellular adhesion in tissue culture—V. Some effects of enzymes on cell detachment. *Exp. Cell Res.*, **30**, 509–20.

WEISS, L. (1963c). Some sorption phenomena at mammalian cell surfaces. *J. Gen. Microbiol.*, **32**, 11–14.

WEISS, L. (1963d). Studies on cellular adhesion in tissue culture—VI. Initial cell contacts and morphogenetic movements. *Exp. Cell Res.*, **31**, 61–9.

WEISS, L. (1964a). Studies on cellular adhesion in tissue culture—VII. Surface activity and cell detachment. *Exp. Cell Res.*, **33**, 277–88.

WEISS, L. (1964b). Cellular locomotive pressure in relation to initial cell contacts. *J. Theor. Biol.*, **6**, 275–81.

WEISS, L. (1965). Studies on cellular adhesion in tissue culture—VIII. Some effects of antisera on cell detachment. *Exp. Cell Res.*, **37**, 540–51.

WEISS, L. and COOMBS, R. R. A. (1963). The demonstration of rupture of cell surfaces by an immunological technique. *Exp. Cell Res.*, **30**, 331–8.

WEISS, P. (1929). Erzwingung elementarer strukturverschiedenheiten am *in vitro* wachsenden gewebe. Die wirkung mechanischer spannung auf richtung und intenstitat des gewebewachstums und ihre analyse. *Arch. f. Entwickmech.*, **116**, 438–554.

WEISS, P. (1934). *In vitro* experiments on the factors determining the course of the outgrowing nerve fiber. *J. Exp. Zool.*, **68**, 393–448.

WEISS, P. (1936). Selectivity controlling the central-peripheral relations in the nervous system. *Biol. Revs.*, **11**, 494–531.

WEISS, P. (1941). Nerve patterns: The mechanics of nerve growth. *Growth.* Suppl., **5**, 163–203.

WEISS, P. (1942). Lid-closure reflex from eyes transplanted to atypical locations in *Triturus torosus*: evidence of a peripheral origin of sensory specificity. *J. Comp. Neurol.*, **77**, 131–69.

WEISS, P. (1945). Experiments on cell and axon orientation *in vitro*: the role of colloidal exudates in tissue organisation. *J. Exp. Zool.*, **100**, 353–86.

WEISS, P. (1947). The problem of specificity in growth and development. *Yale J. Biol. and Med.*, **19**, 235–78.

WEISS, P. (1953). Some introductory remarks on the cellular basis of differentiation. *J. Embryol. Exp. Morph.*, **1**, 181–211.

WEISS, P. (1955). Nervous system. From *Analysis of development*, edited B. H. Willier, P. Weiss and V. Hamburger. Princeton Univ. Press, Princeton, 346–401.

WEISS, P. (1958). Cell contact. *Internat. Rev. Cytol.*, **7**, 391–423.

WEISS, P. (1961). Guiding principles in cell locomotion and cell aggregation. *Exp. Cell Res.*, Suppl., **8**, 260–81.

WEISS, P. and ANDRES, G. M. (1952). Experiments on the fate of embryonic cells (chick) disseminated by the vascular route. *J. Exp. Zool.*, **121**, 449–87.

WEISS, P. and FERRIS, W. (1956). The basement lamella of amphibian skin. Its reconstruction after wounding. *J. Biophys. Biochem. Cytol.*, **2**, Suppl., 275–82.

WEISS, P. and GARBER, B. (1952). Shape and movement of mesenchyme cells as functions of the physical structure of the medium: contributions to a quantitative morphology. *Proc. Nat. Acad. Sci. Wash.*, **38**, 264–80.

WEISS, P. and MOSCONA, A. A. (1958). Type-specific morphogenesis of cartilages developed from dissociated limb and retinal mesenchyme *in vitro*. *J. Embryol. Exp. Morph.*, **6**, 238–46.

WEISS, P. and SCOTT, B. I. H. (1963). Polarization of cell locomotion *in vitro*. *Proc. Nat. Acad. Sci.*, *Wash.*, **50**, 330–6.

WEISS, P. and TAYLOR, A. C. (1944). Further experimental evidence against "neurotropism" in nerve regeneration. *J. Exp. Zool.*, **95**, 23–257.

WEISS, P. and TAYLOR, A. C. (1960). Reconstitution of complete organs from single cell suspensions of chick embryos in advanced stages of differentiation. *Proc. Nat. Acad. Sci., Wash.*, **46**, 1177–85.

WENGER, E. L. (1950). An experimental analysis of relations between parts of the brachial spinal cord of the embryonic chick. *J. Exp. Zool.*, **114**, 51–86.

WESSELS, N. K. (1962). Tissue interactions during skin histodifferentiation. *Devel. Biol.*, **4**, 87–107.

WESSELS, N. (1964a). Tissue interactions and cytodifferentiation. *J. Exp. Zool.*, **157**, 139–52.

WESSELS, N. (1964b). Substrate and nutrient effects upon epidermal basal cell orientation and proliferation. *Proc. Nat. Acad. Sci.*, **52**, 252–9.

WESTON, J. A. (1963). A radioautographic analysis of the migration and localization of trunk neural crest cells in the chick. *Devel. Biol.*, **6**, 279–310.

WETZEL, R. (1929). Untersuchungen am Huhnchen. Die entwicklung des keims wahrend der ersten beiden bruttage. *Arch. f. Entwickmech.*, **119**, 118–321.

WHITEFIELD, F. E. (1964). The use of proteolytic and other enzymes in the separation of slime mould grex. *Exp. Cell Res.*, **36**, 62–72.

WHITTAKER, V. P. and GRAY, E. G. (1962). The synapse. *Brit. Med. Bull.*, **18**, pt. 3, 223–8.

WIGGLESWORTH, V. B. (1959). The role of the epidermal cells in the "migration" of tracheoles in *Rhodnius prolixus* (Hemiptera). *J. Exp. Biol.*, **36**, 632–40.

WILENS, S. (1955). The migration of heart mesoderm and associated areas in *Amblystoma punctatum*. *J. Exp. Zool.*, **129**, 579–605.

WILKINS, D. J., OTTEWILL, R. H. and BANGHAM, A. D. (1962a). On the flocculation of sheep leucocytes—I. Electrophoretic studies. *J. Theor. Biol.*, **2**, 165–75.

WILKINS, D. J., OTTEWILL, R. H. and BANGHAM, A. D. (1962b). On the flocculation of sheep leucocytes—II. Stability studies. *J. Theor. Biol.*, **2**, 176–91.

WILLIER, B. H. (1937). Experimentally produced sterile gonads and the problem of the origin of germ cells in the chick embryo. *Anat. Rec.*, **70**, 89–112.

WILLIER, B. H. (1950). Sterile gonads and the problem of the origin of germ cells in the chick embryo. *Arch. Anat. Micros. Morphol. Exp.*, **39**, 269–70.

WILLIER, B. H. and RAWLES, M. E. (1940). The control of feather color pattern by melanophores grafted from one embryo to another of a different breed of fowl. *Physiol. Zool.*, **13**, 177–99.

WILLMER, E. N. (1958). *Tissue culture: the growth and differentiation of normal tissues in the artificial media.* 3rd edition. Methuen, London, xx + 191.

WILLMER, E. N. (1961). Steroids and cell surfaces. *Biol. Revs.*, **36**, 368–98.

WILLMER, E. N. and JACOBY, F. (1936). Studies on the growth of tissues *in vitro*—IV. On the manner in which growth is stimulated by extracts of embryo tissues. *J. Exp. Biol.*, **13**, 237–48.

WILSON, E. V. (1907). On some phenomena of coalescence and regeneration in sponges. *J. Exp. Zool.*, **5**, 245–58.

WILSON, E. V. (1911). On the behaviour of the dissociated cells in hydroids, *Alcyonaria* and *Asterias*. *J. Exp. Zool.*, **11**, 281–337.

WISCHNITZER, S. (1963). The ultrastructure of the layers enveloping yolk-forming oocytes from *Triturus viridescens*. *Z. f. Zellforsch.*, **60**, 452–62.

WITSCHI, E. (1948). Migration of the germ cells of human embryos from the yolk sac to the primitive gonadal folds. *Contr. Embryology Carn. Inst., Wash.,* **32**, 67–80.

WITSCHI, E. (1950). Genetique et physiologie de la differenciation du sexe. *Arch. Anat. Micros. Morph. Exp.,* **39**, 215–40.

WOLFE, L. S. (1962). Lipid-protein complexes in membranes of nervous tissue. *Canad. J. Biochem. Biophys.,* **40**, 1261–71.

WOLFE, L. S. (1964). Cell membrane constituents concerned with transport mechanisms. *Canad. J. Biochem. Biophys.,* **42**, 971–88.

WOLFF, E. (1954). Potentialités et affinités des tissus, révélées par la culture *in vitro* d'organes en associations hétérogènes et xénoplastique. *Bull. Soc. Zool.,* **79**, 357–67.

WOLFF, E. and HAFFEN, K. (1952). Sur l'intersexualité expérimentale des gonades embryonnaires de canard cultivées *in vitro. Arch. Anat. Micros.,* **41**, 184–207.

WOLFF, E. and MARIN, L. (1957). Sur les mouvements des explants de foie embryonnaire cultivé *in vitro. C.R. Acad. Sci.,* **244**, 2745–7.

WOLFF, E. and SCHNEIDER, N. (1957). La culture d'un sarcome de souris sur des organes de poulet explantés *in vitro. Arch. Anat. Micros.,* **46**, 173–97.

WOLFF, E. and WENIGER, J. P. (1954). Recherches préliminaires sur les chimères d'organes embryonnaires d'Oiseaux et de Mammifères en culture *in vitro. J. Embryol. Exp. Morph.,* **2**, 161–71.

WOLPERT, L. (1960). The mechanics and mechanism of cleavage. *Int. Rev. Cytol.,* **10**, 163–216.

WOLPERT, L. (1966). The mechanical properties of the membrane of the sea-urchin egg during cleavage. *Exp. Cell Res.,* **41**, 385–96.

WOLPERT, L. and GUSTAFSON, T. (1961). Studies on the cellular basis of morphogenesis of the sea-urchin embryo. The formation of the blastula. *Exp. Cell Res.,* **25**, 374–82.

WOLPERT, L. and MERCER, E. H. (1961). An electron microscope study of fertilization of the sea-urchin egg *Psammechinus miliaris. Exp. Cell Res.,* **22**, 45–55.

WOLPERT, L. and MERCER, E. H. (1963). An electron microscope study of the development of the sea-urchin embryo and its radial polarity. *Exp. Cell Res.,* **30**, 280–300.

WOLPERT, L. and O'NEILL, C. H. (1962). Dynamics of the membrane of *Amoeba proteus* studied with labelled specific antibody. *Nature, Lond.,* **196**, 1261–6.

WOLPERT, L., THOMPSON, C. M. and O'NEILL, C. H. (1964). Studies in the isolated membrane and cytoplasm of *Amoeba proteus* in relation to amoeboid movement. From *Primitive motile systems in cell biology,* edited R. D. Allen and N. Kamiya. Princeton Univ. Press, Princeton, 143–72.

WOOD, R. L. (1959). Intercellular attachment in the epithelium of *Hydra* as revealed by electron microscopy. *J. Biophys. Biochem. Cytol.,* **6**, 343–52.

WOOD, S. (1964). Experimental studies of the intravascular dissemination of ascitic V2 carcinoma cells in the rabbit, with special reference to fibrinogen and fibrinolytic agents. *Bull. Swiss Acad. Med. Sci.,* **20**, 92–121.

WOODIN, A. M. and WIENKE, A. A. (1966). The interaction of leucocidin with the cell membrane of the polymorphonuclear leucocyte. *Biochem. J.,* **99**, 479–92.

YAMADA, T. (1962). Mutual adhesiveness of tumor cells in "hepatoma islands" of the rat ascites hepatoma—I and II. Studies on the mechanism of tumor metastasis. *Z. f. Krebsforsch.,* **65**, 75–86.

YAMADA, T. and SATON, H. (1964). Mutual adhesiveness of tumor cells in "hepatoma islands" of the rat ascites hepatoma—III. The influence of nitrogen N-oxide. *Z. f. Krebsforsch.*, **66**, 109–14.

YOUNGNER, J. S. and NOLL, H. (1958). Virus-lipid interactions—I. Concentration and purification of viruses by adsorption to a cholesterol column and studies of the biological properties of lipid-adsorbed virus. *Virology*, **6**, 157–80.

ZEIDMAN, I. (1947). Chemical factors in the mutual adhesiveness of epithelial cells. *Cancer Res.*, **7**, 386–9.

ZEIDMAN, I. (1961). The fate of circulating tumor cells—I. Passage of cells through capillaries. *Cancer Res.*, **21**, 38–9.

ZOTIN, A. E. (1962). механизм перехода жидкосги бластоцеля в полость первичнои кишки у зародьшеи осетра. *Dokl. Akad. Nauk.*, *S.S.S.R.*, **142**, 968–71.

ZWILLING, E. (1954). Dissociation of chick embryo cells by means of a chelating compound. *Science*, **120**, 219.

ZWILLING, E. (1963). Survival and non-sorting of nodal cells following dissociation and reaggregation of definitive streak chick embryos. *Devel. Biol.*, **7**, 642–52.

ADDENDUM

In the period between the completion of the manuscript of this book and the return of the galley proofs to the printers, a number of papers of importance have come to my notice. These papers are listed and briefly described in this addendum, but no critical discussion of their contents will be given. They are described in the order of the chapters to which they are relevant.

Warren *et al.* (1966) have described a new method for isolation of the cell surface from L cells etc. in which various reagents which react with sulphydryl groups are reacted with the surface to stabilise it during isolation. Burger and Goldberg (1967) have identified the antigenic determinant specific to a certain tumour cell surface as containing N-acetylglucosamine, probably in a glycoprotein. Langdon and Sloan (1967), studying the permeation of sugars into erythrocytes etc., suggest that the surface contains lysyl residues involved in the permeation process. Wallach and Kamat (1966) have investigated the contribution which sialic acid makes to the surface charge of the isolated plasmalemma. L. Weiss and Mayhew (1966) suggest that ribonucleic acid is present in the plasmalemma of certain tumour cells, mainly because ribonuclease reduced the surface charge density of these cells. Lenard and Singer (1966) and Wallach and Zahler (1966) have studied respectively the optical rotatory dispersion of red cell ghosts and Ehrlich ascites cell isolated plasma-lemmae, and have identified the presence of protein in their samples; the protein being largely in the α-helix configuration or randomly coiled. Wallach and Zahler also supported this conclusion from the results of I.R. and fluorescence spectroscopy. Chapman *et al.* (1967) have examined the erythrocyte ghost by NMR spectroscopy and have identified choline and sialic acid in the ghost. Hanai *et al.* (1965a, b, c) have studied the capacitance and conductance of artificial "plasmalemmae", and suggest that pores may be present in the live plasmalemma to account for the measured values of capacitance. Similar studies have been carried out by Lauger *et al.* (1967). Leslie and Chapman (1967), Tien and Dawidowicz (1966) and Tien *et al.* (1966) have described new types of artificial lipid membranes. Tsofina *et al.* (1966) have also described new techniques for the production of such membranes. Seufert (1965) has reported transmembrane potentials across artificial lipid membranes containing crude lipids, α-tocopherol and various detergents. Cunningham and Crane (1966) described the results of using negative staining to reveal membrane structure. Finean *et al.* (1966) deduced that red cell ghosts contain bimolecular lipid leaflets while wet using low angle X-ray diffraction. Finean (1966) has produced a masterly review on the structure of the plasmalemma. Rambourg and Leblond (1967) have used a silver stain which they claim reveals the surface coat of cell in electron micrographs: the true plasmalemma was not visible in their micrographs.

Seeman (1967) has described the nature and lifetime of holes in the plasmalemma of haemolysing red cells by following the entry of ferritin etc. tracers into such cells. He found that the holes exist in the main for only 15–25 seconds after the onset of haemolysis by saponin or lysolecithin. Ambrose (1966) has reviewed the techniques and results of cell electrophoresis.

Moscona and Moscona (1967) have published a note criticising the use of EDTA for dispersing tissues. Norrby *et al.* (1966) have investigated the release of DNA into the medium during the trypsinisation of tissues. Moskowitz *et al.* (1966) have described the structure of aggregates. Jones (1966) has reviewed work on aggregation mechanism: unfortunately his description of my work is rather inaccurate.

Vaughan and Trinkaus (1966) have studied the movements of epithelial cells in tissue culture.

Mallette and Anthony (1966) have described the aggregation of adult thyroid cells. Sara (1965) and Sara *et al.* (1966a, b) report that the cells of pairs of many sponge species will form aggregates in which there is no sorting out of the two species types. They also describe mixed aggregates of sponge and anthozoan cells.

Gaze (1967) has published an excellent review on morphogenetic movements, concentrating on the nervous system.

Trinkaus and Lentz (1967) have provided a very detailed analysis of cell movements during the gastrulation of the fish, *Fundulus*, in which they correlate electron microscopy of cell contacts, cell behaviour and morphogenetic movements. Weston and Butler (1966) have continued their work on the emigration of cells from the neural crest of chick embryos and have tested what process if any controls the final siting of the crest cells.

REFERENCES

AMBROSE, E. J. (1966). Electrophoretic behaviour of cells. *Prog. Biophys. and Mol. Biol.*, **16**, 243–65.

BURGER, M. M. and GOLDBERG, A. R. (1967). Identification of a tumor specific determinant on neoplastic cell surfaces. *Proc. Nat. Acad. Sci. Wash.*, **57**, 359–66.

CHAPMAN, D., KAMAT, V. B., GIER, J. DE and PENKETT, S. A. (1967). Nuclear magnetic resonance spectroscopic studies of erythrocyte membranes. *Nature, Lond.*, **213**, 74–5.

CUNNINGHAM, W. P. and CRANE, F. L. (1966). Variation in membrane structure as revealed by negative staining technique. *Exp. Cell Res.*, **44**, 31–45.

FINEAN, J. B. (1966). The molecular organization of cell membrane. *Prog. Biophys. and Mol. Biol.*, **16**, 145–70.

FINEAN, J. B., COLEMAN, R., GREEN, W. G. and LIMBRICK, A. R. (1966). Low-angle X-ray diffraction and electron microscope studies of isolated cell membranes. *J. Cell Sci.*, **1**, 287–95.

GAZE, R. M. (1967). Growth and differentiation. *Ann. Rev. Physiol.*, **29**, 59–86.

HANAI, T., HAYDON, D. A. and TAYLOR, J. (1965a). The influence of lipid composition and of some adsorbed proteins in the capacitance of black hydrocarbon membranes. *J. Theor. Biol.*, **9**, 422–32.

HANAI, T., HAYDON, D. A. and TAYLOR, J. (1965b). The variation of capacitance and conductance of bimolecular lipid membranes with area. *J. Theor. Biol.*, **9**, 433–43.

HANAI, T., HAYDON, D. A. and TAYLOR, J. (1965c). Some further experiments on bimolecular lipid membranes. *J. Gen. Physiol.*, **48**, 59–63.

JONES, B. M. (1966). Invertebrate tissue and organ culture in research. From *Cells and tissues in culture. Methods, biology and physiology*, edited E. N. Willmer. Academic Press, New York.

LANGDON, R. G. and SLOAN, H. R. (1967). Formation of imine bonds between transported sugars and lysyl residues of specific membrane proteins of erythrocytes and fat cells. *Proc. Nat. Acad. Sci. Wash.*, **57**, 401–8.

LAUGER, P., LESSLAUER, W., MARTI, E. and RICHTER, J. (1967). Electrical properties of bimolecular phospholipid membranes. *Biochem. Biophys. Acta*, **135**, 20–32.

LENARD, J. L. and SINGER, S. J. (1966). Protein conformation in cell membrane preparations as studied by optical rotatory dispersion and circular dichroism. *Proc. Nat. Acad. Sci. Wash.*, **56**, 1828–35.

LESLIE, R. B., and CHAPMAN, D. (1967). Artificial phospholipid membranes and bioenergetics. *Chem. Phys. Lipids*, **1**, 143–56.

MALLETTE, J. M. and ANTHONY, A. (1966).   Growth in culture of trypsin dissociated thyroid cells from adult rats. *Exp. Cell Res.*, **41**, 642–51.

MOSCONA, A. A. and MOSCONA, M. H. (1967).   Comparison of aggregation of embryonic cells dissociated with trypsin or versene. *Exp. Cell Res.*, **45**, 239–43.

MOSKOWITZ, M., AMBORSKI, G. F. and WIEKER, G. H. (1966).   Structure development in aggregens. *Nature, Lond.*, **211**, 1047–9.

NORRBY, K., KNUTSON, F. and LUNDIN, P. M. (1966).   On the single cell state in enzymatically produced tumor cell suspensions. *Exp. Cell Res.*, **44**, 421–8.

RAMBOURG, A. and LEBLOND, C. P. (1967).   Electron microscope observations on the carbohydrate-rich cell coat present at the surface of cells in the rat. *J. Cell Biol.*, **32**, 27–53.

SARA, M. (1965).   Aggregazione cellulare interspecifica fra specie diverse di Poriferi e fra Poriferi ed *Anemonia sulcata*. *Boll. Zool.*, **32**, 1067–77.

SARA, M., LIACI, L. and MELONE, N. (1966a).   Mixed cell aggregation between sponges and the Anthozoan *Anemonia Sulcata*. *Nature, Lond.*, **210**, 1168–9.

SARA, M., LIACI, L. and MELONE, N. (1966b).   Bispecific cell aggregation in sponges. *Nature, Lond.*, **210**, 1167–8.

SEEMAN, P. (1967).   Transient holes in the erythrocyte membrane during hypotonic hemolysis and stable holes in the membrane after lysis by saponin and lysolecithin. *J. Cell Biol.*, **32**, 55–70.

SEUFERT, W. D. (1965).   Induced permeability changes in reconstituted cell membrane structure. *Nature, Lond.*, **207**, 174–6.

TIEN, H. T., CARBONE, S. and DAWIDOWICZ, E. A. (1966).   Formation of 'black' lipid membranes by oxidation of cholesterol. *Nature, Lond.*, **212**, 718–19.

TIEN, H. T. and DAWIDOWICZ, E. A. (1966).   Black lipid films in aqueous media; a new type of interfacial phenomenon. *J. Coll. & Interfac. Sci.*, **22**, 438–53.

TRINKAUS, J. P. and LENTZ, T. L. (1967).   Surface specializations of *Fundulus* cells and their relation to cell movements during gastrulation. *J. Cell Biol.*, **32**, 139–53.

TSOFINA, L. M., LIBERMAN, E. A. and BABAKOV, A. V. (1966).   Production of bimolecular protein-lipid membranes in aqueous solutions. *Nature, Lond.*, **212**, 681–3.

VAUGHAN, R. B. and TRINKAUS, J. P. (1966).   Movements of epithelial cell sheets *in vitro*. *J. Cell Sci.*, **1**, 407–13.

WALLACH, D. F. H. and KAMAT, V. B. (1966).   The contribution of sialic acid to the surface charge of fragments of plasma membrane and endoplasmic reticulum. *J. Cell Biol.*, **30**, 660–3.

WALLACH, D. F. H. and ZAHLER, P. H. (1966).   Protein conformation in cellular membranes. *Proc. Nat. Acad. Sci. Wash.*, **56**, 1552–9.

WARREN, L., GLICK, M. C. and NASS, M. K. (1966).   Membranes of animal cells— I. Methods of isolation of the surface membrane. *J. Cell Physiol.*, **68**, 269–88.

WEISS, L. and MAYHEW, E. (1966).   The presence of ribonucleic acid within the peripheral zones of two types of mammalian cell. *J. Cell Physiol.*, **68**, 345–60.

WESTON, J. A. and BUTLER, S. L. (1966).   Temporal factors affecting the localization of neural crest cells in chicken embryos. *Devel. Biol.*, **14**, 246–66.

Index

399

DATE DUE